Plant Biodiversity: Use of Remote Sensing Techniques (Volume 1)

Plant Biodiversity: Use of Remote Sensing Techniques (Volume 1)

Rosanna Mojica

www.callistoreference.com

Callisto Reference,
118-35 Queens Blvd., Suite 400,
Forest Hills, NY 11375, USA

Visit us on the World Wide Web at:
www.callistoreference.com

ISBN: 978-1-64116-747-5 (Hardback)

Cataloging-in-Publication Data

Plant biodiversity : use of remote sensing techniques (Volume 1) / Rosanna Mojica.
p. cm.
Includes bibliographical references and index.
ISBN 978-1-64116-747-5
1. Plant diversity--Remote sensing. 2. Plant diversity--Data processing. 3. Biodiversity.
4. Remote sensing. I. Mojica, Rosanna.

QK46.5.D58 P53 2023
581.7--dc23

Table of Contents

Preface

Over the recent decade, advancements and applications have progressed exponentially. This has led to the increased interest in this field and projects are being conducted to enhance knowledge. The main objective of this book is to present some of the critical challenges and provide insights into possible solutions. This book will answer the varied questions that arise in the field and also provide an increased scope for furthering studies.

The variability and variety of plant life which exists on the Earth is known as plant biodiversity. It faces threats from multiple sources such as extraction of resources, climate change, and human-driven disturbances. Remote sensing is a technology that enables the acquisition of information about an object or phenomenon without making physical contact with the object, in contrast to in-situ or on-site observation. It uses satellites, drones, and air-based sensors to gather information on a specific object or area. The techniques of remote sensing can be utilized for consistent and multi-temporal analysis of plant biodiversity. This book is a detailed explanation of the various applications of remote sensing techniques with respect to plant biodiversity. It will also provide interesting topics for research, which interested readers can take up. Coherent flow of topics, student-friendly language, and extensive use of examples make this book an invaluable source of knowledge.

I hope that this book, with its visionary approach, will be a valuable addition and will promote interest among readers. Each of the authors has provided their extraordinary competence in their specific fields by providing different perspectives as they come from diverse nations and regions. I thank them for their contributions.

Rosanna Mojica

1

An Introduction to Remote Sensing of Plant Biodiversity

Jeannine Cavender-Bares, John A. Gamon, and Philip A. Townsend

1.1 Introduction

Improved detection and monitoring of biodiversity is critical at a time when Earth's biodiversity loss due to human activities is accelerating at an unprecedented rate. We face the largest loss of biodiversity in human history, a loss which has been called the "sixth mass extinction" (Leakey 1996; Kolbert 2014), given that its magnitude is in proportion to past extinction episodes in Earth history detectable from the fossil record. International efforts to conserve biodiversity (United Nations 2011) and to develop an assessment process to document changes in the status and trends of biodiversity globally through the Intergovernmental Science-Policy Platform on Biodiversity and Ecosystem Services (Díaz et al. 2015) have raised awareness about the critical need for continuous monitoring of biodiversity at multiple spatial scales across the globe. Biodiversity itself—the variation in life found among ecosystems and organisms at any level of biological organization—cannot practically be observed everywhere. However, if habitats, functional traits, trait diversity, and the spatial turnover of plant functions can be remotely sensed, the potential exists to globally inventory the diversity of habitats and traits associated with terrestrial biodiversity. To face this challenge, there have been recent calls for

J. Cavender-Bares (✉)
Department of Ecology, Evolution & Behavior, University of Minnesota,
Saint Paul, MN, USA
e-mail: cavender@umn.edu

J. A. Gamon
Department of Earth & Atmospheric Sciences, University of Alberta, Edmonton, AB, Canada

Department of Biological Sciences, University of Alberta, Edmonton, AB, Canada

CALMIT, School of Natural Resources, University of Nebraska – Lincoln, Lincoln, NE, USA

P. A. Townsend
Department of Forest and Wildlife Ecology, University of Wisconsin, Madison, WI, USA

a global biodiversity monitoring system (Jetz et al. 2016; Proença et al. 2017; The National Academy of Sciences 2017). A central theme of this volume is that remote sensing (RS) will play a key role in such a system.

1.2 Why a Focus on Plant Diversity?

Plants and other photosynthetic organisms form the basis of almost all primary productivity on Earth, and their diversity and function underpin virtually all other life on this planet. Plants—collectively called vegetation—regulate the flow of critical biogeochemical cycles, including those for water, carbon, and nitrogen. They affect soil chemistry and other properties, which in turn affect the productivity and structure of ecosystems. Given the importance of plant diversity for providing the ecosystem services on which humans depend—including food production and the regulating services that maintain clean air and freshwater supply (Millennium Ecosystem Assessment 2005; IPBES 2018a)—it is critical that we monitor and understand plant biodiversity from local to global scales, encompassing genetic variation within and among species to the entire plant tree of life (Cavender-Bares et al. 2017; Jetz et al. 2016; Turner 2014).

Of the 340,000 known seed plants on Earth and the 60,000 known tree species (Beech et al. 2017), 1 out of every 5 seed plants and 1 out of every 6 tree species are threatened (Kew Royal Botanic Gardens 2016). Vulnerability to threats ranging from climate change to disease varies among species and lineages because of evolved differences in physiology and spatial proximity to threats. Across all continents, the largest threat to terrestrial biodiversity is land use change due to farming and forestry, while climate change, fragmentation, and disease loom as ever-increasing threats (IPBES 2018a; b; c; d). Many plant species are at risk for extinction due to a combination of global change factors, including drought stress, exotic species, pathogens, land use change, altered disturbance regimes, application of chemicals, and overexploitation.

1.3 The Promise of Remote Sensing to Detect Plant Diversity

Different plants have evolved to synthesize different mixes of chemical compounds arranged in contrasting anatomical forms to support survival and growth. In addition, the structures of plant canopies correspond to different growth strategies in response to climate, environment, or disturbance. Differences among individual plants, populations, and lineages result from contrasting evolutionary histories, genetic backgrounds, and environmental conditions. Because these differences are readily expressed in aboveground physiology, biochemistry, and structure, many of these properties can be detected using spectral reflectance from leaves and plant canopies (Fig. 1.1). Plant pigments absorb strongly electromagnetic radiation in the

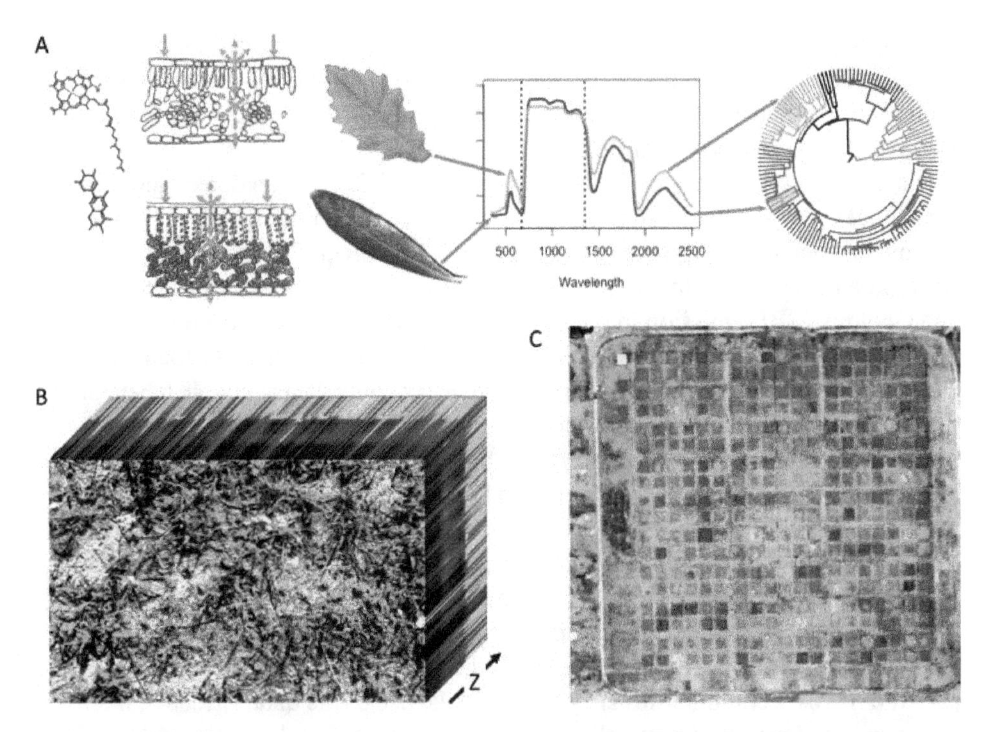

Fig. 1.1 (**a**) The chemical, structural, and anatomical attributes of plants influence the way they interact with electromagnetic energy to generate spectral reflectance profiles that reveal information about plant function and are tightly coupled to their evolutionary origins in the tree of life. (Adapted from Cavender-Bares et al. 2017.) Imaging spectroscopy offers the potential to remotely detect patterns in diversity and chemical composition and vegetation structure that inform our understanding of ecological processes and ecosystem functions. Examples are shown from the Cedar Creek Ecosystem Science Reserve long-term biodiversity experiment. (**b**) The image cube (0.5 m × 1 m) at 1 mm spatial resolution (400–1000 nm) detects sparse vegetation early in the season in which individual plants can be identified. The "Z-dimension" (spectral dimension) illustrates different spectral reflectance properties for different scene elements, including different species. At this spatial resolution, plant diversity is predicted from remotely sensed spectral diversity (Wang et al. 2018). (**c**) AVIRIS NextGen false color image of the full experiment at 1 m spatial resolution (400–2500 nm). Each square is a 9 × 9 m plot with a different plant composition and species richness. Wang et al. (2019) mapped chemical composition and a suite of other functional traits and their uncertainties in all of the experimental plots. By combining spectral data at different scales, proximal and remote imagery can be used to examine the scale dependence of the spectral diversity–biodiversity relationship in detail (e.g., Wang et al. 2018; Gamon et al. Chap. 16)

visible wavelengths (400–700 nm), while other chemical compounds and structural attributes of plants that tend to be conserved through evolutionary history affect longer wavelengths. The patterns of light absorbed, transmitted, and reflected at different wavelengths from vegetation reveal leaf and canopy surface properties, tissue chemistry, and anatomical structures and morphological attributes of leaves, whole plants, and canopies. Thus, technological advances for assessing optical properties of plants provide profound opportunities for detecting functional traits of organisms at different levels of biological organization. These advances are occurring at multiple spatial scales, with technologies ranging from field spectrometers and airborne systems to emerging satellite systems. As a consequence, there is high potential to detect and monitor plant diversity—and other forms of diversity—across a range of spatial scales, and to do so iteratively and continuously, particularly if multiple methods can be properly coordinated.

Calls for a global biodiversity observatory (Fig. 1.2) that can detect and monitor functional plant diversity from space (Jetz et al. 2016; Proença et al. 2017; The National Academy of Sciences 2017; Geller et al., Chap. 20) have been met with widespread support. Forthcoming satellite missions, including the Surface Biology and Geology (SBG) mission in planning stages at the US National Aeronautics and Space Administration (NASA) and related missions in Europe and Japan (Schimel et al., Chap. 19), will make unprecedented spectroscopic data available to scientists, management communities, and decision-makers, but at relatively coarse spatial scales. At the same time, rapid progress is being made with field spectroscopy using unmanned aerial vehicles (UAVs) and other airborne platforms that are offering novel ways to use RS to advance our understanding of the linkages between optical (e.g., spectral or structural) diversity and multiple dimensions of biodiversity (e.g., species, functional, and phylogenetic diversity) at finer spatial scales (Fig. 1.1). These advances present a timely and tremendously important opportunity to detect changes in the Earth's biodiversity over large regions of the planet. One can fairly ask whether user communities are ready to make use of the data. Effective interpretation and application of remotely sensed data to determine the status and trends of plant biodiversity and plant functions across the tree of life—with linkages to all other living organisms—requires integration across vastly different knowledge arenas. Critically, it requires integration with in-situ direct and indirect measures of species distributions, their evolutionary relationships, and their functions. Approaches for integration are the primary focus of this book.

A central requirement to advance monitoring of biodiversity at the global scale is to decipher the sources of variation that contribute to spectral variation, both from a biological perspective and from a physical perspective. Distinct fields of biology have developed a range of methodologies for understanding plant ecological and evolutionary processes that underlie these sources of variation. Similarly, radiative transfer models have been developed largely based on the physics of light interacting with vegetation canopy elements and the atmosphere. These models have yet to capture the full range of plant traits, often preferring to represent "average" vegetation conditions for a region instead of the variation present, so are not yet ready for the task. All of these methods have unique approaches to analyzing complex, multidimensional data sets, and neither the analytical approaches nor the data structures have been brought together in a systematic or comprehensive manner. A common language among disciplines (including biology, physiological ecology, landscape ecology, genetics, phylogenetics, geography, spectroscopy, and radiative transfer) related to the RS of biodiversity is currently lacking. This book provides a framework for how biodiversity, focusing particularly on plants, can be detected using proximally and remotely sensed hyperspectral data (with many contiguous spectral bands) and other tools, such as lidar (with its ability to detect structure). The chapters in this book present a range of perspectives and approaches on how RS can be

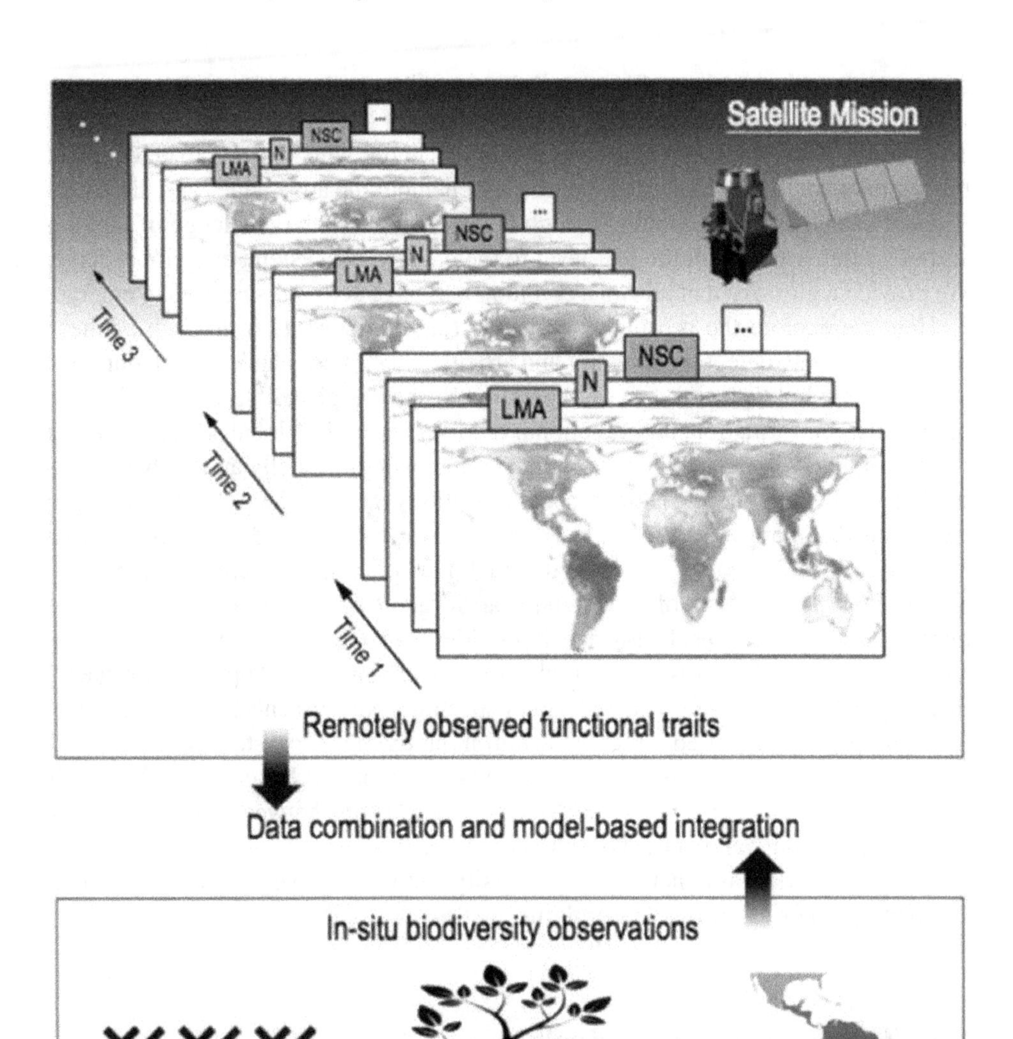

Fig. 1.2 An envisioned global biodiversity observatory in which remotely sensed high spectral resolution spectroscopic data from satellites is integrated with biodiversity observations through natural history studies, phylogenetic systematics, functional trait measurements, and species distribution data. The figure is adapted from Jetz et al. 2016 based on the National Center for Ecological Analysis and Synthesis Working Group "Prospects and priorities for satellite monitoring of global terrestrial biodiversity"

integrated to detect and monitor the status and trends of plant diversity, as well the biodiversity of other organisms and life processes that depend on plants above- and belowground. Biological and computational experts from three disciplines—RS and leaf optics, plant functional biology, and systematics—present insights to advance our understanding of how to link spectral and other kinds of RS data with functional traits, species distributions, and the tree of life for biodiversity detection. The authors detail the approaches and conditions under which efforts to detect plant biodiversity are likely to succeed, being explicit about the advantages and disadvantages of each. A theme running through many chapters is the challenge of moving across spatial scales from the leaf level to the canopy, ecosystem, and global scale. We provide a glossary that allows a common language across disciplines to emerge.

Here we explore the prospects for integrating components from each of these fields to remotely detect biodiversity and articulate the major challenges in our ability to directly link spectral data of vegetation to species diversity, functional traits, phylogenetic information, and functional biodiversity at the global scale. RS offers the potential to fill in data gaps in biodiversity knowledge locally and globally, particularly in remote and difficult-to-access locations, and can help define the larger spatial and temporal background needed for more focused and effective local or regional studies. It also may increase the likelihood of capturing temporal variation, and it allows monitoring of biodiversity at different spatial scales with different platforms and approaches. In essence, it provides the context within which changing global biodiversity patterns can be understood. The concept of "optical surrogacy" (Magurran 2013)—in which the linkage of spectral measurements to associated patterns and processes is used—may be useful in predicting ecosystem processes and characteristics that themselves are not directly observable (Gamon 2008; Madritch et al. 2014; Fig. 1.3). In a broad sense, such relationships between various expressions of biodiversity and optical (spectral) diversity provide a fundamental principle for "why RS works" as a metric of biodiversity and why so many different methods at different scales can provide useful information.

1.4 The Contents of the Book

The first section of the book presents the potential and basis for direct and indirect remote detection of biodiversity.

Cavender-Bares et al. (Chap. 2) present an overview of biodiversity itself, including the in-situ methods and metrics for measuring biodiversity, particularly plant diversity. The chapter provides a layperson's overview of the elements, methods, and metrics for detecting and analyzing biodiversity and points to the potential of spectral data, collected at multiple spatial and biological scales, to enhance the study of biodiversity. In doing so, it bridges an ecological and evolutionary understanding of the diversity of life, considering both its origins and consequences.

Serbin and Townsend (Chap. 3) describe various approaches for measuring plant and ecosystem function using spectroscopy, providing both the historical development of past advances and the potential of these approaches looking forward. The chapter explains why we are able to retrieve functional traits from spectra, which traits can be retrieved, and where spectra show features important for different aspects of plant function. It also raises the challenge of scaling plant function from leaves to canopies and landscapes.

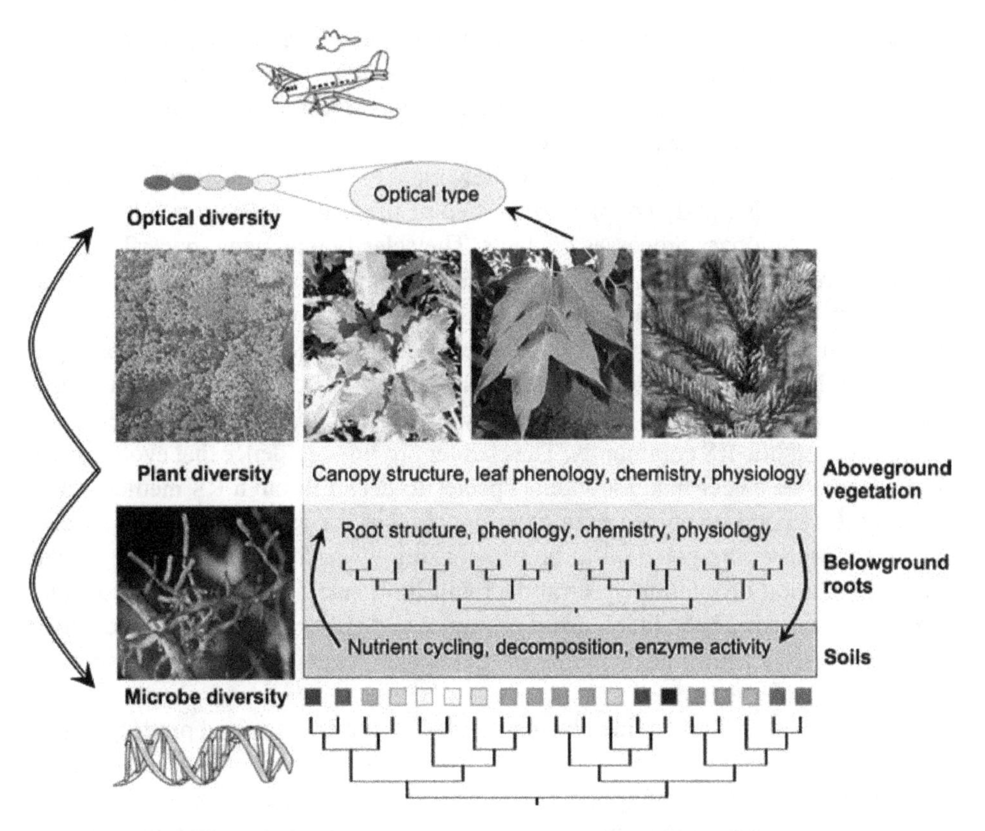

Fig. 1.3 Optical methods for detecting the functional, structural, and chemical components of vegetation, which are tightly coupled to the genetic and phylogenetic backgrounds of plants, are linked to belowground processes and the structure and function of microbial communities

Morsdorf et al. (Chap. 4) then present the Laegeren forest site in Switzerland as a virtual laboratory. They demonstrate how spectroscopy can be operationalized for RS of functional diversity to explain plant biodiversity patterns and ecosystem functions. The Laegeren site is one of the best-studied sites in the world for this purpose and is used as a case study to explain ground truthing and what can be learned from landscape-level detection of functional diversity.

Martin (Chap. 5) summarize the experiences with "spectronomics"—a framework aimed at integrating chemical, phylogenetic, and spectral RS data—using airborne imagery to detect forest composition and function in wet tropical forests. In these vast, largely inaccessible landscapes that harbor enormous taxonomic variation, approaches that rely solely on field-based observations are infeasible, illustrating an essential role for RS. As pioneers in using spectroscopy to detect plant chemistry, function, and biodiversity in tropical forests around the world, these researchers highlight some of the major lessons they have learned.

Pontius et al. (Chap. 6) consider how biodiversity can be protected given current threats to forest and vegetation conditions and present approaches for detailed and accurate detection of forest disturbance and decline. They review current techniques used to assess and monitor forest ecosystem condition and disturbance and outline a general approach for earlier, more detailed and accurate decline assessment. They also discuss the importance of engaging land managers, practitioners, and decision-makers in these efforts to ensure that the products developed can be utilized by stakeholders to maximize their impact.

Meireles et al. (Chap. 7) provide a framework to explain how spectral reflectance data from plants is tightly coupled to the tree of life and demonstrate how spectra can reveal evolutionary processes in plants. They clarify that many spectral features in plants are inherited and are thus very similar among close relatives—in other words, they are highly phylogenetically conserved. Simulations reveal that spectral information of plants appears to follow widely used evolutionary models, making it possible to link plant spectra to the tree of life in a predictive manner. As a consequence, methods developed in evolutionary biology to understand the tree of life can now benefit the RS community. The chapter provides evidence that evolutionary lineages may be easier than individual species to detect through RS methods, particularly if they are combined with other approaches for estimating which species and lineages have the potential to be present in a given location. A caveat is that spatial resolution of satellite spectral data will limit such inferences, but leaf- and canopy-level spectra (obtainable from proximal and airborne sensing) can contribute enormously to our understanding of these fundamental links between spectral patterns and gene sequences.

Madritch et al. (Chap. 8) link aboveground plant biodiversity and productivity to belowground processes. They explain the functional mechanisms—which can be revealed by remotely sensed spectral data—that influence interactions of plant hosts with insects and soil organisms, in turn influencing ecosystem functions, such as decomposition and nutrient cycling. The chapter provides an example of using the concept of surrogacy, in which the biochemical linkage of spectral measurements to associated patterns and processes aboveground is used to provide estimates of soil and microbial processes belowground that are not directly observable via RS.

The next three chapters focus on linking satellite-based remotely sensed data to biodiversity prediction. Pinto-Ledezma and Cavender-Bares (Chap. 9) present an example of how currently available satellite-based RS products can be used to generate next-generation species distribution models to predict where species and lineages are likely occur and the habitats they may have access to and persist in under altered climates in the future. They compare RS-based methods for generating predictive models with widely used approaches that use meteorologically derived climate variables. They demonstrate the advantages of RS-based models in regions where meteorological data is only sparsely available. Such predictive modeling that harnesses species occurrence data and temporal information about the biotic environment may make spectral methods of species and evolutionary lineage detection more tractable.

Building on the availability of satellite RS data with near-global coverage to predict biodiversity, Record et al. (Chap. 10) explore how RS illuminates the relationship between biodiversity and geodiversity—the variety of abiotic features and processes that provide the template for the development of biodiversity. They introduce a variety of globally available geodiversity measures and examine how they can be combined with biodiversity data to understand how biodiversity responds to geodiversity. The authors use the analogy of the "stage" that defines the patterns of life to some degree, often measured as habitat heterogeneity, a key driver of species diversity. They illustrate the approach by examining the relationship between biodiversity and geodiversity with tree biodiversity data from the US Forest Inventory and Analysis Program and geodiversity data from remotely sensed elevation from the Shuttle Radar Topography Mission (SRTM). In doing so, they outline the challenges and opportunities for using RS to link biodiversity to geodiversity.

Paz et al. (Chap. 11) present an approach for using RS data to predict patterns of plant diversity and endemism in the tropics within the Brazilian Atlantic rainforest. They examine how RS environmental data from tropical regions can be used to support biodiversity prediction at multiple spatial, temporal, and taxonomic scales.

Bolch et al. (Chap. 12) summarize the range of approaches that can be used to optimize detection of invasive alien species (IAS), which pose severe threats to biodiversity. These approaches emphasize the ability to detect individual plant species that have distinct functional properties. The chapter presents current RS capabilities to detect and track invasive plant species across terrestrial, riparian, aquatic, and human-modified ecosystems. Each of these systems has a unique set of issues and species assemblages with its own detection requirements. The authors examine how RS data collection in the spectral, spatial, and temporal domains can be optimized for a particular invasive species based on the ecosystem type and image analysis approach. RS approaches are enhancing studies of the invasion processes and enabling managers to monitor invasions and predict the spread of IAS.

The next three chapters of the book explore how components of diversity can be detected spectrally and remotely with a focus on optical detection methods and technical challenges. Lausch et al. (Chap. 13) delve into the complexity of monitoring vegetation diversity and explain how no single monitoring approach is sufficient on its own. The chapter introduces the range of Earth observation (EO) techniques available for assessing vegetation diversity, covering close-range EO platforms, spectral approaches, plant phenomics facilities, ecotrons, wireless sensor networks (WSNs), towers, air- and spaceborne EO platforms, UAVs, and approaches that integrate air- and spaceborne EO data. The chapter presents the challenges with these approaches and concludes with recommendations and future directions for monitoring vegetation diversity using RS.

Ustin and Jacquemoud (Chap. 14) provide the physical basis for detecting the optical properties of leaves based on how they modify the absorption and scattering of energy to reveal variation in function. The chapter provides considerable detail on how the combination of absorption and scattering properties of leaves together creates the shape of their reflectance spectrum. It also reviews and summarizes the most common interactions between leaf properties and light and the physical processes that regulate the outcomes of these interactions.

Schweiger (Chap. 15) describes a set of best practices for planning field campaigns and collecting and processing data, focusing on spectral data of terrestrial plants collected across various levels of measurements, from leaf to canopy to airborne. These approaches also generally apply to RS of aquatic systems, soil, and the atmosphere and to active RS systems, such as lidar, thermal, and satellite data collection. Schweiger discusses how goals for data collection can be broadly classified into model calibration, model validation, and model interpretation.

The final chapters of the book move to the issues of temporal and spectral scale and integration across scales. Gamon et al. (Chap. 16) present a thorough examination of the challenges in spectral methods for detecting biodiversity posed by issues of spatial, temporal, and spectral dimensions of scale. They explain why the size of the organism relative to the pixel size of detection has consequences for spectral detection of different components of biodiversity and draw on a rich history of literature on scaling effects, including geostatistical approaches for sampling across spatial scales. The chapter emphasizes the importance of developing biodiversity monitoring systems that are "scale-aware" as well as the value of an integrated, multi-scale sampling approach.

Schrodt et al. (Chap. 17) outline how environmental and socioeconomic data can be integrated with biodiversity and RS data to expand knowledge of ecosystem functioning and inform biodiversity conservation decisions. They present the concepts, data, and methods necessary to assess plant species and ecosystem properties across spatial and temporal scales and provide a critical discussion of the major challenges.

Fernández et al. (Chap. 18) provide a framework for understanding Essential Biodiversity Variables (EBVs) to integrate in-situ biodiversity observations and RS through modeling. They argue that open and reproducible workflows for data integration are critical to ensure traceability and reproducibility to allow each EBV to be updated as new data and observation systems become available. The chapter makes the case that the development of a globally coordinated system for biodiversity monitoring will require the mobilization of and integration of in-situ biodiversity data not yet publicly available with emerging RS technologies, novel biodiversity models, and informatics infrastructures.

Schimel et al. (Chap. 19) discuss the prospects and pitfalls for RS of biodiversity at the global scale, focusing on imaging spectroscopy and NASA's Surface Biology and Geology mission concept.

Finally, Geller et al. (Chap. 20) provide an epilogue to the book and present a vision for a global biodiversity monitoring system that is flexible and accessible to a range of user communities. Such a system will require a coordinated effort among space agencies, the RS community, and biologists to bring information about the status and trends in biodiversity, ecosystem functions, and ecosystem services together so that different data streams inform each other and can be integrated. The chapter explains that the Group on Earth Observations Biodiversity Observation Network (GEO BON), the International Long Term Ecological Research Site (ILTER) network, the US National Ecological Observatory Network (NEON), and a variety of sponsors and other organizations are working to enhance coordination and to develop guidelines and standards that will serve this vision.

Indeed, a rapidly advancing global movement has emerged with a shared vision to develop the capacity to monitor the status and trends in the Earth's biodiversity. The authors of this book have sought to contribute to that shared vision through their varied perspectives and experiences. Collectively, the chapters present a range of approaches and knowledge that can transform the ability of humanity to detect and interpret the changing functional biodiversity of planet Earth.

1.5 The Origins of the Book

Before closing, we offer a note of acknowledgment on how this book came into existence. The editors, who themselves have contrasting backgrounds spanning several disciplines, were collaboratively funded, starting in 2013, by the US National Science Foundation (NSF) and the NASA Dimensions of Biodiversity program on the project *Linking remotely sensed optical diversity to genetic, phylogenetic and functional diversity to predict ecosystem processes* (DEB-1342872,1342778). We worked together in several field sites for 5 years to advance our own understanding approaches for remote detection of plant biodiversity. The importance and high potential for rapid advances, as well as the need for the involvement of numerous experts, were obvious from the start. With support from the National Institute for Mathematical and Biological Synthesis (NIMBioS) for the working group on Remote Sensing of Biodiversity, we brought together many of the experts represented within the book, including a symposium in the fall of 2018 where authors shared their work and provided feedback to each other. Further interactions were fostered by the National Center for Ecological Analysis and Synthesis (NCEAS), annual meetings of the NASA Biological Diversity and Ecological Forecasting Program, the Keck Institute for Space Studies, the NSF Research Coordination Network on Biodiversity across scales, and bioDISCOVERY, an international research program fostering collaborative interdisciplinary activities on biodiversity and ecosystem science. bioDISCOVERY, which is part of Future Earth and is hosted and supported by the University Zürich's University Research Priority Program on Global Change and Biodiversity, provided generous financial support for the editing process. Additional financial support was provided by the NSF RCN (DEB: 1745562) and the Keck Institute for Space Studies. Mary Hoff served as technical editor for all of the chapters in the book. We express gratitude to all of these institutions and our many colleagues who contributed to the conception of this work along the way.

References

Beech E, Rivers M, Oldfield S, Smith PP (2017) GlobalTreeSearch: the first complete global database of tree species and country distributions. J Sustain For 36:454–489

Cavender-Bares J, Gamon JA, Hobbie SE, Madritch MD, Meireles JE, Schweiger AK, Townsend PA (2017) Harnessing plant spectra to integrate the biodiversity sciences across biological and spatial scales. Am J Bot 104:1–4. https://doi.org/10.3732/ajb.1700061

Díaz S, Demissew S, Carabias J, Joly C, Lonsdale M, Ash N, Larigauderie A, Adhikari JR, Arico S, Báldi A, Bartuska A, Baste IA, Bilgin A, Brondizio E, Chan KMA, Figueroa VE, Duraiappah A, Fischer M, Hill R, Koetz T, Leadley P, Lyver P, Mace GM, Martin-Lopez B, Okumura M, Pacheco D, Pascual U, Pérez ES, Reyers B (2015) The IPBES conceptual framework—connecting nature and people. Curr Opin Environ Sustain 14:1–16

Gamon JA (2008) Tropical remote sensing – opportunities and challenges. In: Kalacska M, Sanchez-Azofeifa GA (eds) Hyperspectral remote sensing of tropical and subtropical forests. CRC Press. Taylor and Francis Group, Boca Raton, pp 297–304

IPBES (2018a) Summary for policymakers of the regional and subregional assessment of biodiversity and ecosystem services for Europe and Central Asia of the Intergovernmental Science-Policy Platform on Biodiversity and Ecosystem Services. IPBES Secretariat, Bonn

IPBES (2018b) Summary for policymakers of the regional assessment report on biodiversity and ecosystem services for Africa of the Intergovernmental Science-Policy Platform on Biodiversity and Ecosystem Services. IPBES Secretariat, Bonn

IPBES (2018c) Summary for policymakers of the regional assessment report on biodiversity and ecosystem services for Asia and the Pacific of the Intergovernmental Science-Policy Platform on Biodiversity and Ecosystem Services. IPBES Secretariat, Bonn

IPBES (2018d) Summary for policymakers of the regional assessment report on biodiversity and ecosystem services for the Americas of the Intergovernmental Science-Policy Platform on Biodiversity and Ecosystem Services. IPBES Secretariat, Bonn

Jetz W, Cavender-Bares J, Pavlick R, Schimel D, Davis FW, Asner GP, Guralnick R, Kattge J, Latimer AM, Moorcroft P, Schaepman ME, Schildhauer MP, Schneider FD, Schrodt F, Stahl U, Ustin SL (2016) Monitoring plant functional diversity from space. Nature Plants 2:16024

Kew Royal Botanic Gardens (2016) State of the World's plants. Board of Trustees of the Royal Botanic Gardens, Kew, London, UK

Kolbert E (2014) The sixth extinction: an unnatural history. Henry Holt & Company, New York

Leakey R, Roger L (1996) The sixth extinction: patterns of life and the future of humankind. Weidenfeld and Nicolson, London

Madritch MD, Kingdon CC, Singh A, Mock KE, Lindroth RL, Townsend PA (2014) Imaging spectroscopy links aspen genotype with below-ground processes at landscape scales. Philos Trans R Soc Lond B Biol Sci 369:20130194

Magurran AE (2013) Measuring biological diversity. Wiley-Blackwell, Malden

Millennium Ecosystem Assessment (2005) Ecosystems and human well-being: current state & trends assessment. Island Press, Washington, D.C.

Proença V, Martin LJ, Pereira HM, Fernandez M, McRae L, Belnap J, Böhm M, Brummitt N, García-Moreno J, Gregory RD, Honrado JP, Jürgens N, Opige M, Schmeller DS, Tiago P, van Swaay CAM (2017) Global biodiversity monitoring: from data sources to essential biodiversity variables. Biol Conserv 213:256–263

Turner W (2014) Sensing biodiversity. Science 346:301

The National Academy of Sciences (2017) Thriving on our changing planet: a decadal strategy for earth observation from space. The National Academies Press, Washington, D.C.

United Nations (2011) Resolution 65/161. Convention on biological diversity

Wang R, Gamon JA, Cavender-Bares J, Townsend PA, Zygielbaum AI (2018) The spatial sensitivity of the spectral diversity–biodiversity relationship: an experimental test in a prairie grassland. Ecol Appl 28:541–556

Wang Z, Townsend PA, Schweiger AK, Couture JJ, Singh A, Hobbie SE, Cavender-Bares J (2019) Mapping foliar functional traits and their uncertainties across three years in a grassland experiment. Remote Sensing of Environment 221:405–416.

2

Detection and Analysis of Biodiversity

Jeannine Cavender-Bares, Anna K. Schweiger, Jesús N. Pinto-Ledezma, and Jose Eduardo Meireles

A treatment of the topic of biodiversity requires consideration of what biodiversity is, how it arises, what drives its current patterns at multiple scales, how it can be measured, and its consequences for ecosystems. Biodiversity science, by virtue of its nature and its importance for humanity, intersects evolution, ecology, conservation biology, economics, and sustainability science. These realms then provide a basis for discussion of how remote detection of biodiversity can advance our understanding of the many ways in which biodiversity is studied and impacts humanity. We start with a discussion of how biodiversity has been defined and the ways it has been quantified. We briefly discuss the nature and patterns of biodiversity and some of the metrics for describing biodiversity, including remotely sensed spectral diversity. We discuss how the historical environmental context at the time lineages evolved has left "evolutionary legacy effects" that link Earth history to the current functions of plants. We end by considering how remote sensing (RS) can inform our understanding of the relationships among ecosystem services and the trade-offs that are often found between biodiversity and provisioning ecosystem services.

J. Cavender-Bares (✉) · J. N. Pinto-Ledezma
Department of Ecology, Evolution and Behavior, University of Minnesota, Saint Paul, MN, USA
e-mail: cavender@umn.edu

A. K. Schweiger
Department of Ecology, Evolution and Behavior, University of Minnesota, Saint Paul, MN, USA

Institut de recherche en biologie végétale, Université de Montréal, Montréal, QC, Canada

J. E. Meireles
Department of Ecology, Evolution and Behavior, University of Minnesota, Saint Paul, MN, USA

School of Biology and Ecology, University of Maine, Orono, ME, USA

2.1 What Is Biodiversity?

Biodiversity encompasses the totality of variation in life on Earth, including its ecosystems, the species generated through evolutionary history across the tree of life, the genetic variation within them, and the vast variety of functions that each organism, species, and ecosystem possess to access and create resources for life to persist. Changes in the Earth's condition, including the actions of humanity, have consequences for the expression of biodiversity and how it is changing through time.

2.2 The Hierarchical Nature of Biodiversity

Since Darwin (1859), we have understood that biodiversity is generated by a process of descent with modification from common ancestors. As a result, biological diversity is organized in a nested hierarchy that recounts the branching history of species (Fig. 2.1a). Individual organisms are nested within populations, which are nested within species and within increasingly deeper clades. This hierarchy ultimately represents the degree to which species are related to each other and often conveys when in time lineages split (Fig. 2.1a).

Evolution results in the accumulation of changes in traits that causes lineages to differ. The degree of trait divergence between taxa is expected to be proportional to the amount of time they have diverged from a common ancestor. As a consequence, distantly related taxa are expected to be phenotypically more dissimilar (Fig. 2.1b).

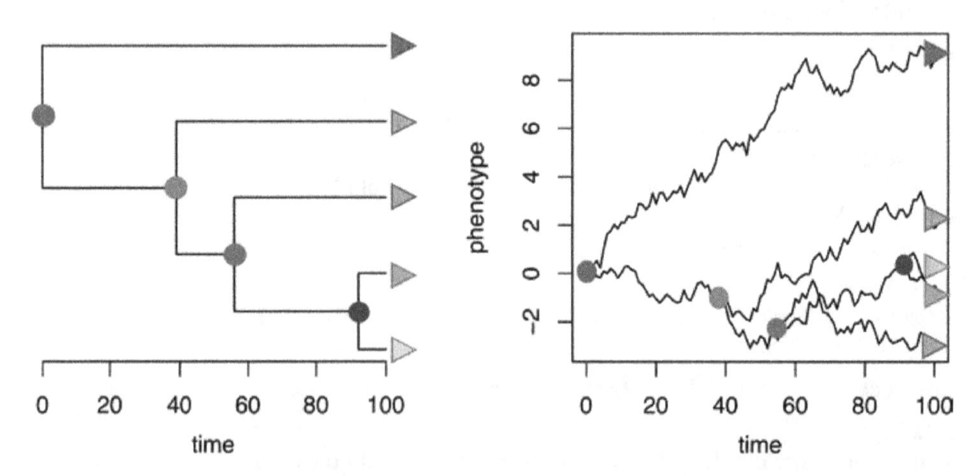

Fig. 2.1 (**a**) The hierarchical organization of biodiversity. Species (triangles) are nested within phylogenetic lineages (clades) due to shared ancestry. All species within a lineage have common ancestor (filled circles). (**b**) Differences in the phenotypes (or trait values) of species (triangles) tend to increase with time since divergence from a common ancestor, shown by the orange circle. The divergence points at which species split are shown in the simulation, and the filled circles indicate the common ancestor of each lineage

Since spectral signals are integrated measures of phenotype, spectra should be more dissimilar among distantly related groups than among close relatives (Cavender-Bares et al. 2016b; McManus et al. 2016; Schweiger et al. 2018). This expectation can be seen from models of evolution in which traits change over time following a random walk (Brownian motion process; Fig. 2.1b) (O'Meara et al. 2006; Meireles et al., Chap. 7). In cases of convergent evolution—where natural selection causes distant relatives to evolve similar functions in similar environments—however, phenotypes can be more similar than expected under Brownian motion.

This hierarchy of life is relevant to RS of plant diversity because certain depths of the tree of life may be more accurately detected than others at different spatial resolutions and geographic regions. For example, it could be easier to detect deeper levels in the hierarchy (such as genera or families) in hyper-diverse communities than in shallow levels (such as species) because deep splits tend to have greater trait divergence. Meireles et al. (Chap. 7) further explain why and how phylogenetic information can be leveraged to detect plant diversity.

2.3 The Making of a Phenotype: Phylogeny, Genes, and the Environment

The phenotype of an organism is the totality of its attributes, and it is quantified in terms of its myriad functions and traits. The phenotype of an organism is a product of the interaction between the information encoded in its genes—the genotype—and the environment over the course of development. Understanding the relative influence of gene combinations, environmental conditions, and ontogenetic stage is an active area of investigation across different disciplines (Diggle 1994; Sultan 2000; Des Marais et al. 2013; Palacio-López et al. 2015).

Although genotypes often play a critical role in determining phenotypic outcomes, many processes can result in mismatches between genotype and phenotype. One of the most well documented of these processes is known as phenotypic plasticity—when organisms with the same genotype display different phenotypes, usually in response to different environmental conditions (Bradshaw 1965; Scheiner 1993; Des Marais et al. 2013). Plasticity can also result in distinct genotypes developing similar phenotypes when growing under the same environmental conditions.

A similar story can be told about the relationship of phenotypic similarity and phylogenetic relatedness. As we have seen earlier, closely related taxa are expected to be more similar to each other than distantly related taxa. However, convergent evolution can lead to plants from different branches of the tree of life to evolve very similar traits—such as succulents, which are found within both euphorbia and very distantly related cacti taxa.

The fact that phenotypes can be, but not necessarily are, directly related to specific genotypes and phylogenetic history should be considered when remotely sensing biodiversity. Only phenotypes can be remotely sensed directly. Genetic and phylogenetic information can only be inferred from spectra to the degree that absorption features of plant chemical or structural characteristics at specific wavelengths relate to phenotypic information. However, the effects plant traits have on spectra are only partially understood. Identifying the regions of the spectrum that are influenced by specific traits is complicated by overlapping absorption features and subtle differences in plant chemical, structural, morphological, and anatomical characteristics that simultaneously influence the shape of the spectral response (Ustin and Jacquemoud, Chap. 14).

2.4 Patterns in Plant Diversity

One of the most intensively studied patterns in biodiversity is the latitudinal gradient, in which low-latitude tropical regions harbor more species, genera, and families of organisms than high-latitude regions. In particular, wet tropical areas tend to reveal higher diversity of organisms than colder and drier climates (Fig. 2.2). Humboldt (1817) documented these patterns quite clearly for plant diversity. Naturalists since then have sought to explain these patterns.

Tropical biomes have existed longer than more recent biomes, such as deserts, Mediterranean climates, and tundra, which expanded as the climate began to cool some 35 million years ago. Tropical biomes also cover more land surface area than other biomes. Tropical species thus have had more time and area (integrated over the time since their first appearance) for species to evolve and maintain viable populations (Fine and Ree 2006). Lineages that originally evolved in the tropics may also have been less able to disperse out of the tropics and to evolve new attributes adapted to cold or dry climates—due to phylogenetic conservatism—restricting their ability to diversify (Wiens and Donoghue 2004). However, not all lineages follow this latitudinal gradient. Ectomycorrhizal fungi, for example, show higher diversity at temperate latitudes, where they likely have higher tree host density (Tedersoo and Nara 2010). Moreover, other measures of diversity do not necessarily follow these patterns. Variation in functional attributes of species, for example, follow different patterns depending on the trait (Cavender-Bares et al. 2018; Echeverría-Londoño et al. 2018; Pinto-Ledezma et al. 2018b). Specific leaf area, one of the functional traits that is highly aligned with the leaf economic spectrum (discussed below), shows higher variation at high latitudes than low latitudes across the Americas. In contrast, seed size shows higher variation at low latitudes (Fig. 2.2b).

At regional scales, variation in the environment, as discussed by Record et al. (Chap. 10), sets the stage for variation in biodiversity because species have evolved to inhabit and can adapt to different environments, which allows them to partition resources and occupy different niches created by environmental variation. Thus, habitat diversity begets biodiversity, and remotely sensed measures of environmental variation have long been known to predict biodiversity patterns (Kerr et al. 2001).

Land area is another long-observed predictor of species diversity, first described for species within certain guilds on island archipelagoes (Diamond and Mayr 1976).

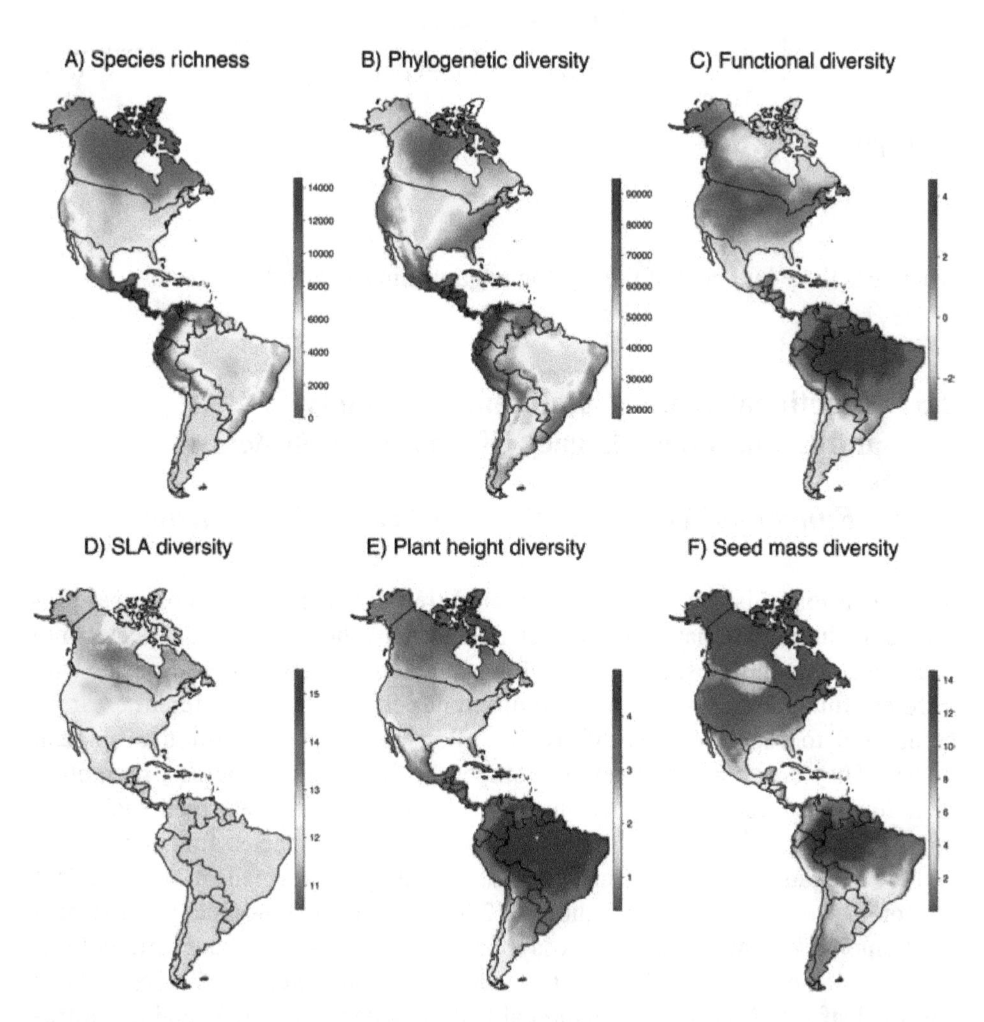

Fig. 2.2 Map of the Americas showing plant species richness, phylogenetic diversity, and functional diversity. Species richness and phylogenetic and functional diversities were estimated based on available information from the Botanical Information and Ecology Network (BIEN) database (Enquist et al. 2016, https://peerj.com/preprints/2615/). Distribution of functional diversity (trait mean) for three functional traits (**d–f**) was log-transformed for plotting purposes. (**a**) Species richness; (**b**) phylogenetic diversity; (**c**) first principal component of functional trait means; (**d**) specific leaf area (mm^2/mg); (**e**) plant height (m); and (**f**) seed mass (mg). Diversity metrics were calculated from an estimated presence-absence matrix (PAM) for all vascular plant species at 1 degree spatial resolution (PAM dimension = 5353 pixels × 98,291 species) using range maps and predicted distributions. Functional diversity is based on the first principal component of a principal component analysis (PCA) of species means for the three functional traits

These observations led to the generalization that richness (number of species, S) increases with available land area (A), giving rise to the well-known species-area relationships, in which the log of species number increases linearly with the log of the area available:

$$\log(S) = z \times \log(A) + c, \quad (2.1)$$

or simply,

$$S = cA^z, \quad (2.2)$$

where c is the y-intercept of the log-log relationship and z is the slope.

2.5 Functional Traits, Community Assembly, and Evolutionary Legacy Effects on Ecosystems

2.5.1 Functional Traits and the Leaf Economic Spectrum

There is a long history of using functional traits to understand ecological processes, including the nature of species interactions, the assembly of species into ecological communities, and the resulting functions of ecosystems. Species with different functions are likely to have different performance in different environments and to use resources differently, allowing them to partition ecological niches. They are thus less likely to compete for the same resources, promoting their long-term coexistence. An increased focus on trait-based methodological approaches to understanding the relationship between species functional traits and the habitats or ecological niches was spurred by the formalization of the leaf economic spectrum (LES) (Wright et al. 2004). The LES shows that relationships exist among several key traits across a broad range of species and different climates (Reich et al. 1997; Wright et al. 2004) and that simple predictors, such as specific leaf area (SLA, or its reciprocal leaf mass per area, LMA) and leaf nitrogen content, represent a major axis of life history variation. This axis ranges from slow-growing ("conservative") species that tolerate low-resource environments to fast-growing ("acquisitive") species that perform well in high-resource environments (Reich 2014). Variations in relatively easy-to-measure plant traits are tightly coupled to hard-to-measure functions, such as leaf lifespan and growth rate, which reveal more about how a plant invests and allocates resources over time to survive in different kinds of environments. High correlations of functional traits provide strong evidence for trait coordination across the tree of life. The variation in plant function across all of its diversity is relatively constrained and can be explained by a few major axes of trait information (Díaz et al. 2015). Conveniently, traits such as SLA and N are readily detectable via spectroscopy. Other traits—such as leaf lifespan or photosynthetic rates—that are harder to measure but are correlated with these readily detectable traits can thus be inferred, permitting greater insight into ecological processes.

2.5.2 *Plant Traits, Community Assembly, and Ecosystem Function*

Considerable evidence supports the perspective that plant traits influence how species sort along environmental gradients and are linked to abiotic environmental filters that prevent species without the appropriate traits from persisting in a given location. Traits thus influence the assembly of species in communities—and consequently, the composition, structure, and function of ecosystems. Variation in traits among individual plants and species within communities indicates differences in resource use strategies of plants, which have consequences for ecosystem functions, such as productivity and resistance to disturbance, disease, and extreme environmental conditions. Moreover, the distribution of plant traits within communities influences resource availability for other trophic levels, above- and belowground, which affects community structure and population dynamics in other trophic levels. A major goal of functional ecology is to develop predictive rules for the assembly of communities based on an understanding of which traits or trait combinations (e.g., the leaf-height-seed (LHS) plant ecology strategy, sensu Westoby 1998) are important in a given environment, how traits are distributed within and among species, and how those traits relate to mechanisms driving community dynamics and ecosystem function (Shipley et al. 2017). This predictive framework requires selecting relevant traits; describing trait variation and incorporating this variation into models; and scaling trait data to community- and ecosystem-level processes (Funk et al. 2017). Selecting functional traits for ecological studies is not trivial. Depending on the question, individual traits or trait combinations can be selected that contribute to a mechanistic understanding of the critical processes examined. One can distinguish *response traits*, which influence a species response to its environment, and *effect traits*, which influence ecosystem function (Lavorel and Garnier 2002). These may or may not be different traits. Disturbance or global change factors that influence whether a species can persist within a habitat or community based on its response traits may impact ecosystem functions in complex ways (Díaz et al. 2013). Plant traits are at the heart of understanding how the evolutionary past influences ongoing community assembly processes and ecosystem function (Fig. 2.3). Traits also influence species interactions, which contribute to continuing evolution. Remotely sensed plant traits, if detected and mapped (Serbin and Townsend, Chap. 3; Morsdorf et al., Chap. 4) at the appropriate pixel size and spatial extent (Gamon et al., Chap. 16), can provide a great deal of insight into these different processes (Fig. 2.4).

2.5.3 *Phylogenetic, Functional, and Spectral Dispersion in Communities*

The rise of phylogenetics in community ecology was based on the idea that functional similarity due to shared ancestry should be predictive of environmental sorting and limiting similarity. These processes depend on physiological tolerances

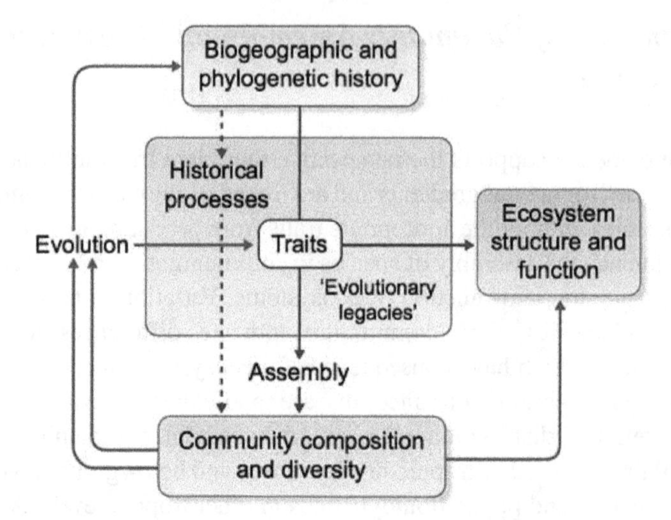

Fig. 2.3 Plant traits that have evolved over time influence how plants assemble into communities, which shapes ecosystem structure and function. Traits reflect biogeographic and environmental legacies and evolve in response to changing environments. They play a central role in ecological processes influencing the distribution of organisms and community assembly. A range of traits influence the way plants reflect light, such that many traits can be mapped continuously across large spatial extents with imaging spectroscopy. The remote detection of plant traits provides incredible potential to observe and understand patterns that reveal information about community assembly, changes in ecosystem function, and how legacies from the past shape community structure and ecosystem processes today. (Reprinted from Cavender-Bares et al. 2019, with permission)

in relation to environmental gradients and intensity of competition as a consequence of shared resource requirements (Webb 2000a, 2002). The underlying conceptual framework was formalized in terms of functional traits in individual case studies (Cavender-Bares et al. 2004; Verdu and Pausas 2007). The tendency to oversimplify the interpretation of phylogenetic patterns in communities, whereby phylogenetic overdispersion was equated with the outcome of competitive exclusion and phylogenetic clustering was interpreted as evidence for environmental sorting, led to a series of studies investigating the importance of scale (Cavender-Bares et al. 2006; Swenson et al. 2006) and the role of Janzen-Connell-type mechanisms, i.e., density-dependent mortality due to pathogens and predators (Gilbert and Webb 2007; Parker et al. 2015). Further developments revealed that the relationship between patterns and ecological processes is context-dependent—in particular, with respect to spatial scale (Emerson and Gillespie 2008; Cavender-Bares et al. 2009; Gerhold et al. 2015). Later studies revisited assumptions about the nature of competition and expected evolutionary and ecological outcomes (Mayfield and Levine 2010). Likewise, interpreting spectral dispersion will depend on the spatial resolution and pixel (grain) size of remotely sensed imagery relative to plant size (Marconi et al. 2019) as well as on the consideration of specific spectral regions and their functional importance. When traits and spectral regions are highly phylogenetically conserved (see Meireles et al., Chap. 7), trait, phylogenetic, and spectral data provide equivalent information. However, when some traits and spectral regions are

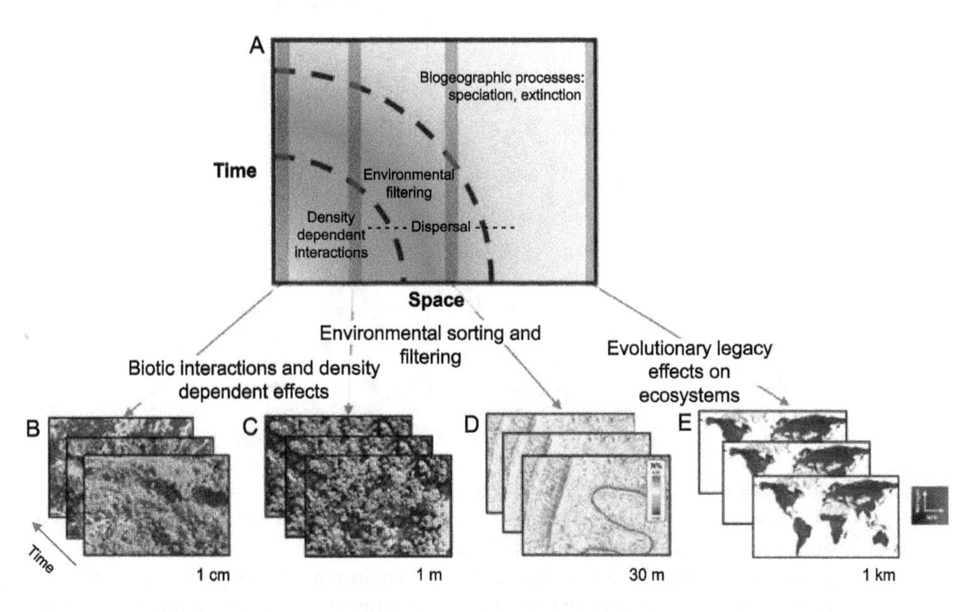

Fig. 2.4 (**a**) Biological processes change with spatial and temporal scale as do the patterns they give rise to. (Adapted from Cavender-Bares et al. 2009.) Detection and interpretation of those patterns will shift with spatial resolution (pixel size) and extent (**b–e**). (**a**) At high spatial resolutions (1 cm pixel size)—that allow detection of individual herbaceous plants and their interactions—and relatively restricted spatial extents in which the abiotic environment is fairly homogeneous, spectral dissimilarity among pixels may indicate complementarity of contrasting functional types. (**b**) The grain size sufficient to detect species interactions is likely to shift with plant size. For example, the interactions of trees in the Minnesota oak savanna and their vulnerability to density-dependent diseases, such as oak wilt (*Bretziella fagacearum*), can be studied at a 1 m pixel size. (**c**) At somewhat larger spatial resolution (30 m pixel sizes) and extent, environmental sorting—which includes interactions of species with both the biotic and abiotic environments—may be detected by comparing spectral similarity of neighbors and comparing mapped functional traits to environmental variation. Images adapted from Singh et al. (2015). The ability to detect change through time may be especially important in understanding species interactions and ecological sorting processes in relation to the biotic and abiotic environment. (**d**) At the global scale, it may be possible to detect the evolutionary legacy effects. For instance, regions with similar climate and geology can differ in vegetation composition and ecosystem function as a consequence of differences in which lineages evolved in a given biogeographic region and their historical migration patterns. Shown are mapped values of %N and NPP based on Moderate Resolution Imaging Spectroradiometer (MODIS) data. (Adapted from Cavender-Bares et al. 2016a)

conserved, but others related to species interactions or with the abiotic environment vary considerably among close relatives, there is the potential to tease apart spectral signals that may relate to species interactions.

Spatial patterns of spectral similarity and dissimilarity also have the potential to provide meaningful information about ecological processes and the forces that dominate community assembly at a particular scale. For example, to the extent that spectral similarity of neighboring plants can be determined, high spectral similarity might indicate that functionally and/or phylogenetically similar individuals are sorting into the same environment, while spectral dissimilarity might indicate that quite distinct individuals are able to coexist if they exhibit complementarity by

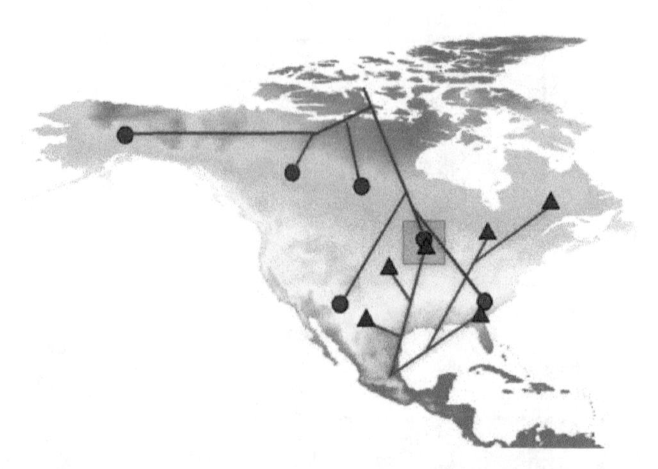

Fig. 2.5 Evolutionary legacy effects as a consequence of biogeographic origin. Two lineages are shown that have contrasting origins, one from the tropics and one from high latitudes. Both diversified and expanded to colonize intermediate latitudes such that their descendants sometimes co-occur. Lineages with ancestors from contrasting climatic environments likely differ in functional traits that reflect their origins and thus may assemble in contrasting microenvironments within the local communities where they co-occur. (Adapted from Cavender-Bares et al. 2016a)

partitioning resources. Approaches that use spectral detection of patterns that might be interpreted within this framework will need to pay close attention to the pixel-to-plant size ratio—or the grain size at which biological diversity varies (Gamon et al., Chap. 16; Serbin and Townsend, Chap. 3; Schimel et al., Chap. 19)—as well as to the spatial extent at which density-dependent processes and environmental sorting pressures are strongest. Often these processes are expected to dominate at different spatial scales, such that competition and Janzen-Connell-type mechanisms operate at very local scales, while environmental sorting may be more important at landscape scales. Other factors, such as the geographic locations and environmental conditions under which lineages diversified, may impact spectral patterns of phylogenetic, functional trait, and spectral similarity at continental scales (Fig. 2.5). At the same time, spectral similarity will be driven by similar ecological forces, since both genetic and phylogenetic compositions, as well as environmental factors, drive phenotypic variation that can be spectrally detected.

2.6 Evolutionary Legacy Effects on Ecosystems

Ecological communities are formed by resident species (incumbents) and colonizer species. Incumbents may have originated in the study region (or at least have had considerable time to adapt to their biotic and abiotic environment), whereas colonizers evolved elsewhere and subsequently dispersed into the region. However, the processes that determine species distributions and the assembly of ecological communities are complex. Species within communities experience unique combinations of evolutionary constraints and innovations due to legacies of their biogeographic origins and the environmental conditions in which they evolved (Cavender-Bares et al. 2016a; Pinto-Ledezma et al. 2018a). Historical contingencies play a role in which lineages can take advantage of opportunities to diversify following climate change or other disturbances and environmental transitions. The rate of species range expansion and contraction and the evolution of species functional traits that allow species to establish and persist in some regions or under particular environmental conditions but not elsewhere are shaped by biogeographic history (Moore et al. 2018). For example, when species from two distinct lineages—one that evolved in tropical climates and the other that evolved in temperate climates—colonize a new environment, they are predicted to persist in contrasting microhabitats as a consequence of niche conservatism (Ackerly 2003; Harrison 2010; Cavender-Bares et al. 2016a). These evolutionary legacies—collectively referred to as "historical factors" (Ricklefs and Schluter 1993)—operate at different spatial and temporal scales that leave their imprints on species current functional attributes and distributions and consequently on ecosystem function itself (Fig. 2.4c, Cavender-Bares et al. 2016a). RS approaches can help reveal how the deep past has influenced current biodiversity patterns and ecosystem function by decoupling climate and geological setting from ecosystem function. Current and forthcoming RS instruments (Lausch et al., Chap. 13; Schimel et al., Chap. 19) enable the monitoring of plant productivity, dynamics of vegetation growth, seasonal changes in chemical composition, and other ecosystem properties independently of climate and geology. These technologies thus provide opportunities to detect how biodiversity is sorted across the globe and to determine how variable ecosystem functions can be in the same geological and environmental setting. Both are important for developing robust predictive models of how lineages respond to current and future environmental conditions with important consequences for managing ecosystems in the Anthropocene.

2.7 Quantifying Multiple Dimensions of Biodiversity

Several major dimensions of biodiversity have emerged in the literature that capture different aspects of the variation of life. *Taxonomic diversity* focuses on differences between species or between higher-order clades, such as genera or families. Estimating the numbers and/or abundances of different taxa across units of area captures this variation. *Phylogenetic diversity* captures the evolutionary distances between species or individuals, represented in terms of millions of years since divergence from a common ancestor or molecular distances based on accumulated mutations since divergence. *Functional diversity* focuses on the variation among species as a consequence of measured differences in their functional traits, frequently calculated as a multivariate metric but also calculated for individual trait variation. *Spectral diversity* captures the variability in spectral reflectance from vegetation (or from other surfaces), either measured and calculated among individual plants or, more commonly, calculated among pixels or among other meaningful spatial units.

Biodiversity metrics can have different components, including (1) taxonomic units; (2) abundance, frequency, or biomass of those units and their degree of evenness; and (3) the dispersion or distances between those units in trait, evolutionary, or spectral space. Myriad metrics quantify the major dimensions and components of diversity. Here we briefly describe several frequently used metrics; the equation for each metric and the source citation that provides the full details are given in Table 2.1.

Table 2.1 A brief summary of metrics that are commonly used to estimate the diversity of different facets/dimensions of plant diversity

Metric	Equation	Definition	Reference
Whittaker alpha	α	Number of species found in a sample or particular area, generally expressed as species richness	Whittaker (1960)
Whittaker beta	γ/α	Variation of species composition between two samples. Can be interpreted as the effective number of distinct compositional units in the region	Whittaker (1960)
Whittaker gamma	γ	Overall diversity (number of species) within a region	Whittaker (1960)
Shannon's H	$-\sum_{i=1}^{S} p_i \ln p_i$	Metric that characterizes species diversity in a sample. Assumes that all species are represented in the sample and that individuals within species were sampled randomly	Shannon (1948)
Simpson's D	$\frac{1}{\sum_{i=1}^{S} P_i^2}$	Metric that characterizes species diversity in a sample. Contrary to Shannon's H, Simpson's D captures the variance of the species abundance distribution	Simpson (1949)
Faith's PD (phylogenetic diversity)	$\sum_{e \in z(T)} \lambda e$	Sum of the lengths of all phylogenetic branches (from the root to the tip) spanned by a set of species	Faith (1992)
PSV (phylogenetic species variability)	$\frac{ntrC - \Sigma C}{n(n-1)} = 1 - \bar{c},$	Measures the variability in an unmeasured neutral trait or the relative amount of unshared edge length	Helmus (2007)
PSR (phylogenetic species richness)	nPSV	The deviation from species richness, penalized by close relatives	Helmus (2007)

(continued)

Table 2.1 (continued)

Metric	Equation	Definition	Reference
PSE (phylogenetic species evenness)	$\frac{mdiag(C)' M - M'CM}{m^2 - \bar{m}_i m}$	PSV metric modified to account for relative species abundance or simply abundance-weighted PSV	Helmus (2007)
${}^{q}PD(T)$ (phylogenetic branch diversity)	$\left[\sum_{i=1}^{B} L_i x \left(\frac{a_i}{\bar{T}}\right)^q\right]^{\frac{1}{1-q}}$	Hill number (the effective total branch length) of the average time of a tree's generalized entropy over evolutionary time intervals	Chao et al. (2010)
${}^{q}D(T)$ (phylogenetic Hill numbers)	$\left(\frac{{}^{q}PD(\bar{T})}{\bar{T}}\right)$	Effective number of species or lineages	Chao et al. (2010)
FRic (functional richness)	Quickhull algorithm	Quantity of functional space filled by the sample. The number of species within the sample must be higher than the number of functional traits	Barber et al. (1996), Villeger et al. (2008)
FEve (functional evenness)	$\frac{\sum_{i=1}^{S-1} \min\left(PEW_i, \frac{1}{S-1}\right) - \frac{1}{S-1}}{1 - \frac{1}{S-1}}$	Quantifies the abundance distribution in functional trait space	Villeger et al. (2008)
FDiv (functional divergence)	$\frac{\Delta d + \overline{dG}}{\Delta \lvert d \rvert + \overline{dG}}$	Metric that measures the spread of species abundance across trait space	Villeger et al. (2008)
${}^{q}D(TM)$ (functional trait dispersion)	$1 + (S - 1) \times {}^{q}E(T) \times M'$	Metric that quantify the effective number of functionally distinct species for a given level of species dispersion	Scheiner et al. (2017)
βsor (Sørensen pairwise dissimilarity)	$\frac{b+c}{2a+b+c}$	Compares the shared species relative to the mean number of species in a sample. Bray-Curtis dissimilarity is a special case of Sørensen dissimilarity that accounts for species abundance	Sørensen (1948), Baselga (2010)
βsim (Simpson pairwise dissimilarity)	$\frac{\min(b,c)}{a+\min(b,c)}$	Similar to Sørensen dissimilarity but independent of species richness	Simpson (1943), Lennon et al. (2001), Baselga (2010)
βjac (Jaccard index)	$\frac{a}{a+b+c}$	Metric that compares the shared species to the total number of species in all samples	Jaccard (1900)

2.7.1 *The Spatial Scale of Diversity: Alpha, Beta, and Gamma Diversity*

Diversity metrics are designed to capture biological variation at different spatial extents. Alpha diversity (α) represents the diversity within local communities, which are usually spatial subunits within a region or landscape. Whittaker first defined beta diversity (β) as the variation in biodiversity among local communities and gamma diversity (γ) as the total biodiversity in a region or a region's species pool (Whittaker 1960).

$$\beta = \frac{\gamma}{\alpha}, \tag{2.3}$$

where β is beta diversity, γ gamma diversity, and α alpha diversity.

Other authors have defined beta diversity differently (see Tuomisto 2010), including using variance partitioning methods (Legendre and De Cáceres 2013).

2.7.2 *Taxonomic Diversity*

Species richness is the number of species for a given area. It does not include abundance of individuals within species. However, the relative abundances, frequency, and biomass of species within a community matter in terms of capture rarity and evenness. Abundance-weighted metric, such as Simpson's diversity index (D), incorporates both richness and evenness. A set of indices based on Hill numbers—a unified standardization method for quantifying and comparing species diversity across samples, originally presented by Mark Hill (1973)—were refined by Chao et al. (2005, 2010). These are generalizable to all of the dimensions of diversity and consider the number of species and their relative abundances within a local community. Hill numbers require the specification of the diversity order (q), which determines the sensitivity of the metric to species relative abundance. Different orders of q result in different diversity measures; for example, $q = 0$ is simply species richness, $q = 1$ gives the exponential of Shannon's entropy index, and $q = 2$ gives the inverse of Simpson's concentration index.

2.7.3 *Phylogenetic Diversity*

Phylogenetic diversity (PD) considers the extent of shared ancestry among species (Felsenstein 1985). For example, a plant community composed of two species that diverged from a common ancestor more recently is less phylogenetically diverse than a community of two species that diverged less recently. Faith's (1992) metric of PD sums the branch lengths among species within a community (from the root of the phylogeny to the tip). One feature of this metric is that it scales with species richness because as new species are added into the community, new branch lengths are also added. Other metrics were subsequently developed that calculate the mean evolutionary distances among species independently of the number of species [e.g., mean phylogenetic distance (MPD, Webb 2000b; Webb et al. 2002) or phylogenetic species variability (PSV), Helmus 2007]. Helmus (2007) developed two more phylogenetic diversity metrics that scale either with richness or by incorporating species abundances. Phylogenetic species richness (PSR) increases with the number of species, but reduces the effect of species richness proportionally to their degree of shared ancestry. Phylogenetic species evenness (PSE) is similar to PSV but includes abundances by adding individuals as additional tips descending from a single species node, with branch lengths of 0. Chao et al. (2010) defined the phylogenetic Hill number, ${}^{q}D(T)$, as the effective number of equally abundant and equally distinct lineages and phylogenetic branch diversity, ${}^{q}PD(T)$, as the effective total lineage length from the root node (i.e., the total evolutionary history of an assemblage) (Chao et al. 2014).

Phylogenetic endemism is another aspect of biodiversity that can be estimated from phylogenetic information and range maps of species (Faith et al. 2004). Phylogenetic endemism can be simply defined as the quantity of PD restricted to a given geographic area. This metric thus focuses on geographic areas, rather than on species, to discern areas of high endemism based on evolutionary history for conservation purposes.

2.7.4 Functional Diversity

Widely used metrics of functional diversity consider the area or volume of trait space occupied by a community of species, the distances of each species to the center of gravity of those traits, and the trait distances between species (Mouillot et al. 2013). Functional attribute diversity (FAD) is a simple multivariate metric calculated as the sum of species pairwise distances of all measured continuous functional traits (Walker et al. 1999). Villeger et al. (2008) developed a series of functional diversity metrics that incorporate trait dispersion and distance among species as well as species abundances, including functional richness (FRic), functional divergence (FDiv), and functional evenness (FEve). Building on the framework of Villeger et al. (2008), Laliberté and Legendre (2010) developed functional dispersion (FDis), a functional diversity metric that is independent of species richness and can include species relative abundances (Table 2.1).

Scheiner's functional trait dispersion [${}^{q}D(TM)$, Scheiner et al. 2017] calculates the effective number of species (or units) that are as distinct as the most distinct species (or unit) in that community. ${}^{q}D(TM)$ decomposes diversity estimates into three components: the number of units (S), functional evenness [${}^{q}E(T)$, the extent to which units are equally dispersed], and mean dispersion [M', the average distance or the distinctiveness of these units]. Functional diversity measured as ${}^{q}D(TM)$ is maximized when there are more units in a community that are more equitably distributed (or less clumped) and more dispersed (or positioned further apart) in space. Like Chao's approach, ${}^{q}D(TM)$ includes Hill numbers (^{q}), which allow weighting of abundances: small and large q values emphasize rare and common species, respectively. Like many other biodiversity metrics, ${}^{q}D(TM)$ can be calculated from pairwise distances among species or individuals; thus, the metric can be applied to estimate different dimensions of biodiversity, including functional, phylogenetic (Scheiner 2012; Presley et al. 2014) and spectral components (Schweiger et al. 2018).

Briefly, functional trait dispersion [${}^{q}D(TM)$] is calculated as:

$$ {}^{q}D(TM) = 1 + (S-1) \times {}^{q}E(T) \times M' \quad (2.4)$$

where:

S = species richness
$E(T)$ = trait evenness
M' = trait dispersion
q = Hill number

2.7.5 *Spectral Diversity*

Like taxonomic, functional, and phylogenetic diversity, spectral diversity can be calculated in many different ways. Spectral alpha diversity metrics include the coefficient of variation of spectral indices (Oindo and Skidmore 2002) or spectral bands among pixels (Hall et al. 2010; Gholizadeh et al. 2018, 2019; Wang et al. 2018, the convex hull volume (Dahlin 2016) and the convex hull area (Gholizadeh et al. 2018) of pixels in spectral feature space, the mean distance of pixels from the spectral centroid (Rocchini et al. 2010), the number of spectrally distinct clusters or "spectral species" in ordination space (Féret and Asner 2014), and spectral variance (Laliberté et al. 2019). Schweiger et al. (2018) applied ${}^{q}D(TM)$ to species mean spectra and to individual pixels extracted at random from high-resolution proximal RS data. The second approach is independent of species identity and uses the same number of pixels per community for analysis. In this manner, the problem of diversity scaling with the number of species in a community is eliminated, and greater differences in reflectance spectra among pixels result in increased spectral diversity. Conceptually, spectral diversity metrics are versatile and can be tailored to match taxonomic or phylogenetic units, e.g., by using mean spectra for focal taxa, or to resemble functional diversity by selecting spectral bands that align with known absorption features for specific chemical traits or spectral indices that capture plant characteristics of known ecological importance. If measured at the appropriate scale (see Gamon et al. Chap. 16), spectral diversity can integrate the variation captured by other metrics of diversity and similarly predicts ecosystem function (Fig. 2.6).

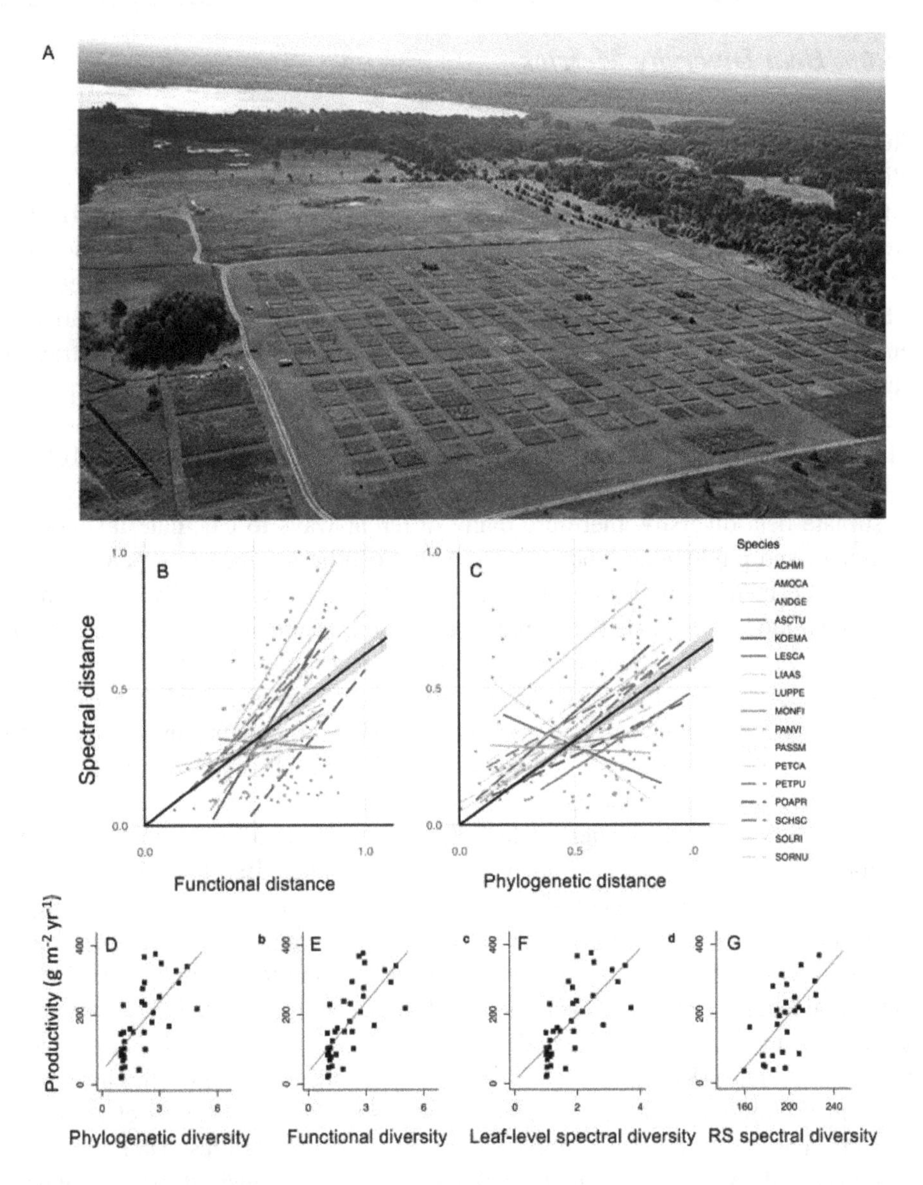

Fig. 2.6 (**a**) Aerial photo of the Cedar Creek long-term biodiversity experiment (BioDIV) (Courtesy of Cedar Creek Ecosystem Science Reserve). (**b**) Pairwise phylogenetic and (**c**) functional distances for the 17 most abundant prairie-grassland species in BioDIV are well-predicted by their spectral distances based on leaf-level spectral profiles (400–2500 nm). (**d**) Phylogenetic, (**e**) functional, and (**f**) leaf-level spectral diversities based on Scheiner's ${}^{q}D(TM)$ metric all predict ecosystem productivity in BioDIV. (**g**) Independent of information about species identities or their abundances, remotely sensed spectral diversity detected at high spatial resolution (1 mm) also predicts productivity. All graphs are redrawn from Schweiger et al. 2018. Species abbreviations in **b** and **c** are as follows: ACHMI = *Achillea millefolium* L., AMOCA = *Amorpha canescens* Pursh, ANDGE = *Andropogon gerardii* Vitman, ASCTU = *Asclepias tuberosa* L., KOEMA = *Koeleria macrantha* (Ledeb.) Schult., LESCA = *Lespedeza capitata* Michx., LIAAS = *Liatris aspera* Michx., LUPPE = *Lupinus perennis* L., MONFI = *Monarda fistulosa* L., PANVI = *Panicum virgatum* L., PASSMI = *Pascopyrum smithii* (Rydb.) Á. Löve, PETCA = *Petalostemum candidum* (Willd.), PETPU = *Petalostemum purpureum* (Vent.) Rydb., POAPR = *Poa pratensis* L., SCHSC = *Schizachyrium scoparium* Michx., SOLRI = *Solidago rigida* L., SORNU = *Sorghastrum nutans* (L.) Nash

2.7.6 *Beta Diversity Metrics*

Whittaker's 1960 definition of beta diversity (Eq. 2.3) quantified the degree of differentiation among communities in relation to environmental gradients. Under this definition, beta diversity is defined as the ratio between regional (gamma) and local (alpha) diversities (Eq. 2.3) and measures the number of different communities in a region and the degree of differentiation between them (Whittaker 1960; Jost 2007). Indices such as Bray-Curtis dissimilarity and Jaccard and Sørensen indices evaluate similarity of communities based on the presence or abundance of species within them. Metrics of similarity used for species have been adapted for phylogenetic and functional trait distances (Bryant et al. 2008; Graham and Fine 2008; Kembel et al. 2010; Cardoso et al. 2014) and can equally be applied to spectral information (Gamon et al., Chap. 16).

While the ratio between regional and local communities provides a simple means to estimate beta diversity, there are many different ways to calculate taxonomic, functional, and phylogenetic beta diversity that can be grouped into pairwise and multiple-site metrics (reviewed in Baselga 2010). Notably, beta diversity can be partitioned into components that capture species replacement—the "turnover component"—caused by the exchange of species among communities and differences in the number of species, the "nestedness component," caused by differences in the number of species among communities. The turnover component can be interpreted as the difference between two community assemblages that contain contrasting subsets of species from a regional source pool, while the nestedness component represents the difference in species composition between two communities due to attrition of species in one assemblage relative to the other (Baselga 2010; Cardoso et al. 2014). Examining these different components of beta diversity for multiple dimensions of plant diversity provides a means to discern the role of historical and ongoing environmental sorting processes in the distribution of plant diversity at continental extents (Pinto-Ledezma et al. 2018b). In contrast to traditional diversity metrics, spectral diversity (alpha and beta) is only beginning to receive attention in biodiversity studies (Rocchini et al. 2018). Although different approaches have been proposed (Schmidtlein et al. 2007; Féret and Asner 2014; Rocchini et al. 2018; Laliberté et al. 2019), the estimation and mapping of dissimilarities in spectral composition (i.e., the variation among pixels) is similar to traditional estimations of beta diversity. For example, Laliberté et al. (2019) adapted the total community composition variance approach (Legendre and De Cáceres 2013) to estimate spectral diversity as spectral variance, partitioning the spectral diversity of a region (gamma diversity) into additive alpha and beta diversity components.

2.8 Links Between Plant Diversity, Other Trophic Levels, and Ecosystem Functions

Plant diversity has consequences for other trophic levels, sometimes reducing herbivory on focal species (Castagneyrol et al. 2014), but also increasing the diversity of insects and their predators in an ecosystem (Dinnage et al. 2012; Lind et al. 2015).

The distribution of plant traits within communities influences resource availability for other trophic levels above- and belowground, which affects community assembly and population dynamics across trophic levels. Diversity of neighbors surrounding focal trees can both increase and decrease pathogen and herbivore pressure on them (Grossman et al. 2019). Thus, while we know that plant diversity impacts other trophic levels, consistent rules across the globe that explain how and why these impacts occur remain elusive. An increasing number of studies reveal that plant diversity influences belowground microbial diversity and composition (Madritch et al. 2014; Cline et al. 2018). While these relationships are significant, they may explain limited variation given the number of other factors that influence microbial diversity and potentially due to a mismatch in sampling scales. Ultimately, it appears that chemical composition and productivity of aboveground components of ecosystems that can be remotely sensed are critical drivers of belowground processes, including microbial diversity (Madritch et al., Chap. 8).

Biodiversity loss is known to substantially decrease ecosystem functioning and ecosystem stability (Cardinale et al. 2011; O'Connor et al. 2017). Yet, the nature and scale of biodiversity-ecosystem function relationships remains a central question in biodiversity science. The issue is one that is ready to be tackled across scales using RS technology. The long-term biodiversity experiment at Cedar Creek Ecosystem Science Reserve (Tilman 1997) (Fig. 2.6), for example, has revealed the increasing effects of biodiversity on productivity over time (Reich et al. 2012) and that phylogenetic and functional diversity are highly predictive of productivity (Cadotte et al. 2008; Cadotte et al. 2009). Remotely sensed spectral diversity also predicts productivity (Sect. 2.9). Increased stability has also been linked to both higher plant richness (Tilman et al. 2006) and phylogenetic diversity (Cadotte et al. 2012) in this experiment. Tree diversity experiments show similar effects of increasing productivity with diversity (Tobner et al. 2016; Grossman et al. 2017) (Fig. 2.7), and these same trends emerge as the dominant pattern in forest plots globally

Fig. 2.7 The Forest and Biodiversity (FAB) experiment at the Cedar Creek Ecosystem Science Reserve shows overyielding (**a**)—greater productivity than expected in species-rich communities compared to monocultures—also called the net biodiversity effect (NBE). Curves show 90% predictions from multiple linear regression models (yellow 2013–2014; blue 2014–2015). (Redrawn from Grossman et al. 2018.) Photos (**b, c**) show juvenile trees grown in mixtures with varying neighborhood composition. The first phase of the experiment, shown here, includes three 600 m^2 blocks, each consisting of 49 plots (9.25 m^2) planted in a grid with 0.5 m spacing

(Liang et al. 2016). Hundreds of rigorous biodiversity experiments have been designed and conducted to tease apart effects of changing numbers of species (richness) from effects of changing identities of species (composition) (O'Connor et al. 2017; Grossman et al. 2018; Isbell et al. 2018). Complementarity among diverse plant species that vary in their functional attributes and capture and respond to resources differently is the primary explanation for increasing productivity with diversity (Williams et al. 2017). Nevertheless, both the nature of biodiversity-ecosystem function (BEF) relationships and their causal mechanisms remain variable and scale dependent in natural systems. In the Nutrient Network global grassland experiments, in which communities have assembled naturally, the relationship between diversity and productivity is variable (Adler et al. 2011). In tropical forest plots around the globe, at spatial extents of 0.04 ha or less, the biodiversity-productivity relationship is strong. However, as scales increase to 0.25 or 1.0 ha, the relationship is no longer consistently positive and can frequently be negative (Chisholm et al. 2013). These varied relationships at contrasting spatial scales may result from nonlinear, hump-shaped relationships between biodiversity and ecosystem function across resource availability gradients as the nature of species interactions and their level of complementarity shift (Jaillard et al. 2014). RS methods—including imaging spectroscopy and LiDAR—that can detect both the diversity and the structure and function of ecosystems (Martin, Chap. 5; Atkins et al. 2018) can discern these relationships across spatial extents and biomes in natural systems. They thus have high potential to enhance our understanding of the scale and context dependence of linkages between biodiversity and ecosystem function (Grossman et al. 2018).

2.9 Incorporating Spectra into Relationships Between Biodiversity and Ecosystem Function

Detection of spectral diversity, in particular, offers the potential to contribute to the quantification of BEF relationships at large scales (Schweiger et al. 2018) and is thus worth discussing in more detail. The variability captured by spectral diversity in a given ecosystem depends on the way the spectral diversity is calculated, as well as its spatial and spectral resolution (Sect. 2.7.5; Gamon et al., Chap. 16). From a functional perspective, spectral profiles measured at the leaf level depend on the chemical, structural, morphological, and anatomical characteristics of leaves (Ustin and Jacquemoud, Chap. 14). Variation in spectra and spectral diversity can be used to test hypotheses about how specific traits influence ecosystem function, community composition, and other characteristics of ecosystems, when using spectral bands or spectral indices with known associations with specific plant traits (Serbin and Townsend, Chap. 3). Moreover, spectral bands and indices can be weighted based on prior information about the relative contribution of individual traits to specific ecosystem characteristics. However, while the absorption features of some chemical traits are known, the effects of other, particularly nonchemical, plant traits on spectra are less well understood, in part due to overlapping spectral features and challenges associated with accurately describing nonchemical traits (Ustin and Jacquemoud, Chap. 14). Using the full spectral profile of plants in spectral diversity calculations provides a means to integrate chemical, structural, morphological, and anatomical variation and to acknowledge the many ways plants differ from one another.

It is certainly more complicated to decipher the biological meaning of spectral diversity calculated from spectral profiles than from measures of biodiversity that are based on a specific set of plant traits or spectral bands or indices with known links to specific traits. However, the variance that is explained by models based on spectral profiles can be partitioned into known and unknown sources of variation. This provides a means to assess the relative contribution of traits with known spectral characteristics and traits that are less well understood spectrally or that are of yet-unrecognized importance. At the canopy level, when spectra are measured from a distance, the question of what spectra and spectral diversity represent is further complicated by the influences that plant architecture, soil, and other materials have on the spectral characteristics of image pixels (Wang et al. 2018; Gholizadeh et al. 2018). Again, the degree to which these characteristics matter for a particular ecosystem needs to be evaluated in the particular context of the study. Some ecosystem components such as shade, soil, rock, or debris, which influence remotely sensed spectra, are biologically meaningful because they influence light availability and microclimate and provide resources for other trophic levels.

The association between plant spectra and traits can be illustrated by plotting spectral distances against functional distances or dissimilarity, as illustrated using species from the Cedar Creek biodiversity experiment (Fig. 2.6d). Given that functional differences among species are expected to increase with evolutionary divergence time (Fig. 2.1b), positive relationships are also expected among spectral and phylogenetic distances. The observed associations among spectral, functional, and phylogenetic dissimilarity (Fig. 2.6a, b) allow biodiversity metrics based on any of these dimensions of biodiversity to explain a similar proportion of the total variability in aboveground productivity (Fig. 2.6c–e), which is known to increase with the functional diversity of the plant community in this system (Cadotte et al. 2009). The species in the biodiversity experiment at Cedar Creek are relatively functionally dissimilar and distantly related, such that spectral, functional, and phylogenetic diversity also predict species richness (not shown). One advantage of spectral diversity is that the metric can be calculated from remotely sensed image pixels without depending on information about the distribution and abundance of species in an area, their functional traits, or phylogenetic relationships (Schweiger et al. 2018). By extracting a random number of high-resolution image pixels in each plant community, Schweiger et al. (2018) found that remotely sensed spectral diversity explained the biodiversity effect on aboveground productivity about as well as spectral diversity calculated using leaf-level spectra (Fig. 2.6).

2.10 Links Between Biodiversity and Ecosystem Services

Humans benefit from ecosystem functions and biodiversity. The benefits we derive from nature, often called ecosystem services, are a product of the biodiversity—assembled over millions of years—and ecosystem properties of a given region, or the whole Earth (Daily 1997). Daily (1997) defines ecosystem services as "the conditions and processes through which natural ecosystems, and the species that make them up, sustain and fulfill human life." Ecosystem services, referred to by the Intergovernmental Science-Policy Platform on Biodiversity and Ecosystem Services (IPBES) as "nature's contributions to people" (Díaz et al. 2018), are a socioecological concept that emerged from the Millennium Ecosystem Assessment (2005) and include provisioning, regulating, supporting, and cultural services. Some ecosystem service categories include direct benefits of biodiversity—through the use and spiritual values that humans establish with elements of biodiversity and ecosystems—and indirect benefits through the contributions of biodiversity to critical regulating ecosystem functions. The diversities of functional traits of plants make up the primary productivity of life on Earth and are essential to the ecosystem services on which all life depends. Assessment of ecosystem services depends on understanding both the ecosystem functions on which ecosystem services are derived and how services are valued by humans (Schrodt et al., Chap. 17). Modeling efforts that incorporate remotely sensed data can be used to describe ecosystem functions and quantify the services they generate (Sharp et al. 2018). (For modeling tools that enable mapping and valuing ecosystem services, see https://naturalcapitalproject.stanford.edu/invest/.)

2.11 Trade-Offs Between Biodiversity and Ecosystem Services

Biodiversity—as well as many regulating services to which biodiversity contributes and upon which it depends—frequently shows a negative trade-off with provisioning ecosystem services, such as agricultural production (Haines-Young and Potschin 2009). The nature of these trade-offs depends on the biophysical context, including the climate, soils, hydrology, and geology, and will differ among regions. A trade-off curve represents the limits set by these biophysical constraints and can be thought of in economic terms as an "efficiency frontier" that sets the boundaries on possible combinations of biodiversity (or regulating services) and provisioning services (Polasky et al. 2008). Combinations above the curve are not possible; outcomes beneath the curve provide fewer total benefits than what is actually possible from the environment. Quantifying the biodiversity and ecosystem service potential from land and how they trade off are critical to efficient management of ecosystems. Current RS tools and forthcoming technologies are well-poised to decrease uncertainty in estimates of biodiversity—ecosystem service trade-offs—and can

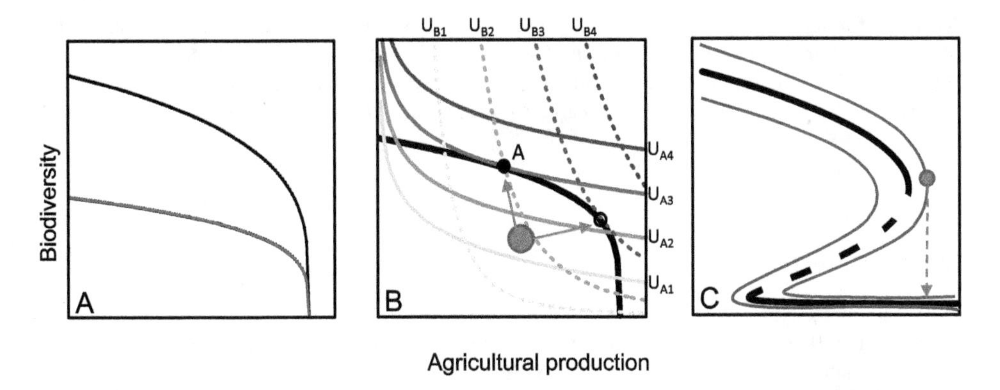

Fig. 2.8 RS technologies that enable detection of biodiversity and ecosystem functions aid in modeling their trade-offs. (**a**) The "efficiency frontier," or biophysical constraints that limit biodiversity and crop production, depends on the specific climatic, historical, and resource context of the land area and on the growth or replenishment rate of the natural system. These constraints can vary among ecosystems (red vs black curves). (**b**) Where along the efficiency frontier we want to manage for depends on human values. The superimposed curves show isolines of equal utility (U_{A1-4} or U_{B1-4}) for two different stakeholders (A and B) who have sharply different willingness to give up natural habitat for crop production and vice versa. Utility—or benefits to each stakeholder—increases moving from yellow to dark red. The two points at which the highest utility curve for each stakeholder intersects with the efficiency frontier represent the greatest feasible benefit to the stakeholder (points A and B). Often ecosystems are managed well below the efficiency frontier (green circle). RS may enable detection of components of biodiversity and ecosystem services at relevant spatial scales that can inform stakeholders about improved outcomes and aid negotiation among stakeholders. (**c**) Some trade-offs can have thresholds and tipping points that, once traversed, may result in a degraded alternative state. RS approaches that can aid in predicting uncertainties and temporal variability in trade-offs, indicated by thin blue lines, can help maximize ecosystem service benefits without overshooting thresholds that risk pushing the system into a degraded state. (Adapted from Cavender-Bares et al. 2015b)

contribute meaningfully to decision-making and resource management (Chaplin-Kramer et al. 2015; de Araujo Barbosa et al. 2015; Schrodt et al., Chap. 17).

Where along the efficiency frontier we wish to target our management efforts depends on human preferences. These can differ strongly among different stakeholders that have contrasting priorities (Cavender-Bares et al. 2015a, b). Distinguishing the biophysical limits of ecosystems from contrasting stakeholder preferences for what they want from ecosystems is a critical contribution to participatory processes that enable dialogue and progress toward sustainability (Cavender-Bares et al. 2015b; King et al. 2015). RS technologies that can enhance detection of biodiversity as well as both regulating and provisioning ecosystem services—and changes in these at multiple scales—can thus increase clarity in decision-making processes in the face of rapid global change (Fig. 2.8).

Acknowledgments We thank Shan Kothari for providing valuable feedback on an earlier version of the manuscript. The authors acknowledge funding from a NSF and NASA Dimensions grant (DEB 1342778), as well as the Cedar Creek NSF Long-Term Ecological Research grant (DEB

1234162) for support of some of the data presented here. We also thank the National Institute for Mathematical and Biological Synthesis (NIMBioS) working group on "Remotely Sensing Biodiversity," the Keck Institute for Space Studies working group "Unlocking a New Era in Biodiversity Science," and the NSF Research Coordination Network "Biodiversity Across Scales" (DEB 1745562).

References

Ackerly DD (2003) Community assembly, niche conservatism and adaptive evolution in changing environments. Int J Plant Sci 164:S165–S184

Adler PB, Seabloom EW, Borer ET, Hillebrand H, Hautier Y, Hector A, Harpole WS, O'Halloran LR, Grace JB, Anderson TM, Bakker JD, Biederman LA, Brown CS, Buckley YM, Calabrese LB, Chu C-J, Cleland EE, Collins SL, Cottingham KL, Crawley MJ, Damschen EI, Davies KF, DeCrappeo NM, Fay PA, Firn J, Frater P, Gasarch EI, Gruner DS, Hagenah N, Hille Ris Lambers J, Humphries H, Jin VL, Kay AD, Kirkman KP, Klein JA, Knops JMH, La Pierre KJ, Lambrinos JG, Li W, MacDougall AS, McCulley RL, Melbourne BA, Mitchell CE, Moore JL, Morgan JW, Mortensen B, Orrock JL, Prober SM, Pyke DA, Risch AC, Schuetz M, Smith MD, Stevens CJ, Sullivan LL, Wang G, Wragg PD, Wright JP, Yang LH (2011) Productivity is a poor predictor of plant species richness. Science 333:1750–1753

Atkins JW, Fahey RT, Hardiman BS, Gough CM (2018) Forest canopy structural complexity and light absorption relationships at the subcontinental scale. J Geophys Res Biogeosci 123:1387–1405

Barber CB, Dobkin DP, Huhdanpaa H (1996) The quickhull algorithm for convex hulls. ACM Trans Math Softw 22(4):469–483

Baselga A (2010) Partitioning the turnover and nestedness components of beta diversity: partitioning beta diversity. Glob Ecol Biogeogr 19:134–143

Bradshaw A (1965) Evolutionary significance of phenotypic plasticity in plants. Adv Genet 13:115–155

Bryant JA, Lamanna C, Morlon H, Kerkhoff AJ, Enquist BJ, Green JL (2008) Microbes on mountainsides: contrasting elevational patterns of bacterial and plant diversity. Proc Natl Acad Sci 105:11505–11511

Cadotte MW, Cardinale BJ, Oakley TH (2008) Evolutionary history and the effect of biodiversity on plant productivity. Proc Natl Acad Sci 105:17012–17017

Cadotte MW, Cavender-Bares J, Tilman D, Oakley TH (2009) Using phylogenetic, functional and trait diversity to understand patterns of plant community productivity. PLoS One 4:e5695

Cadotte MW, Dinnage R, Tilman D (2012) Phylogenetic diversity promotes ecosystem stability. Ecology 93:S223–S233

Cardinale BJ (2011) Biodiversity improves water quality through niche partitioning. Nature 472:86–89

Cardoso P, Rigal F, Carvalho JC, Fortelius M, Borges PAV, Podani J, Schmera D (2014) Partitioning taxon, phylogenetic and functional beta diversity into replacement and richness difference components. J Biogeogr 41:749–761

Castagneyrol B, Jactel H, Vacher C, Brockerhoff EG, Koricheva J (2014) Effects of plant phylogenetic diversity on herbivory depend on herbivore specialization. J Appl Ecol 51:134–141

Cavender-Bares J, Ackerly DD, Baum DA, Bazzaz FA (2004) Phylogenetic overdispersion in Floridian oak communities. Am Nat 163:823–843

Cavender-Bares J, Keen A, Miles B (2006) Phylogenetic structure of floridian plant communities depends on taxonomic and spatial scale. Ecology 87:S109–S122

Cavender-Bares J, Kozak KH, Fine PVA, Kembel SW (2009) The merging of community ecology and phylogenetic biology. Ecol Lett 12:693–715

Cavender-Bares J, Balvanera P, King E, Polasky S (2015a) Ecosystem service trade-offs across global contexts and scales. Ecol Soc 20:22
Cavender-Bares J, Polasky S, King E, Balvanera P (2015b) A sustainability framework for assessing trade-offs in ecosystem services. Ecol Soc 20:17
Cavender-Bares J, Ackerly DD, Hobbie SE, Townsend PA (2016a) Evolutionary legacy effects on ecosystems: biogeographic origins, plant traits, and implications for management in the era of global change. Annu Rev Ecol Evol Syst 47:433–462
Cavender-Bares J, Meireles JE, Couture JJ, Kaproth MA, Kingdon CC, Singh A, Serbin SP, Center A, Zuniga E, Pilz G, Townsend PA (2016b) Associations of leaf spectra with genetic and phylogenetic variation in oaks: prospects for remote detection of biodiversity. Remote Sens 8:221. https://doi.org/10.3390/rs8030221
Cavender-Bares J, Arroyo MTK, Abell R, Ackerly D, Ackerman D, Arim M, Belnap J, Moya FC, Dee L, Estrada-Carmona N, Gobin J, Isbell F, Köhler G, Koops M, Kraft N, Macfarlane N, Mora A, Piñeiro G, Martínez-Garza C, Metzger J-P, Oatham M, Paglia A, Peri PL, Randall R, Weis J (2018) Status and trends of biodiversity and ecosystem functions underpinning nature's benefit to people. In: Regional and subregional assessments of biodiversity and ecosystem services: regional and subregional assessment for the Americas. IPBES Secretariat, Bonn
Cavender-Bares J (2019) Diversification, adaptation, and community assembly of the American oaks (*Quercus*), a model clade for integrating ecology and evolution. New Phytol 221:669–692
Chao A, Chazdon RL, Colwell RK, Shen T-J (2005) A new statistical approach for assessing similarity of species composition with incidence and abundance data. Ecol Lett 8:148–159
Chao A, Chiu C-H, Jost L (2010) Phylogenetic diversity measures based on Hill numbers. Philos Trans R Soc Lond Ser B Biol Sci 365:3599–3609
Chao A, Chiu C-H, Jost L (2014) Unifying species diversity, phylogenetic diversity, functional diversity, and related similarity and differentiation measures through Hill numbers. Annu Rev Ecol Evol Syst 45:297–324
Chaplin-Kramer R, Sharp RP, Mandle L, Sim S, Johnson J, Butnar I, Milà i Canals L, Eichelberger BA, Ramler I, Mueller C, McLachlan N, Yousefi A, King H, Kareiva PM (2015) Spatial patterns of agricultural expansion determine impacts on biodiversity and carbon storage. Proc Natl Acad Sci 112:7402
Chisholm RA, Muller-Landau HC, Abdul Rahman K, Bebber DP, Bin Y, Bohlman SA, Bourg NA, Brinks J, Bunyavejchewin S, Butt N, Cao H, Cao M, Cárdenas D, Chang L-W, Chiang J-M, Chuyong G, Condit R, Dattaraja HS, Davies S, Duque A, Fletcher C, Gunatilleke N, Gunatilleke S, Hao Z, Harrison RD, Howe R, Hsieh C-F, Hubbell SP, Itoh A, Kenfack D, Kiratiprayoon S, Larson AJ, Lian J, Lin D, Liu H, Lutz JA, Ma K, Malhi Y, McMahon S, McShea W, Meegaskumbura M, Razman SM, Morecroft MD, Nytch CJ, Oliveira A, Parker GG, Pulla S, Punchi-Manage R, Romero-Saltos H, Sang W, Schurman J, Su S-H, Sukumar R, Sun IF, Suresh HS, Tan S, Thomas D, Thomas S, Thompson J, Valencia R, Wolf A, Yap S, Ye W, Yuan Z, Zimmerman JK (2013) Scale-dependent relationships between tree species richness and ecosystem function in forests. J Ecol 101:1214–1224
Cline LC, Hobbie SE, Madritch MD, Buyarski CR, Tilman D, Cavender-Bares JM (2018) Resource availability underlies the plant-fungal diversity relationship in a grassland ecosystem. Ecology 99:204–216
Dahlin KM (2016) Spectral diversity area relationships for assessing biodiversity in a wildland-agriculture matrix. Ecol Appl 26(8):2758–2768
Daily GC (1997) Nature's services. Island Press, Washington, D.C.
Darwin C (1859) On the origin of species. Murray, London
de Araujo Barbosa CC, Atkinson PM, Dearing JA (2015) Remote sensing of ecosystem services: a systematic review. Ecol Indic 52:430–443
Des Marais D, Hernandez K, Juenger T (2013) Genotype-by-environment interaction and plasticity: exploring genomic responses of plants to the abiotic environment. Annu Rev Ecol Evol Syst 44:5–29
Diamond JM, Mayr E (1976) Species-area relation for birds of the Solomon Archipelago. Proc Natl Acad Sci U S A 73:262–266

Díaz S, Purvis A, Cornelissen JHC, Mace GM, Donoghue MJ, Ewers RM, Jordano P, Pearse WD (2013) Functional traits, the phylogeny of function, and ecosystem service vulnerability. Ecol Evol 3:2958–2975

Díaz S, Kattge J, Cornelissen JHC, Wright IJ, Lavorel S, Dray S, Reu B, Kleyer M, Wirth C, Colin Prentice I, Garnier E, Bönisch G, Westoby M, Poorter H, Reich PB, Moles AT, Dickie J, Gillison AN, Zanne AE, Chave J, Joseph Wright S, Sheremet'ev SN, Jactel H, Baraloto C, Cerabolini B, Pierce S, Shipley B, Kirkup D, Casanoves F, Joswig JS, Günther A, Falczuk V, Rüger N, Mahecha MD, Gorné LD (2015) The global spectrum of plant form and function. Nature 529:167

Díaz S, Pascual U, Stenseke M, Martín-López B, Watson RT, Molnár Z, Hill R, Chan KMA, Baste IA, Brauman KA, Polasky S, Church A, Lonsdale M, Larigauderie A, Leadley PW, van Oudenhoven APE, van der Plaat F, Schröter M, Lavorel S, Aumeeruddy-Thomas Y, Bukvareva E, Davies K, Demissew S, Erpul G, Failler P, Guerra CA, Hewitt CL, Keune H, Lindley S, Shirayama Y (2018) Assessing nature's contributions to people. Science 359:270

Diggle P (1994) The expression of andromonoecy in *Solanum hirtum* (Solanaceae)—phenotypic plasticity and ontogenetic contingency. Am J Bot 81:1354–1365

Dinnage R, Cadotte MW, Haddad NM, Crutsinger GM, Tilman D (2012) Diversity of plant evolutionary lineages promotes arthropod diversity. Ecol Lett 15:1308–1317

Echeverría-Londoño S, Enquist BJ, Neves DM, Violle C, Boyle B, Kraft NJB, Maitner BS, McGill B, Peet RK, Sandel B, Smith SA, Svenning J-C, Wiser SK, Kerkhoff AJ (2018) Plant functional diversity and the biogeography of biomes in North and South America. Front Ecol Evol 6:219

Emerson BC, Gillespie RG (2008) Phylogenetic analysis of community assembly and structure over space and time. Trends Ecol Evol 23:619–630

Enquist BJ, Condit R, Peet RK, Schildhauer M, Thiers BM (2016) Cyberinfrastructure for an integrated botanical information network to investigate the ecological impacts of global climate change on plant biodiversity. https://doi.org/10.7287/peerj.preprints.2615v2

Faith DP (1992) Conservation evaluation and phylogenetic diversity. Biol Conserv 61:1–10

Faith DP, Reid CAM, Hunter J (2004) Integrating phylogenetic diversity, complementarity, and endemism for conservation assessment. Conserv Biol 18:255–261

Felsenstein J (1985) Phylogenies and the comparative method. Am Nat 125:1–15

Féret J-B, Asner GP (2014) Mapping tropical forest canopy diversity using high-fidelity imaging spectroscopy. Ecol Appl 24:1289–1296

Fine P, Ree R (2006) Evidence for a time integrated species area effect on the latitudinal gradient in tree diversity. Am Nat 168:796–804

Funk JL, Larson JE, Ames GM, Butterfield BJ, Cavender-Bares J, Firn J, Laughlin DC, Sutton-Grier AE, Williams L, Wright J (2017) Revisiting the Holy Grail: using plant functional traits to understand ecological processes: plant functional traits. Biol Rev 92:1156–1173

Gerhold P, Cahill JF, Winter M, Bartish IV, Prinzing A (2015) Phylogenetic patterns are not proxies of community assembly mechanisms (they are far better). Funct Ecol 29:600–614

Gholizadeh H, Gamon JA, Zygielbaum AI, Wang R, Schweiger AK, Cavender-Bares J (2018) Remote sensing of biodiversity: soil correction and data dimension reduction methods improve assessment of α-diversity (species richness) in prairie ecosystems. Remote Sens Environ 206:240–253

Gholizadeh H, Gamon JA, Townsend PA, Zygielbaum AI, Helzer CJ, Hmimina GY, Yu R, Moore RM, Schweiger AK, Cavender-Bares J (2019) Detecting prairie biodiversity with airborne remote sensing. Remote Sens Environ 221:38–49

Gilbert GS, Webb CO (2007) Phylogenetic signal in plant pathogen-host range. Proc Natl Acad Sci U S A 104:4979–4983

Graham C, Fine P (2008) Phylogenetic beta diversity: linking ecological and evolutionary processes across space and time. Ecol Lett 11:1265–1277

Grossman JJ, Cavender-Bares J, Hobbie SE, Reich PB, Montgomery RA (2017) Species richness and traits predict overyielding in stem growth in an early-successional tree diversity experiment. Ecology 98:2601–2614

Grossman JJ, Vanhellemont M, Barsoum N, Bauhus J, Bruelheide H, Castagneyrol B, Cavender-Bares J, Eisenhauer N, Ferlian O, Gravel D, Hector A, Jactel H, Kreft H, Mereu S, Messier C, Muys B, Nock C, Paquette A, Parker J, Perring MP, Ponette Q, Reich PB, Schuldt A, Staab M, Weih M, Zemp DC, Scherer-Lorenzen M, Verheyen K (2018) Synthesis and future research directions linking tree diversity to growth, survival, and damage in a global network of tree diversity experiments. Environ Exp Bot 152:68

Grossman JJ, Cavender-Bares J, Reich PB, Montgomery RA, Hobbie SE (2019) Neighborhood diversity simultaneously increased and decreased susceptibility to contrasting herbivores in an early stage forest diversity experiment. J Ecol 107:1492–1505

Haines-Young R, Potschin M (2009) The links between biodiversity, ecosystem services and human well-being. In: Raffaelli D (ed) Ecosystem ecology: a new synthesis. Cambridge University Press, Cambridge

Hall K, Johansson LJ, Sykes MT, Reitalu T, Larsson K, Prentice HC (2010) Inventorying management status and plant species richness in semi-natural grasslands using high spatial resolution imagery. Appl Veg Sci 13(2):221–233

Harrison S, Damschen EI, Grace JB (2010) Ecological contingency in the effects of climatic warming on forest herb communities. Proc Natl Acad Sci USA 107:19362–19367

Helmus MR (2007) Phylogenetic measures of biodiversity. Am Nat 169:E68–E83

Hill MO (1973) Diversity and evenness: a unifying notation and its consequences. Ecology 54:427–432

Humboldt, A (1817) *De distributione geographica plantarum secundum coeli temperiem et altitudinem montium: prolegomena* (On the Distribution of Plants) CE, Lutetiae Parisiorum, In: Libraria Graeco-Latino-Germanica

Isbell F, Cowles J, Dee LE, Loreau M, Reich PB, Gonzalez A, Hector A, Schmid B (2018) Quantifying effects of biodiversity on ecosystem functioning across times and places. Ecol Lett 21:763–778

Jaccard P (1900) Contribution au problème de l'immigration post-glaciaire de la flore alpine. Bull Soc Vaud Sci Nat 36:87–130

Jaillard B, Rapaport A, Harmand J, Brauman A, Nunan N (2014) Community assembly effects shape the biodiversity-ecosystem functioning relationships. Funct Ecol 28:1523–1533

Jost L (2007) Partitioning diversity into independent alpha and beta components. Ecology 88:2427–2439

Kembel SW, Ackerly DD, Blomberg SP, Cornwell WK, Cowan PD, Helmus MR, Morlon H, Webb CO (2010) Picante: R tools for integrating phylogenies and ecology. Bioinformatics 26:1463–1464

Kerr JT, Southwood TRE, Cihlar J (2001) Remotely sensed habitat diversity predicts butterfly species richness and community similarity in Canada. Proc Natl Acad Sci 98:11365–11370

King E, Cavender-Bares J, Balvanera P, Mwampamba TH, Polasky S (2015) Trade-offs in ecosystem services and varying stakeholder preferences: evaluating conflicts, obstacles, and opportunities. Ecol Soc 20:25

Laliberté E, Legendre P (2010) A distance-based framework for measuring functional diversity from multiple traits. Ecology 91(1):299–305

Legendre P, De Cáceres M (2013) Beta diversity as the variance of community data: dissimilarity coefficients and partitioning. Ecol Lett 16(8):951–963

Laliberté E, Schweiger AK, Legendre P (2019) Partitioning plant spectral diversity into alpha and beta components. Ecol Lett https://doi.org/10.1111/ele.13429

Lavorel S, Garnier E (2002) Predicting changes in community composition and ecosystem functioning from plant traits: revisiting the Holy Grail. Funct Ecol 16:545–556

Lennon JJ, Koleff P, Greenwood JJD, Gaston KJ (2001) The geographical structure of British bird distributions: diversity, spatial turnover and scale. J Anim Ecol 70:966–979

Liang J, Crowther TW, Picard N, Wiser S, Zhou M, Alberti G, Schulze E-D, McGuire AD, Bozzato F, Pretzsch H, de-Miguel S, Paquette A, Hérault B, Scherer-Lorenzen M, Barrett CB, Glick HB, Hengeveld GM, Nabuurs G-J, Pfautsch S, Viana H, Vibrans AC, Ammer C, Schall P,

Verbyla D, Tchebakova N, Fischer M, Watson JV, Chen HYH, Lei X, Schelhaas M-J, Lu H, Gianelle D, Parfenova EI, Salas C, Lee E, Lee B, Kim HS, Bruelheide H, Coomes DA, Piotto D, Sunderland T, Schmid B, Gourlet-Fleury S, Sonké B, Tavani R, Zhu J, Brandl S, Vayreda J, Kitahara F, Searle EB, Neldner VJ, Ngugi MR, Baraloto C, Frizzera L, Bałazy R, Oleksyn J, Zawiła-Niedźwiecki T, Bouriaud O, Bussotti F, Finér L, Jaroszewicz B, Jucker T, Valladares F, Jagodzinski AM, Peri PL, Gonmadje C, Marthy W, O'Brien T, Martin EH, Marshall AR, Rovero F, Bitariho R, Niklaus PA, Alvarez-Loayza P, Chamuya N, Valencia R, Mortier F, Wortel V, Engone-Obiang NL, Ferreira LV, Odeke DE, Vasquez RM, Lewis SL, Reich PB (2016) Positive biodiversity-productivity relationship predominant in global forests. Science 354:aaf8957

Lind EM, Vincent JB, Weiblen GD, Cavender-Bares JM, Borer ET (2015) Trophic phylogenetics: evolutionary influences on body size, feeding, and species associations in grassland arthropods. Ecology 96:998–1009

Madritch MD, Kingdon CC, Singh A, Mock KE, Lindroth RL, Townsend PA (2014) Imaging spectroscopy links aspen genotype with below-ground processes at landscape scales. Philos Trans R Soc Lond B Biol Sci 369:20130194

Marconi S, Graves S, Weinstein B, Bohlman S, White E (2019) Rethinking the fundamental unit of ecological remote sensing: Estimating individual level plant traits at scale. bioRxiv:556472

Mayfield MM, Levine JM (2010) Opposing effects of competitive exclusion on the phylogenetic structure of communities. Ecol Lett 13:1085–1093

McManus MK, Asner PG, Martin ER, Dexter GK, Kress JW, Field BC (2016) Phylogenetic structure of foliar spectral traits in tropical forest canopies. Remote Sens 8:196. https://doi.org/10.3390/rs8030196

Millennium Ecosystem Assessment (2005) Ecosystems and human well-being: biodiversity synthesis. World Resources Institute, Washington, D.C.

Moore TE, Schlichting CD, Aiello-Lammens ME, Mocko K, Jones CS (2018) Divergent trait and environment relationships among parallel radiations in Pelargonium (Geraniaceae): a role for evolutionary legacy? New Phytol 219:794–807

Mouillot D, Graham NAJ, Villéger S, Mason NWH, Bellwood DR (2013) A functional approach reveals community responses to disturbances. Trends Ecol Evol 28:167–177

Oindo BO, Skidmore AK (2010) Interannual variability of NDVI and species richness in Kenya. Int J Remote Sens 23(2):285–298

O'Connor MI, Gonzalez A, Byrnes JEK, Cardinale BJ, Duffy JE, Gamfeldt L, Griffin JN, Hooper D, Hungate BA, Paquette A, Thompson PL, Dee LE, Dolan KL (2017) A general biodiversity–function relationship is mediated by trophic level. Oikos 126:18–31

O'Meara BC, Ané C, Sanderson MJ, Wainwright PC (2006) Testing for different rates of continuous trait evolution using likelihood. Evolution 60:922–933

Palacio-López K, Beckage B, Scheiner S, Molofsky J (2015) The ubiquity of phenotypic plasticity in plants: a synthesis. Ecol Evol 5:3389–4000

Parker IM, Saunders M, Bontrager M, Weitz AP, Hendricks R, Magarey R, Suiter K, Gilbert GS (2015) Phylogenetic structure and host abundance drive disease pressure in communities. Nature 520:542–544

Pinto-Ledezma JN, Jahn AE, Cueto VR, Diniz-Filho JAF, Villalobos F (2018a) Drivers of phylogenetic assemblage structure of the Furnariides, a widespread clade of lowland neotropical birds. Am Nat 193:E41–E56

Pinto-Ledezma JN, Larkin DJ, Cavender-Bares J (2018b) Patterns of beta diversity of vascular plants and their correspondence with biome boundaries across North America. Front Ecol Evol 6:194

Polasky S, Nelson E, Camm J, Csuti B, Fackler P, Lonsdorf E, Montgomery C, White D, Arthur J, Garber-Yonts B, Haight R, Kagan J, Starfield A, Tobalske C (2008) Where to put things? Spatial land management to sustain biodiversity and economic returns. Biol Conserv 141:1505–1524

Presley SJ, Scheiner SM, Willig MR (2014) Evaluation of an integrated framework for biodiversity with a new metric for functional dispersion. PLoS One 9:e105818

Reich PB (2014) The world-wide 'fast-slow' plant economics spectrum: a traits manifesto. J Ecol 102:275–301
Reich PB, Walters MB, Ellsworth DS (1997) From tropics to tundra: global convergence in plant functioning. Proc Natl Acad Sci 94:13730–13734
Reich PB, Tilman D, Isbell F, Mueller K, Hobbie SE, Flynn DFB, Eisenhauer N (2012) Impacts of biodiversity loss escalate through time as redundancy fades. Science 336:589–592
Ricklefs RE, Schluter D (1993) In: Ricklefs RE, Schluter D (eds) Species diversity: regional and historical influences. University of Chicago Press, Chicago, pp 350–363
Rocchini D, Balkenhol N, Carter GA, Foody GM, Gillespie TW, He KS, Kark S, Levin N, Lucas K, Luoto M, Nagendra H, Oldeland J, Ricotta C, Southworth J, Neteler M (2010) Remotely sensed spectral heterogeneity as a proxy of species diversity: Recent advances and open challenges. Ecol Inform 5(5):318–329
Rocchini D, Luque S, Pettorelli N, Bastin L, Doktor D, Faedi N, Feilhauer H, Féret JB, Foody GM, Gavish Y, Godinho S, Kunin WE, Lausch A, Leitão PJ, Marcantonio M, Neteler M Ricotta C, Schmidtlein S, Vihervaara P, Wegmann M, Nagendra H (2018) Measuring β-diversity by remote sensing: A challenge for biodiversity monitoring. Methods Ecol Evol 9(8):1787–1798
Scheiner SM (1993) Genetics and evolution of phenotypic plasticity. Annu Rev Ecol Syst 24:35–68
Scheiner SM (2012) A metric of biodiversity that integrates abundance, phylogeny, and function. Oikos 121:1191–1202
Scheiner SM, Kosman E, Presley SJ, Willig MR (2017) Decomposing functional diversity. Methods Ecol Evol 8:809–820
Schmidtlein S, Zimmermann P, Schüpferling R, Weiß C (2007) Mapping the floristic continuum: Ordination space position estimated from imaging spectroscopy. J Veg Sci 18(1):131–140
Schweiger AK, Cavender-Bares J, Townsend PA, Hobbie SE, Madritch MD, Wang R, Tilman D, Gamon JA (2018) Plant spectral diversity integrates functional and phylogenetic components of biodiversity and predicts ecosystem function. Nat Ecol Evol 2:976–982
Simpson GG (1943) Mammals and the nature of continents. Am J Sci 241:1–31
Simpson EH (1949) Measurement of diversity. Nature 163(4148):688–688
Singh A, Serbin SP, McNeil BE, Kingdon CC, Townsend PA (2015) Imaging spectroscopy algorithms for mapping canopy foliar chemical and morphological traits and their uncertainties. Ecol Appl 25:2180–2197
Shannon CE (1948) A mathematical theory of communication. Bell Syst Tech J 27(3):379–423
Sharp R, Tallis HT, Ricketts T, Guerry AD, Wood SA, Chaplin-Kramer R, Nelson E, Ennaanay D, Wolny S, Olwero N, Vigerstol K, Pennington D, Mendoza G, Aukema J, Foster J, Forrest J, Cameron D, Arkema K, Lonsdorf E, Kennedy C, Verutes G, Kim CK, Guannel G, Papenfus M, Toft J, Marsik M, Bernhardt J, Griffin R, Glowinski K, Chaumont N, Perelman A, Lacayo M, Mandle L, Hamel P, Vogl AL, Rogers L, Bierbower W, Denu D, Douglass J (2018) In: T. N. C. Project (ed) InVEST 3.5.0. User's Guide. Stanford University, University of Minnesota, The Nature Conservancy, and World Wildlife Fund
Shipley B, Belluau M, Kühn I, Soudzilovskaia NA, Bahn M, Penuelas J, Kattge J, Sack L, Cavender-Bares J, Ozinga WA, Blonder B, van Bodegom PM, Manning P, Hickler T, Sosinski E, Pillar VDP, Onipchenko V, Poschlod P (2017) Predicting habitat affinities of plant species using commonly measured functional traits. J Veg Sci 28:1082–1095
Sørensen TA (1948) A method of establishing groups of equal amplitude in plant sociology based on similarity of species content, and its application to analyses of the vegetation on Danish commons. K Dan Vidensk Selsk Biol Skr 5:1–34
Sultan S (2000) Phenotypic plasticity for plant development, function and life history. Trends Plant Sci 5:537–542
Swenson NG, Enquist BJ, Pither J, Thompson J, Zimmerman JK (2006) The problem and promise of scale dependency in community phylogenetics. Ecology 87:2418–2424
Tedersoo L, Nara K (2010) General latitudinal gradient of biodiversity is reversed in ectomycorrhizal fungi. New Phytol 185:351–354
Tilman D (1997) Community invasibility, recruitment limitation, and grassland biodiversity. Ecology 78:81–92

Tilman D, Reich PB, Knops JMH (2006) Biodiversity and ecosystem stability in a decade-long grassland experiment. Nature 441:629–632

Tobner CM, Paquette A, Gravel D, Reich PB, Williams LJ, Messier C (2016) Functional identity is the main driver of diversity effects in young tree communities. Ecol Lett 19:638–647

Tuomisto H (2010) A diversity of beta diversities: straightening up a concept gone awry. Part 1. Defining beta diversity as a function of alpha and gamma diversity. Ecography 33:2–22

Verdu M, Pausas JG (2007) Fire drives phylogenetic clustering in Mediterranean Basin woody plant communities. J Ecol 95:1316–1323

Villeger S, Mason NWH, Mouillot D (2008) New multidimensional functional diversity indices for a multifaceted framework in functional ecology. Ecology 89:2290–2301

Walker B, Kinzig A, Langridge J (1999) Plant attribute diversity, resilience, and ecosystem function: the nature and significance of dominant and minor species. Ecosystems 2:95–113

Wang R, Gamon JA, Cavender-Bares J, Townsend PA, Zygielbaum AI (2018) The spatial sensitivity of the spectral diversity–biodiversity relationship: an experimental test in a prairie grassland. Ecol Appl 28:541–556

Webb CO (2000a) Exploring the phylogenetic structure of ecological communities: an example for rain forest trees. Am Nat 156:145–155

Webb CO (2000b) Exploring the phylogenetic structure of ecological communities: an example for rain forest trees. Am Nat 156:145–155

Webb CO, Ackerly DD, McPeek MA, Donoghue MJ (2002) Phylogenies and community ecology. Annu Rev Ecol Syst 33:475–505

Westoby M (1998) A leaf-height-seed (LHS) plant ecology strategy scheme. Plant Soil 199:213–227

Whittaker RH (1960) Vegetation of the Siskiyou Mountains, Oregon and California. Ecol Monogr 30:279–338

Wiens JJ, Donoghue MJ (2004) Historical biogeography, ecology and species richness. Trends Ecol Evol 19:639–644

Williams LJ, Paquette A, Cavender-Bares J, Messier C, Reich PB (2017) Spatial complementarity in tree crowns explains overyielding in species mixtures. Nat Ecol Evol 1:0063

Wright IJ, Reich PB, Westoby M, Ackerly DD, Baruch Z, Bongers F, Cavender-Bares J, Chapin T, Cornelissen JH, Diemer M (2004) The worldwide leaf economics spectrum. Nature 428:821–827

3

Plant Functional Traits: Scaling and Mapping

Shawn P. Serbin and Philip A. Townsend

3.1 Introduction

Fossil energy use and land use change are the dominant drivers of the accelerating increase in atmospheric CO_2 concentration and the principal causes of global climate change (IPCC 2018; IPBES 2018). Many of the observed and projected impacts of rising CO_2 concentration and increased anthropogenic pressures on natural resources portend increasing risks to global terrestrial biomes, including direct impacts on biodiversity, yet the uncertainty surrounding the forecasting of biodiversity change, future climate, and the fate of terrestrial ecosystems by biodiversity and Earth system models (ESMs) is unacceptably high, hindering informed policy decisions at national and international levels (Jetz et al. 2007; Friedlingstein et al. 2014; Rice et al. 2018). As such, the impact of our changing climate and altered disturbance regimes on terrestrial ecosystems is a major focus of a number of disciplines, including the biodiversity, remote sensing (RS), and global change research communities.

Here we provide an overview of approaches to scale and map plant functional traits and diversity across landscapes with a focus on current approaches, leveraging on best practices provided by Schweiger (Chap. 15), benefits and issues with general techniques for linking and scaling traits and spectra, and other key considerations that need to be addressed when utilizing RS observations to infer plant functional traits across diverse landscapes.

S. P. Serbin (✉)
Brookhaven National Laboratory, Environmental and Climate Sciences Department, Upton, NY, USA
e-mail: sserbin@bnl.gov

P. A. Townsend
Department of Forest and Wildlife Ecology, University of Wisconsin, Madison, WI, USA

3.1.1 Plant Traits and Functional Diversity

The importance of characterizing leaf and plant functional traits across scales is tied to the crucial role these traits play in mediating ecosystem structure, functioning, and resilience or response to perturbations (Lavorel and Garnier 2002; Reich et al. 2003; Wright et al. 2004; Reich 2014; Funk et al. 2017). The structural, biochemical, physiological, and phenological properties of plants regulate the growth and performance or fitness of plants and their ability to propagate or survive in diverse environments. As such, these traits are used to characterize the axes of variation that define broad plant functional types (PFTs), which in turn describe global vegetation patterns and properties (Ustin and Gamon 2010; Díaz et al. 2015), particularly in ESMs (Bonan et al. 2002; Wullschleger et al. 2014). Our focus here will be on leaf traits related to nutrition and defense that broadly fit within the concept of the leaf economics spectrum (LES, Wright et al. 2004), because these are most amendable to measurements using spectral methods. Other traits relating to reproductive strategies, hydraulics, physiology (though see Serbin et al. 2015), wood characteristics, etc. may be inferred from the traits described here, especially when combined with climate, soils, topography, or other data that generally are not directly detectable using RS.

Leaf nutritional properties and morphology are strong predictors of the photosynthetic capacity, plant growth, and biogeochemical cycling of terrestrial ecosystems (Aber and Melillo 1982; Green et al. 2003; Wright et al. 2004; Díaz et al. 2015). With respect to litter turnover and nutrient cycling, leaf traits that correspond to the distribution and magnitude of structural carbon and chemical compounds such as lignin and cellulose are used to infer the recalcitrant characteristics of canopy foliage (Madritch et al., Chap. 8). Capturing the spatial variation in these traits can therefore provide critical information on the nutrient cycling potential of ecosystems (Ollinger et al. 2002). On the other hand, leaf mass per area (LMA)—the ratio of a leaf's dry mass to its surface area—and its reciprocal, specific leaf area (SLA), correspond to a fundamental trade-off of leaf construction costs versus light-harvesting potential (Niinemets 2007; Poorter et al. 2009). The amount of foliar nitrogen within a leaf, on a mass (N_{mass}, %) or area (N_{area}, g/m^2) basis, strongly regulates the photosynthetic capacity of leaves given its fundamental role in the light-harvesting pigments of leaves (chlorophyll a and b) and photosynthetic machinery, namely, the enzyme RuBisCo (Field and Mooney 1986; Evans and Clarke 2018). Other traits, such as the concentration or content of water and accessory pigments, are important indicators of plant health and stress (Ustin et al. 2009). Moreover, the covariation of traits is also a primary focus of ecological and biodiversity research given strong trade-offs defining different leaf form and function (Díaz et al. 2015). For example, across the spectrum of plant functional diversity (Wright et al. 2004), foliar nitrogen and LMA form a key axis of variation that describes end-members between "cheap" thinner, low-LMA leaves with high leaf nitrogen, higher photosynthetic rates and faster turnover versus thick, expensive leaves with high LMA, low nitrogen, slower turnover, and longer leaf life spans. Other traits with strong evidence for detection in the literature relate to plant allocation strategies (e.g., starch and sugar content) or defense compounds, such as phenolics (e.g., Asner et al. 2015; Kokaly and Skidmore 2015; Couture et al. 2016; Ely et al. 2019).

Despite the importance of characterizing leaf and plant functional traits across global biomes, the plasticity and high functional diversity of these traits makes this apparently simple goal extremely challenging (Reich et al. 1997; Wu et al. 2017; Osnas et al. 2018), and as such global coverage has been historically limited to specific biomes (Schimel et al. 2015). Leaf traits can vary strongly within and across species (Serbin et al. 2014; Osnas et al. 2018) and are strongly mediated by an array of biotic and abiotic factors (Díaz et al. 2015; Neyret et al. 2016; Butler et al. 2017). Within a canopy, for example, functional traits typically show high variation with average light condition and quality (Niinemets 2007; Neyret et al. 2016) where lower canopy leaves tend to be thinner and have lower photosynthetic rates and altered pigment pools to account for the lower light quality. Plant traits can also change across local resource gradients, including with variations in water, nutrient availability, and disturbance legacy (Singh et al. 2015; Butler et al. 2017; Enquist et al. 2017). Importantly, this pattern can be confounded by species composition, which is generally the strongest driver of trait variation.

RS has provided new avenues to explore trait variation at larger scales and continuously across landscapes (Fig. 3.1). For example, Dahlin et al. (2013) observed that leaf functional traits were more strongly mediated by plant community composition than environment across a water-limited Mediterranean ecosystem, explaining 46–61% of the variation on the landscape. Likewise, McNeil et al. (2008) found that 93% of variation in nutrient cycling in northern hardwood forests of the US Adirondacks could be explained by species identity. Yet the presence or absence of specific plant species is, in part, a consequence of habitat sorting processes and the adaptive mechanisms of plants that influence the environments in which they can persist, including their modification of traits in response to local conditions (Reich et al. 2003). Mapping species or communities to infer traits is impractical at anything other than the local scale due to the presence of more than 200,000 plant species on Earth. Dispersal and other stochastic processes also play a role. Across broad environmental gradients, traits display much larger variation, where climate, topography, and edaphic conditions drive changes in plant community composition and structure, which, in turn, drive the patterns of potential and realized plant traits in any one location (Díaz et al. 2015; Butler et al. 2017). Finally, factors such as convergent evolution may make some species spectrally similar, while phenology and phenotypic variation may make the same species look different across locations.

Temporal regulation of traits is a key factor driving changes in functional properties and the resulting functioning of the ecosystem. Seasonal changes in traits can be significant (e.g., Yang et al. 2016) and can strongly regulate vegetation functioning (e.g., Wong and Gamon 2015). Moreover, during the lifetime of a leaf, traits can change significantly (e.g., Wilson et al. 2001; Niinemets 2016), and in evergreen species, leaf age has been shown to be a strong covariate with functional trait values (e.g., Chavana-Bryant et al. 2017; Wu et al. 2017). Age-dependent and phenological changes in leaf traits can, in turn, have significant impacts on ecosystem functioning

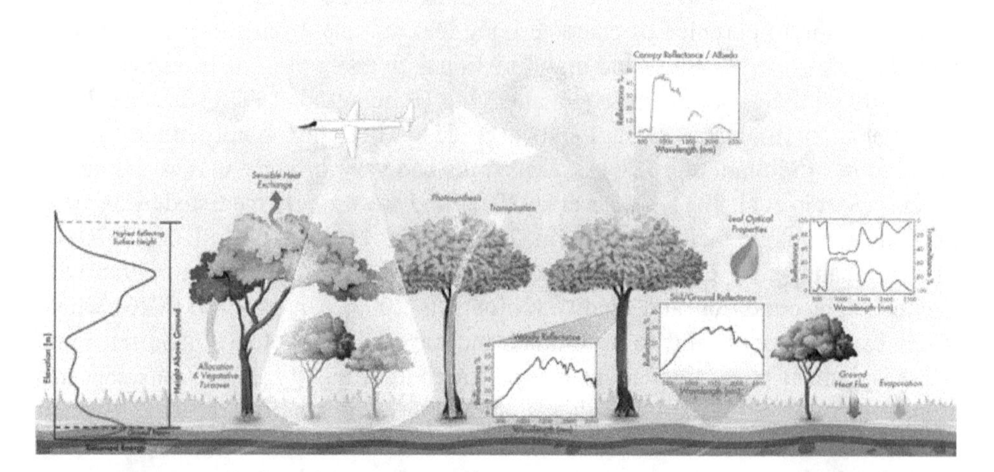

Fig. 3.1 There is a strong coupling between vegetation composition, structure and function, and the signatures observed by remote sensing instrumentation. Passive optical, thermal, and active sensing systems can be used to identify and map a range of phenomena, including minor to major variation in vegetation properties, health, and status across a landscape. Specifically, high spectral resolution imaging spectroscopy data can be used to infer functional traits of the vegetation through the measurement of canopy-scale optical properties which are driven by variation in leaf biochemistry and morphology, as well as overall canopy structure

(Wu et al. 2016). Given the role plant traits play in community assembly, characterizing the distribution, spatial patterns, and seasonality of traits is crucial for improved prediction of biodiversity change and ecosystem responses to global change.

Numerous plant trait databases have been developed to store information on the variation in functional traits across space and time (e.g., Wright et al. 2004; Kattge et al. 2011; LeBauer et al. 2018) needed to inform biodiversity and ecological modeling research. However, repeated direct measurement of plant traits is logistically challenging, which limits the geographic and temporal coverage of trait variation in these databases. Moreover, capturing plant trait variation through time is critical, but currently lacking from most observations (but with notable exceptions, e.g., Stylinski et al. 2002; Yang et al. 2016) given a host of additional technical and monetary challenges. In particular, efforts to collect direct, repeat samples of functional traits in remote areas, such as high-latitude ecosystems and the remote tropics, can be severely hindered by access and other logistical considerations.

On the other hand, RS can provide the critical unifying observations to link in-situ measurements of plant traits to the larger spatial and temporal scales needed to improve our understanding of global functional and plant biodiversity (Fig. 3.1, Table 3.1). As such, a strong interest in the use of RS to characterize foliar functional traits and their diversity has emerged from three key areas: research in RS of leaf optical properties (Jacquemoud et al. 2009), the concept of the leaf

Table 3.1 List of key foliar functional traits that can be estimated from imaging spectroscopy

Functional characterization[a]	Trait	Example of functional role	Example Citations
Primary	Foliar N (% dry mass or area based)	Critical to primary metabolism (e.g., Rubisco).	Johnson et al. (1994), Gastellu-Etchegorry et al. (1995), Mirik et al. (2005), Martin et al. (2008), Gil-Pérez et al. (2010), Gökkaya et al. (2015), Kalacska et al. (2015), Singh et al. (2015)
	Foliar P (% dry mass)	DNA, ATP synthesis	Mirik et al. (2015), Mutanga and Kumar (2007), Gil-Pérez et al. (2010), Asner et al. (2015)
	Sugar (% dry mass)	Carbon soiree	Asner and Martin (2015)
	Starch (% dry mass)	Storage compound, carbon reserve	Matson et al. (1994)
	Chlorophyll-total (ng g^{-1})	Light-harvesting capability	Johnson et al. (1994), Zarco-Tejada et al., (2000a); Moorthy et al., (2008); Gil-Pérez et al. (2010), Zhang et al. (2008)
	Carotenoids (ng g^{-1})	Light harvesting, antioxidants	Datt (1998), Zarco-Tejada et al. (2000a)
	Other pigments (e.g., anthocyanins; ng g^{-1})	Photoprotection. NPQ	van den Berg and Perkins (2005)
	Water content (% fresh mass)	Plant water status	Gao and Goetz (1995), Gao (1996), Serrano et al., (2000), Asner et al. (2015)
Physical	Leaf mass per area (g m^{-2})	Measure of plant resource allocation strategies	Asner et al. (2015), Singh et al. (2015)
	Fiber (% dry mass)	Structure, decomposition	Mirik et al. (2005), Singh et al. (2015)
	Cellulose (% dry mass)	Structure, decomposition	Gastellu-Etchegorry et al. (1995), Thulin et al. (2014), Singh et al. (2015)
	Lignin (% dry mass)	Structure, decomposition	Singh et al. (2015)
Metabolism	Vcmax (μmol m^{-2} s^{-1})	Rubisco-limited photosynthetic capacity	Serbin et al. (2015)
	Photochemical Reflectance Index (PRI)	Indicator of non-photochemical quenching (NPQ) and photosynthetic efficiency, xanthophyll cycle	Gamon et al. (1992), Asner et al. (2004)
	Fv/Fm	Photosynthetic capacity	Zarco-Tejada et al. (2000c)

(continued)

Table 3.1 (continued)

Functional characterization[a]	Trait	Example of functional role	Example Citations
Secondary	Bulk phenolics (% dry mass)	Stress responses	Asner et al. (2015)
	Tannins (% dry mass)	Defenses, nutrient cycling, stress responses	Asner et al. (2015)

[a]Categories of functional characterization are for organizational purposes only: *Primary* refers to compounds that are critical to photosynthetic metabolism; *Physical* refers to non-metabolic attributes that are also important indicators of photosynthetic activity and plant resource allocation; *Metabolism* refers to measurements used to describe rate limits on photosynthesis; and *Secondary* refers compounds that are not directly related to plant growth, but indirectly related to plant function through associations with nutrient cycling, decomposition, community dynamics, and stress responses

economics spectrum (Wright et al. 2004), and the development of global-scale foliar trait databases (Kattge et al. 2011). Within the signals observed by passive optical, thermal, and active sensing systems, such as light detection and ranging (lidar) platforms, is a whole host of underlying leaf chemical, physiological, and plant structure information that drives the spatial and temporal variation in RS observations (Ollinger 2011; Figs. 3.1, 3.2, and 3.3). As a result, RS provides the only truly practical approach to observing spatial and temporal variation in plant traits, canopy structure, ecosystem functioning, and biodiversity in absence of being able to map all species or communities everywhere (Schimel et al. 2015; Jetz et al. 2016). RS observations can provide the synoptic view of terrestrial ecosystems and capture changes on the landscape from disturbances and necessary temporal coverage via multiple repeat passes or targeted collection at specific phenological stages, yielding information needed to fill critical gaps in trait observations across global biomes (Cavender-Bares et al. 2017; Schimel et al., Chap. 19).

3.1.2 Historical Advances in Remote Sensing of Vegetation

Over the last four-plus decades, passive optical RS has been used as a key tool for characterizing and monitoring the composition, structure, and functioning of terrestrial ecosystems across space and time. For example, spectral vegetation indices (SVIs), such as the normalized difference vegetation index (NDVI), have been used to capture broad-scale plant seasonality or phenology and changes in composition, monitor plant pigmentation and stress, and track changes in productivity through time and in response to environmental change (e.g., Goward and Huemmrich 1992; Kasischke et al. 1993; Myneni and Williams 1994; Gamon et al. 1995; Ahl et al. 2006; Mand et al. 2010). Platforms, such as the Advanced Very High Resolution Radiometer (AVHRR), originally designed for atmospheric research, have been

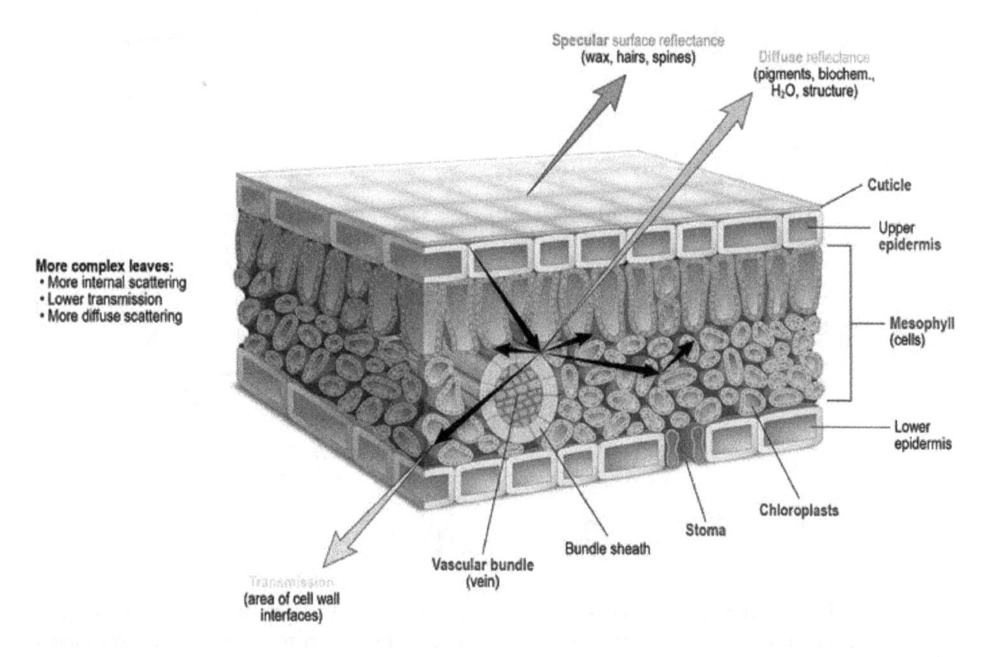

Fig. 3.2 The internal structure and biochemistry of leaves within a canopy control the optical signatures observed by remote sensing instrumentation. The amount of incident radiation that is reflected by, transmitted through, or absorbed by leaves within a canopy is regulated by these structural and biochemical properties of leaves. For example, leaf properties such as a thick cuticle layer, high wax, and/or a large amount of leaf hairs can significantly influence the amount of first-surface reflectance (that is the reflected light directly off the outer leaf layer that does not interact with the leaf interior), causing less solar radiation to penetrate into the leaf. The thickness of the mesophyll layer associated with other properties, such as thicker leaves, can cause higher degree of internal leaf scattering, less transmittance through the leaf, and higher absorption in some wavelengths. Importantly, the diffuse reflectance out of the leaf is that modified by internal leaf properties and contains useful for mapping functional traits

leveraged to capture changes in plant "greenness" based on the ratio of red absorption in leaves (signal of pigmentation levels and change) to near-infrared reflectance (tied to internal cellular structure and water content) to monitor changes in plant vigor and change (e.g., Tucker et al. 2001; Zhou et al. 2001; Goetz et al. 2005; Goetz et al. 2006). With the advent of focused Earth-observing (EO) sensors, such as the Landsat constellation, the science and use of optical RS observations for monitoring plant properties and functioning increased substantially (e.g., Chen and Cihlar 1996; Turner et al. 1999; Townsend 2002; Jones et al. 2007; Sonnentag et al. 2007; Drolet et al. 2008; Foster et al. 2008; Peckham et al. 2008; Yilmaz et al. 2008). Since the earliest uses, optical RS observations from the leaf to suborbital to satellite EO platforms have been heavily leveraged in the plant sciences, RS, and biodiversity communities (e.g., Jacquemoud et al. 1995; Roberts et al. 2004; Ustin et al. 2004; Gitelson et al. 2006; Hilker et al. 2008; Pettorelli et al. 2016; Cavender-Bares et al. 2017).

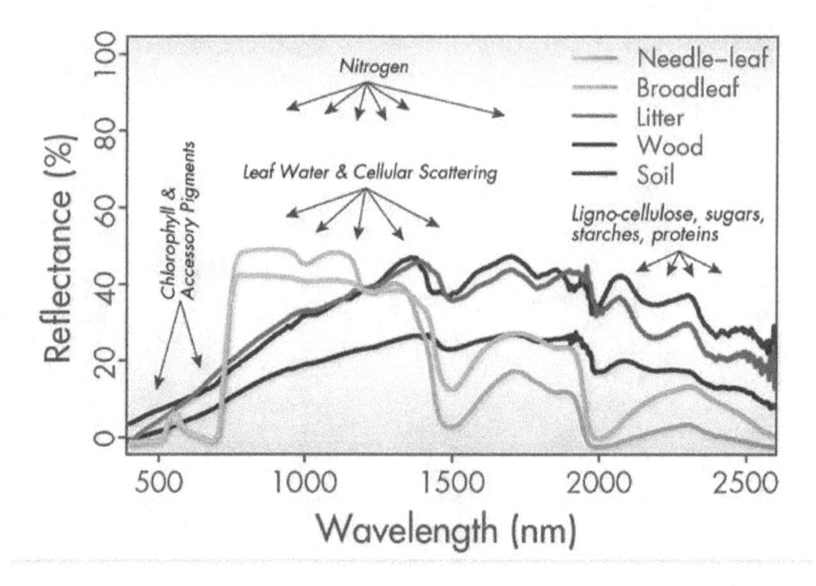

Fig. 3.3 High spectral resolution measurements of leaves and plant canopies enable the indirect, non-contact measurement of key structural and chemical absorption features that are associated with the physiological and biochemical properties of plants

3.1.3 Remote Sensing as a Tool for Scaling and Mapping Plant Traits

The use of leaf-level spectroscopy to understand plant functioning via biochemistry dates to the early twentieth century with papers describing light absorption and reflectance (Shull 1929; McNicholas 1931; Rabideau et al. 1946; Clark 1946; Krinov 1953). Billings and Morris (1951) made a direct linkage to differing ecological strategies of plants, in particular demonstrating that visible and near-infrared reflectance of species growing in different environments is directly linked to strategies associated with thermoregulation. Similarly, Gates et al. (1965) connected the interaction of light with leaves to internal leaf pigments and leaf structure (Fig. 3.2.) and how this relates to larger ecological processes.

By the 1970s, work with spectrophotometers at the US Department of Agriculture (USDA) led to the use of spectral methods for constituent characterization—near-infrared spectroscopy (NIRS) to predict moisture, protein, fat, and carbohydrate content of feed (Norris and Hart 1965; Norris et al. 1976; Shenk et al. 1981; Davies 1998; Workman and Weyer 2012), generally using linear regression on dry samples. In the 1980s and 1990s, field and laboratory studies used these earlier spectrometer systems to develop relationships and approaches to link leaf optical properties and underlying biochemical and structural properties, including variations in leaf moisture condition (Hunt and Rock 1989). For example, Elvidge (1990) utilized spectroscopy to describe optical properties of dried plant materials in the 0.4–2.5 micron range that enable detection of plant biochemistry from spectroscopy. Similarly, Curran (1989) summarized spectral features across this same spectral range that could be used in RS of plants, identifying not just the specific absorption features associated with pigments but also features related to harmonics and overtones related to molecular bonds of hydrogen (H) with carbon (C), nitrogen (N), and oxygen (O) in organic compounds (e.g., Fig. 3.3). In addition, by the late 1980s, researchers began to utilize novel, experimental airborne imaging spectrometer systems to map vegetation canopy chemistry in diverse landscapes. Using an early-generation NASA imaging spectrometer, the airborne imaging spectrometer (AIS, Vane and Goetz 1988), these studies illustrated the capacity to map landscape variation in foliar biochemical properties, including nitrogen and lignin (Peterson et al. 1988; Wessman et al. 1988; Wessman et al. 1989). AIS was the precursor to the Airborne Visible/Infrared Imaging Spectrometer (AVIRIS, Vane 1987). Following on this work, several others explored the impacts of leaf functional traits on reflectance properties of plant canopies and the ability to retrieve canopy chemistry, leveraging several important airborne campaigns including the Oregon Transect Ecosystem Research (OTTER) project and the Accelerated Canopy Chemistry Program (ACCP) (e.g., Card et al. 1988; Peterson et al. 1988; Matson et al. 1994; Bolster et al. 1996; Martin and Aber 1997).

These early studies became the basis for studies using imaging spectrometry to infer nutrient use and cycling in natural ecosystems (e.g., Martin and Aber 1997; Ollinger et al. 2002; Ollinger and Smith 2005). By the 1990s, the promise of spectroscopy for ecological characterization led to the increased use of handheld portable spectrometers in the field (e.g., instruments from Analytical Spectral Devices, GER, Spectra Vista Corporation, Spectral Evolution, Ocean Optics, LiCor, and PP Systems), as well as research that led to the use of narrowband SVIs for characterizing rapid changes in leaf function in response to the environment and leaf physiology (e.g., photochemical reflectance index, PRI, Gamon et al. 1992; Penuelas et al. 1995; Gamon et al. 1997). The review by Cotrozzi et al. (2018) provides a more detailed summary of the history of spectroscopy for plant studies, while Table 3.1 provides a summary of the key functional traits observable with spectroscopic RS approaches. As a consequence of studies at the leaf level and using early imaging spectrometers, a host of airborne sensor systems emerged, such as AVIRIS (Green et al. 1998), HyMap (Cocks et al. 1998), Airborne Prism Experiment (APEX, Schaepman et al. 2015), the Carnegie Airborne Observatory (CAO, Asner et al. 2012), AVIRIS-Next Generation (Miller et al. 2018; Thompson et al. 2018), and the US National Ecological Observatory Network (NEON) imaging spectrometer (Kampe et al. 2010) in the twenty-first century. The NASA prototype satellite EO-1 (Middleton et al. 2013) included the Hyperion sensor as an early test of the capacity to make hyperspectral measurements from space, leading to the development of a number of spaceborne missions planned for the early 2020s (Schimel et al., Chap. 19).

3.1.4 Key Considerations for the Use of Imaging Spectroscopy Data for Scaling and Mapping Plant Functional Traits

One of the chief challenges to effectively using imaging spectroscopy has been the acquisition of data of sufficient resolution, quality, and consistency for broad application in vegetation studies (Table 3.2). This necessitates measurements in the shortwave infrared (SWIR, 1100–2500 nm) in addition to the visible and near infrared (VNIR, 400–1100 nm). While VNIR wavelengths are most sensitive to pigments and overall canopy health, longer wavelengths are required to retrieve many biochemicals and LMA (Serbin et al. 2014; Kokaly and Skidmore 2015; Serbin et al. 2015; Singh et al. 2015). Spectral resolution is critical as well, with 10 nm band spacing and 10 nm full-width half maximum (FWHM) generally considered essential to identify traits detected based on narrow absorption features. Even finer resolution is required to detect spectral features that rely on narrow (<0.5 nm) atmospheric windows, such as solar-induced fluorescence (SIF, Yang et al. 2018). Other key considerations include sufficient signal-to-noise ratio (SNR) to identify important spectral features, accounting for both coherent and random noise related to detector sensitivity, dark current, and stray light. Additional sensor characteristics important to using imaging spectroscopy include spectral distortion. Most sensors are push-broom sensors, in which an image is constructed via the forward movement of the platform. Spatial samples are measured in the X-dimension (pixels) of the detector array and spectral wavelengths in the Y-dimension. Nonuniformity may arise due to differences in detectors in both dimensions, meaning that different detectors in the X-dimension see different central wavelengths (smile) and offsets in the Y-dimension lead to band-to-band misregistration (keystone). All of these effects can influence the ability to detect traits reliably within one scene or across multiple scenes using common algorithms. Full understanding of detector (and thus image) uniformity as well as the measurement point-spread function in 3-D (spatial X [detector X], spatial Y [platform movement], and spectral [detector Y]) is critical to accurate retrievals.

All RS data require some level of post-processing. Imaging spectroscopy is no different; prior to implementing algorithms for trait retrieval (Sect. 3.2.2), additional efforts must be undertaken to ensure consistent measurements in consistent units such that retrievals from imagery from multiple sources, dates, locations, etc. can be compared. Minimally, pixel measurements should be converted to radiances (w m^{-2} sr^{-1} nm^{-1}) based on laboratory calibrations and regular vicarious measurements of stable targets. With proper instrument characterization, keystone, smile, and other radiometric artifacts can be reduced. Subsequently, atmospheric corrections to convert radiance to reflectance (percent) are essential for cross-site studies. The approaches to atmospheric correction are numerous and tailored to particular environments, e.g., terrestrial vs. aquatic systems. Even within terrestrial applications, approaches differ among airborne data products (e.g., NASA's AVIRIS-Classic and AVIRIS-NG sensors vs. NEON AOP) and do not necessarily yield consistent reflectance imagery. Finally, new approaches that take advantage of

Table 3.2 Traceability matrix for a global imaging spectroscopy misson for terrestrail ecosystem functioning and biogeochemical processes

Science target	Science objectives	Functional characterization	Trait	Spectal range and sampling	Other measurement characteristics	Example citations
Theme III: Marine and Terrestrial Ecosystems and Natural Resource Management New essential measurements of the biochemical, physiological and functional attributes of the Earth's terrestrial vegetation	O1. To deliver new quantification of biogeochemical cycles, ecosystem functioning and factors that influence vegetation health and ecosystem services O2. To advance Earth system modes with improved process representation and quantification.	Primary biochemical content	Foliar N (% dry mass or area based)	450–2450 nm @ ≤15 nm	Seasonal cloud free measurement for ≤ 80% terrestrial vegetation areas. Radiometric range and sampling to capture range of vegetation signals from tropical to high latitude summers. Signals-to-Noise Ratio consistent with tropical to high latitude vegetation (e.g., red region, >500:1). At least three years of measurement to capture inter-annual variability and seasonally as robust baseline for ≥80 of the terrestrial ecosystems.	Johnson et al. (1994), Gastellu-Etchegorry et al. (1995), Mirik et al. (2005), Martin et al. (2008), Gil-Pérez et al. (2010), Gökkaya et al. (2015), Kalacska et al. (2015), Singh et al. (2015)
			Foliar P (% dry mass)	450–2450 nm @ ≤15 nm		Mirik et al. (2005), Mutangao and Kumar (2007), Gil-Pérez et al. (2010), Asner et al. (2015)

(continued)

Table 3.2 (continued)

Science target	Science objectives	Functional characterization	Trait	Spectal range and sampling	Other measurement characteristics	Example citations
			Sugar (% dry mass)	1500–2400 nm @ ≤15 nm		Asner and Martin (2015)
			Starch (% dry mass)	1500–2400 nm @ ≤15 nm		Matson et al. (1994)
			Chlorophyll-total (mg g^{-1})	450–740 nm @ ≤ 10 nm		Johnson et al. (1994), Zarco-Tejada et al. (2000a), Gil-Perez et al. (2010), Zhang et al. (2008), Kalacska et al. (2015)
			Carotenoids (mg g^{-1})	450–740 nm @ ≤10 nm		Datt (1998), Zarco-Tejada et al. (2000a)
			Other pigments (e.g., anthocyanins; mg g^{-1})	980 nm ± 40, 1140 ± 50 @ ≤20 nm		van den Berg and Perkins (2005)
			Water content (% fresh mass)	1100–2400 nm @ ≤20 nm		Gao and Goetz (1995), Gao (1996), Thompson et al. (2015), Asner et al. (2016)
			Leaf mass per area (g m^{-2})	1500–2400 nm @ ≤20 nm		Asner et al. (2015), Singh et al. (2015)
			Fiber (% dry mass)	1500–2400 nm @ ≤20 nm		Mirik et al. (2005), Singh et al. (2015)

		Physical/ structural content	Cellulose (% dry mass)	1500–2400 nm @ ≤20 nm		Gastellu-Etchegorry et al. (1995), Thulin et al. (2014), Singh et al. (2015)
			Lignin (% dry mass)	1500–2400 nm @ ≤15 nm		Johnson et al. (1994), Gastellu-Etchegorry et al. (2014), Thulin et al. (2014), Singh et al. (2015)
		Metabolism	Vcmax (μmol m^{-2} s^{-1})	450–2450 nm @ ≤15 nm		Serbin et al. (2015)
			Photochemical Reflectance Index (PRI).	450 to 650 nm @ ≤10 nm		Gamon et al. (1992), Asner et al. (2004)
			Fraction of absorbed photosynthetically active radiation by chlororphyll, fAPARchl.	450 to 800 nm @ ≤20 nm		
		Secondary biochemcial content	Bulk phoeolics (% dry mass)	1100–2400 nm @ ≤10 nm		Asner et al. (2015)
			Tannins (% dry mass)	1100–2400 nm @ 10 nm		Asner et al. (2015)
		Required for atmosopheric correction	Water vapor Cirrus clouds Aerosos	980 nm ± 50, 1140 ± 50 @ ≤20 nm		Thompson et al. (2015), Gao et al. (1993)
				940 nm ± 30, 1140 ± 40 @ ≤20 nm		
				450–1200 nm @ ≤20 nm		

advances in computing capacities and newer optimal estimation (OE) approaches for radiative transfer retrieval of atmospheric parameters are poised to transform atmospheric correction in the 2020s (Thompson et al. 2018).

Following atmospheric correction, scene-dependent corrections are often required, including corrections for different illumination and reflectance due to sun-target-sensor geometry, i.e., the bidirectional reflectance distribution function (BRDF). Current methods to correct for across-track (and along-track) illumination variation account for differences in vegetation structure and density, either through continuous functions (Schläpfer et al. 2015; Weyermann et al. 2015) or using land-cover stratification (Jensen et al. 2018). However, BRDF corrections are also rapidly changing and likely will be improved by new OE methods. As well, methods requiring land cover stratification are generally limited to local studies, whereas broad-scale implementation across biomes and through time will be most stable as long as scene-specific stratification is not required.

In addition to BRDF, corrections for topographic illumination are required (Singh et al. 2015). However, such corrections can result in poor performance for highly shaded slopes; they enhance noise on shaded slopes while suppressing signal on illuminated slopes. In addition, differential illumination may still remain in images due to multiple sensor artifacts as well as effects of vegetation structure (Knyazikhin et al. 2013). These effects can be effectively addressed using vector normalization (Feilhauer et al. 2010; Serbin et al. 2015) or continuum removal (e.g., Dahlin et al. 2013). Such approaches largely address structure-induced reflectance effects of broadleaf and graminoid canopies, with minor variances remaining in conifers. The residual effect of canopy structure on trait mapping largely relates to an inability to fully account for within-canopy scattering of diffuse radiation, especially in conifer forests.

Finally, when integrating data from multiple sources to map canopy traits, users must address wavelength calibrations. Different sensors may have different band centers, and these may change (on airborne devices) as they are recalibrated from time to time. This requires image resampling, which is data and processing intensive and—to be done precisely—requires good knowledge of spectral response functions or model recalibration to new wavelengths.

3.2 Linking Plant Functional Traits to Remote Sensing Signatures

All materials interact with light energy in different and characteristic ways. With respect to terrestrial ecosystems, spectroscopic RS leverages spectroradiometers, which measure the intensity of light energy reflected from or transmitted through leaves, plant canopies, or other materials (e.g., wood, soil, Fig. 3.3). The absorbing and scattering properties of the individual elements (e.g., leaves, twigs, stems) within the canopy or surface (soil) are defined by their physical and 3-D structure as well as chemical constituents or bonds (Figs. 3.2 and 3.3), which drives the vari-

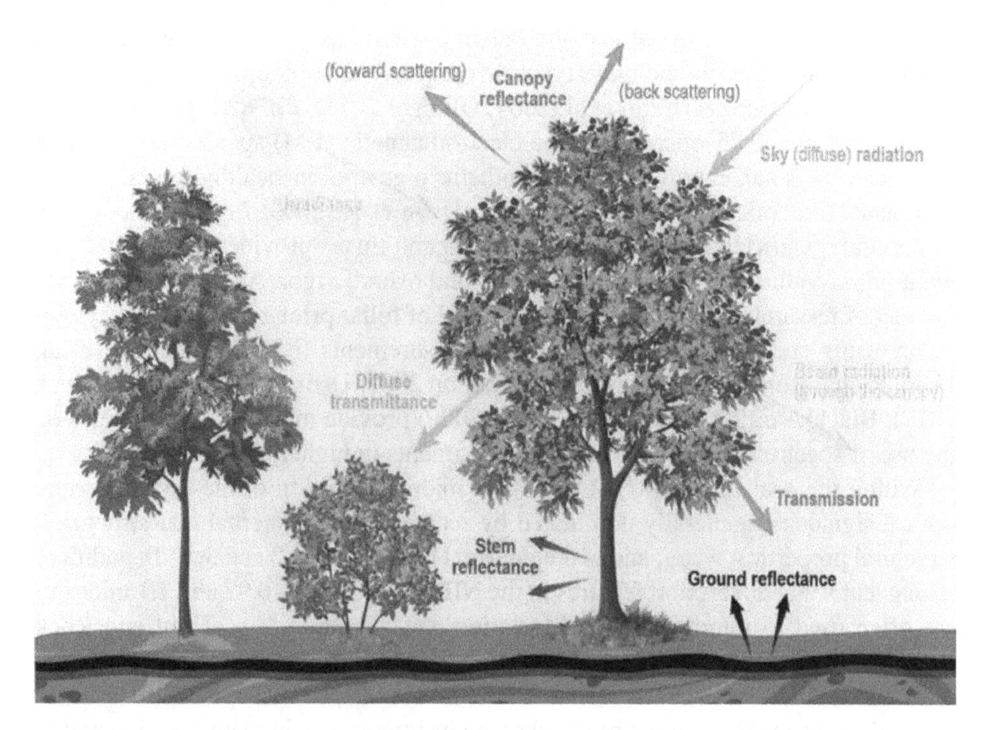

Fig. 3.4. Similar to those of a leaf, the properties of vegetation canopies strongly control the optical signatures observed by passive remote sensing instrumentation (Ollinger 2011). Specifically, the height and three-dimensional shape of the individual plants comprising the canopy as well as their leaf area index (LAI), leaf optical properties and stem and soil optical properties regulate the amount of incident radiation that reflects back from and transmits through a canopy. In addition, canopy properties and sun-sensor geometry can modify the shape and strength of the reflectance signature of vegetation canopies, which requires careful consideration when developing methods to map leaf functional traits

ability observed in reflectance spectra (Figs. 3.1 and 3.4). Thus, the underlying variation in plant canopy structure, function, and leaf traits in turn drives the optical properties and spectral signatures detected by RS platforms (Ollinger 2011). As such, the capacity to infer plant health, status, stress, and leaf and plant functional traits with optical RS observations is tied to the physical principle that plant physiological properties, structure, and distribution of foliage within plant canopies are reflected in the RS signatures of leaves within a canopy (Curran 1989; Kokaly et al. 2009; Ollinger 2011).

3.2.1 Spectroscopy and Plant Functional Traits

With the advent of laboratory and field spectrometer instrumentation, the leaf to landscape-scale RS of vegetation traits and functional properties began in earnest in the early 1980s (Sect. 3.1.3). As stated in Sect. 3.1.4, there are a host of important considerations with the use of leaf and imaging spectroscopy for scaling plant functional traits. In addition, the underlying drivers of vegetation optical properties are complex and numerous (Ustin et al. 2004; Ollinger 2011). For example, in the visible range (~0.4–0.75 microns) of the electromagnetic (EM) spectrum, the strong absorption of solar energy by photosynthetic pigments in healthy, green foliage dominates the optical properties of leaves (Ustin et al. 2009; Figs. 3.2. and 3.3). Importantly, knowledge of leaf pigment pools and fluxes provides key insight into plant photosynthesis, environmental stress, and overall vigor. As such a significant amount of research has focused on the retrieval of foliar primary and accessory pigments using spectroscopic and other RS measurements (e.g., Jacquemoud et al. 1996; Richardson et al. 2002; Sims and Gamon 2002; Ustin et al. 2009; Féret et al. 2017). Blackburn (2007) and Ustin et al. (2009) provide more detailed reviews on the use of spectroscopy to remotely sense pigments in higher plants.

Within the near-infrared (NIR, ~0.8–1.2 microns) portion of the EM spectrum, optical signals are generally dominated by scattering from internal leaf structures, structural properties, water, and leaf epidermal layer (Figs. 3.2 and 3.3). In addition, strong leaf water absorption features in the NIR, centered on ~0.97 and 1.1 microns, are often used to remotely sense vegetation water content (e.g., Hunt and Rock 1989; Gao and Goetz 1995; Sims and Gamon 2003; Stimson et al. 2005; Colombo et al. 2008). Much of the early research into the use of spectroscopic RS focused on leaf and canopy water content retrieval given its importance in plant function and as an important indicator of moisture (Fig. 3.5.) and other stress. In attached, fresh leaf material, water also dominates the spectral absorption features of the SWIR (1.3–2.5 micron) portion of the EM (Hunt and Rock 1989; Sims and Gamon 2003); as a result, spectral optical properties are strongly regulated by leaf and canopy water content in this region (Fig. 3.5). Along with water absorption, a number of other biochemical and structural trait absorption features exist in the SWIR wavelength region (Fig. 3.3), including cellulose, lignin, structural carbon, and nutrients and proteins (Curran 1989; Elvidge 1990; Kokaly et al. 2009; Ollinger 2011; Ely et al., 2019). Removal of water from leaf materials can sometimes enhance the detection of these absorption features (e.g., see Serbin et al. 2014 and references within; Fig. 3.5). However, at the canopy scale, a number of studies have demonstrated the capacity to retrieve these foliar biochemical properties in the SWIR region (e.g., Wessman et al. 1988; Martin and Aber 1997; Townsend et al. 2003; Kokaly et al. 2009; Asner et al. 2015; Singh et al. 2015), perhaps because of the increased signal due to multiple scattering within canopies (Baret et al. 1994).

In addition to the underlying leaf biochemical and structural characteristics, leaf orientation, display, and distribution in a canopy are also strong drivers of plant optical properties (Ollinger, 2011; Fig. 3.4). Decreasing the leaf area of a canopy generally results in a higher reflectance signal from elements deeper within the canopy, including twigs, branches, stems, and soil/litter layer (Asner 1998; Asner et al. 2000; Ollinger 2011). Canopies with flat, horizontal leaves tend to have higher NIR reflectance than those with more erect, vertical leaves, depending on the sun-sensor geometry. Leaf anatomy and average leaf angle vary widely across species

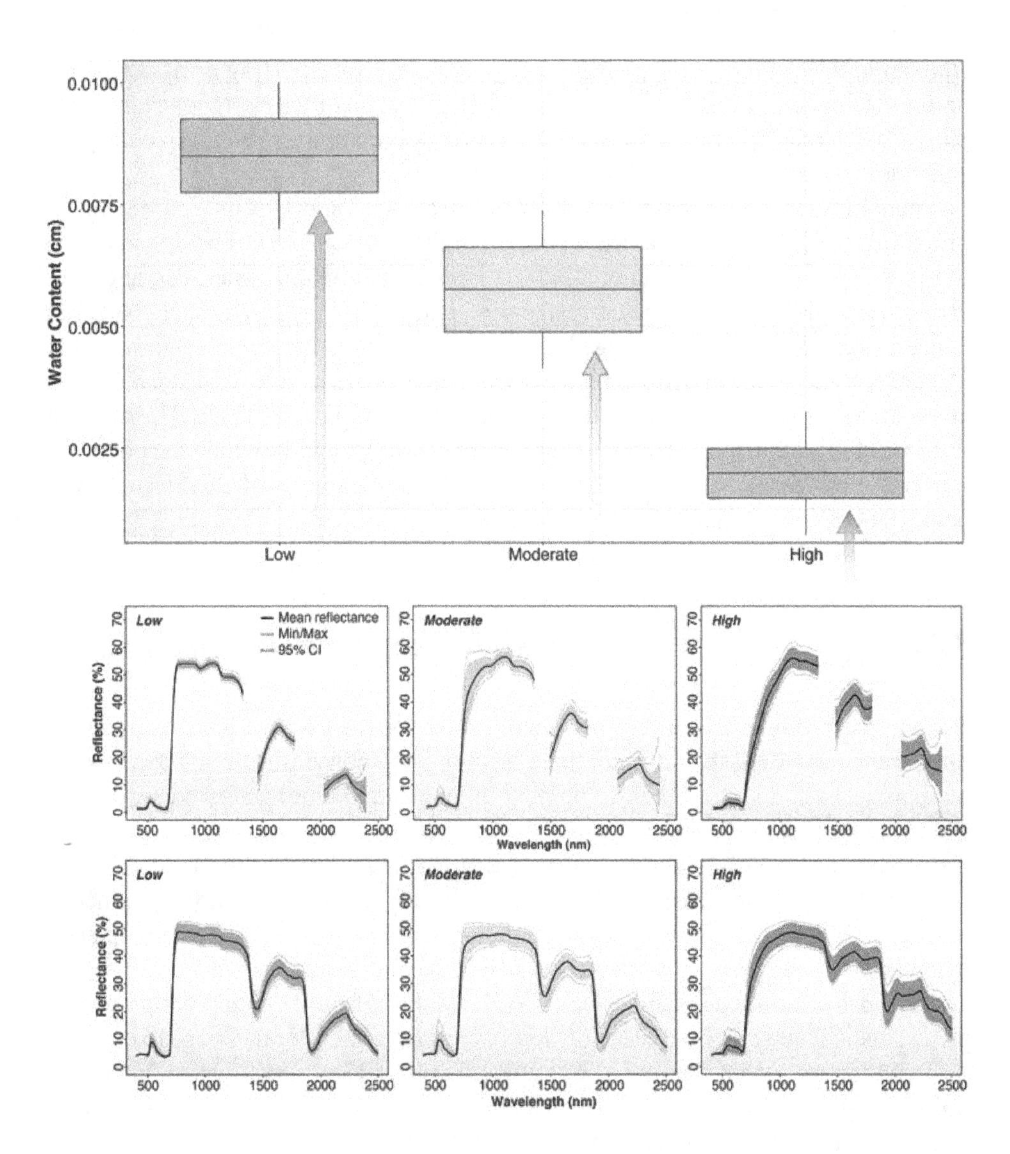

Fig. 3.5. Together, leaf optical properties and canopy architecture regulate the remote sensing signatures observed in remote sensing data. In addition, changes in leaf internal biochemistry or structure (i.e., functional traits) as a result of biotic or abiotic factors can change these signatures over space and time. For example, a prolonged drought can cause changes in leaf internal water content and potentially a redistribution of internal pigmentation. We can simulate the potential changes in optical signatures associated with a drought using a leaf and canopy-scale radiative transfer models (RTM), in this case PROSPECT-5b (Féret JB et al. 2008) and SAIL (Verhoef and Bach 2007), to illustrate the changes in leaf an canopy spectra over the course of a low, moderate, and high drought event. Here we modified pigment and water content from low to high for a range of canopies, as represented by different LAIs, and for canopy-scale reflectance, we incorporated the sensor characteristics of AVIRIS-classic (Green et al. 1998) to illustrate what the canopy reflectance might look like from that sensor. (For illustration purposes only)

(Falster and Westoby 2003), with consequences for interpreting optical RS signatures (Ollinger 2011). Thus, when considering the use of RS approaches for mapping leaf traits, careful consideration of vegetation structure, collection characteristics, and sensor design is important.

Phenology, leaf seasonality, and leaf age are also important drivers of optical properties for a number of reasons. First, leaf traits can change significantly over the lifetime of a leaf (e.g., Wilson et al. 2001; Niinemets 2016; Chavana-Bryant et al. 2017; Wu et al. 2017), and the corresponding leaf optical properties will change in concert (Yang et al. 2016). Average leaf angle distribution can also change with leaf age or seasonally from younger, recently expanded leaves to fully expanded (Raabe et al. 2015), which can have significant impacts on canopy reflectance (Huemmrich 2013). Finally, atmospheric, insect, or other stressors typically change the chemical makeup of leaves and so their optical properties (e.g., Couture et al. 2013; Ainsworth et al. 2014; Cotrozzi et al. 2018).

3.2.2 Approaches for Linking Traits and Spectral Signatures

Despite the promise and utility of spectroscopy for the retrieval and mapping of plant traits across space and time, there has not been consensus or standardization of approaches and algorithm development in the RS and biodiversity communities. This is not entirely unexpected given the complexity of connecting traits and RS observations across the various scales of interest, from leaves to individual trees, communities, and landscapes (Schweiger, Chap. 15). In addition, early approaches (e.g., Peterson et al. 1988) were often later deemed inappropriate and often replaced by other techniques (e.g., Grossman et al. 1996). Access to more powerful, improved, and cheaper computing resources has also allowed for the exploration of more complex statistical and machine-learning approaches (see Schweiger, Chap. 15).

Two primary approaches have been utilized to link RS observations to functional traits—empirical, statistically based techniques and radiative transfer modeling (RTM; see also Meireles et al., Chap. 7; Ustin, Chap. 14).

3.2.2.1 Empirical Scaling Approaches

With respect to empirical techniques, the use of SVIs was one of the earliest methods to explore the capacity to link a range of plant functional traits to vegetation spectra. Typically, with this approach a single SVI is linked with a trait of interest, such as leaf pigments or water content, to develop a simple statistical relationship between the trait of interest and corresponding variation in optical properties (e.g., Sims and Gamon 2003; Gitelson 2004; Colombo et al. 2008; Feret et al. 2011). The derived model is then used to estimate trait values for new leaves using only spectral measurements. This approach typically assumes the researcher has an *a priori* understanding of the links between the trait and resulting variation in the electromagnetic spectrum and thus selects specific wavelengths, and therefore SVI, for their analysis. An alternative approach is to explore the spectra and trait space to identify new or previously unknown SVIs that maximize the correspondence between optical properties and traits of interest (e.g., Inoue et al. 2008), akin to a data mining exercise. A challenge of this approach can be interpretation of the selected SVIs, where the resulting vegetation indices may not contain wavelengths with known absorption features relating to the trait of interest. The same general approach can also leverage multiple SVIs, provided the research avoids highly correlated portions of the spectrum (Grossman et al., 1996), to attempt to capture how variation in the trait of interest is reflected in various portions of the EM spectrum to other sites and plant species. However, a limitation to the use of SVIs has been the ability to generalize across broad canopy architectures, species, and environments due to the often site-specific modeling results or potential signal saturation issues with some SVIs (Shabanov et al. 2005; Glenn et al. 2008).

Continuous spectral wavelet transforms have been used to reduce the dimensionality of spectral data prior to developing simple statistical models (e.g., Blackburn and Ferwerda 2008). Wavelets are functions that are used to decompose a full, complex signal into simpler component sub-signals. When used with spectral data, the full reflectance signature can be decomposed in a way that allows the resulting wavelet coefficients assigned to each sub-signal to be related to concentrations of chemical constituents or other traits of interest, through standard statistical modeling approaches (e.g., linear regression). Previous studies have explored the use of wavelet methods to retrieve a host of functional traits, including pigments, water, and nitrogen content (e.g., Blackburn and Ferwerda 2008; Cheng et al. 2011; Li et al. 2018; Wang et al. 2018). Continuum removal together with band-depth analysis (Kokaly and Clark 1999) has also been utilized as a means to retrieve the chemical composition of leaves. In this approach, continuum removal lines are fit through the absorption features of interest based on those regions not in the areas of interest, then the original spectra are divided by corresponding values of the continuum removal line. The band centers can then be found by finding the minimum of the continuum-removed spectra. Normalization of the band centers is often used to standardize the values across samples. These data are then used to develop models to predict functional traits at the leaf and canopy scales, including foliar nitrogen and recalcitrant properties, such as the amount of lignin and cellulose (Kokaly et al. 2009).

In addition to the empirical SVI approach, as discussed in Schweiger (Chap. 15), partial least-squares regression (PLSR) modeling has been used extensively in the development of spectra-trait models for measuring, scaling, and mapping plant functional traits (e.g., Ollinger et al. 2002; Townsend et al. 2003; Asner and Martin 2008; Martin et al. 2008; Dahlin et al. 2013; Singh et al. 2015; Ely et al. 2019). A key attribute of PLSR is the capacity to utilize the entire measured portion of the EM spectrum as predictors (i.e., X matrix) without requiring a priori selection of wavelengths or SVIs (Wold et al. 1984; Geladi and Kowalski 1986; Wold et al. 2001). PLSR avoids collinearity (i.e., spectral autocorrelation across wavelengths) in the predictor variables (i.e., reflectance wavelengths), even if predictors exceed the number of observations (Geladi and Kowalski 1986; Wold et al. 2001; Carrascal et al. 2009). This is done through singular value decomposition (SVD), which reduces the X matrix down to relatively few non-correlated latent components. While PLSR was originally used in chemometrics, the features and benefits of PLSR also fit well within the goals of connecting spectral signatures to leaf functional traits. PLSR leverages the fact that different portions of the EM spectrum change in concert with various nutritional, structural, and morphological properties of leaves and canopies—in other words, leveraging the known covariance between variations in leaf optical properties and leaf traits (Ollinger 2011). Importantly, PLSR also allows for univariate or multivariate modeling where multiple predictands (i.e., Y matrix) can be modeled simultaneously with the same spectral matrix to account for the covariance between X and Y but also among the various Y (response) variables (Wold et al. 1984; Geladi and Kowalski 1986; Wold et al. 2001). Wolter et al. (2008) review of the use of PLSR in RS research, and Carrascal et al. (2009) summarize its use in ecology, as well as key features of PLSR.

Several approaches and implementations of PLSR have been used within the overarching "plant trait mapping" paradigm, including various spectral transformations and the use of prescreening of wavelengths or down-selection of suitable of pixels (e.g., Townsend et al. 2003; Feilhauer et al. 2010; Schweiger, Chap. 15; Asner et al. 2015). In a typical PLSR implementation (e.g., Fig. 3.6), foliar samples are first collected from vegetation canopies and processed to obtain the functional traits of interest. For leaf-scale algorithms, the optical properties of the leaves are typically measured in situ or within a small window (2–4 hours) prior to further

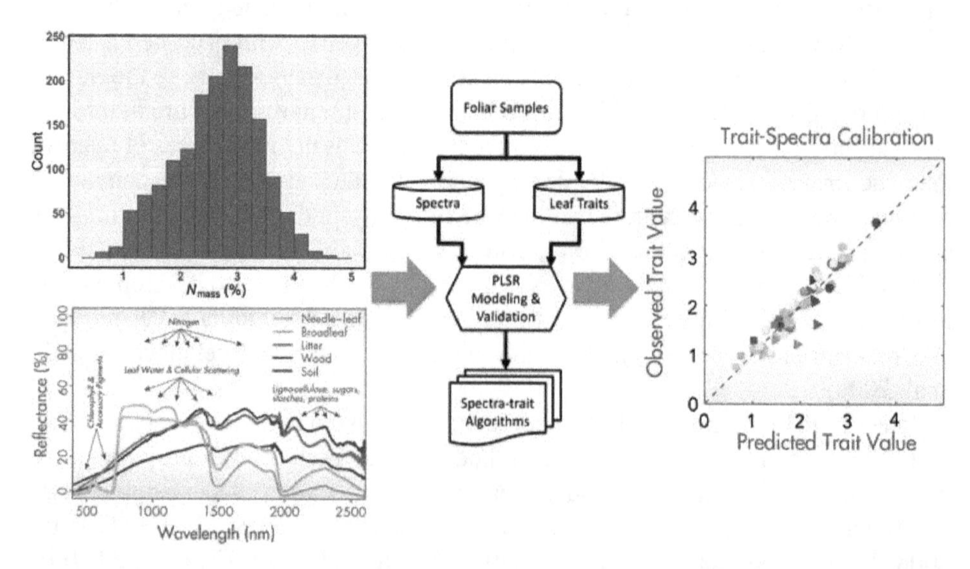

Fig. 3.6. A simple example illustrating how leaf functional traits and optical properties (e.g. reflectance) are combined in an empirical partial least-squares regression (PLSR) modeling approach to develop spectra-trait algorithms. The input traits and reflectance spectra are combined and used to train and test a PLSR model, using either cross-validation and/or independent validation (e.g., Serbin et al. 2014), and the resulting model can then be applied to other spectral measurements to estimate the traits of interest

processing. Leaf and/or image spectra for the pixel containing the plots or sample locations are then linked with these functional trait measurements to develop the PLSR algorithm. Typically, for models utilizing imaging spectroscopy data, plot-scale estimates of traits are derived using measurements of basal area, leaf area by species, or other means to produce a weighted average of each trait by dominant species within given ground area (e.g., McNeil et al. 2008; Singh et al. 2015). The algorithm is evaluated using internal validation during model development (e.g., cross-validation) and/or using a set of training and validation data to build and test the model predictive capacity across a range of similar samples and optical properties. Some approaches utilize additional steps to characterize the uncertainties associated with the sample collection, measurements, and other issues (e.g., instrument noise) in the PLSR modeling step. For example, Serbin et al. (2014) and Singh et al. (2015) introduced a novel PLSR approach that can account for uncertainty in the prediction of trait values, which has later been used by other groups (Asner et al. 2015). Image-scale algorithms are often used to derive functional trait maps (e.g., Fig. 3.7) to explore the spatial and/or temporal patterns of traits across the landscapes of interest (e.g., Ollinger et al. 2002; McNeil et al. 2008).

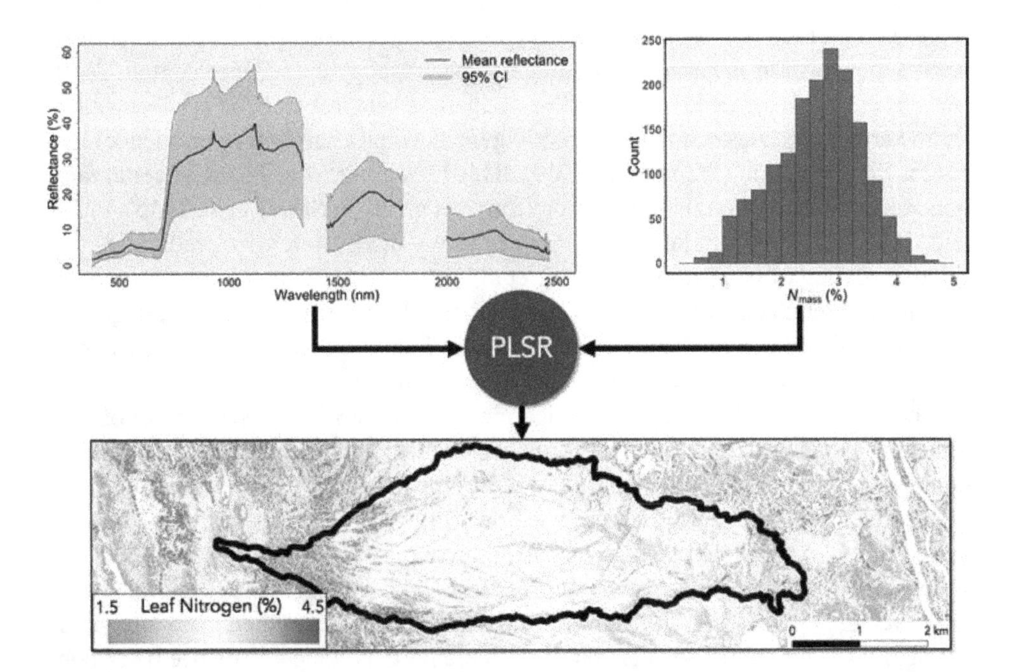

Fig. 3.7. Much like developing a leaf-scale PLSR model for estimating leaf functional traits, such as leaf nitrogen concentration (Fig. 3.6), we can also utilize high spectral resolution imaging spectroscopy data, such as that from NASA AVIRIS to build models applicable at the canopy to landscape scales (e.g., Dahlin et al. 2013; Singh et al. 2015). Here we show a simple illustration of the linkage between functional traits scaled to the canopy, for example based on a weighted average of the dominant species in the plot, connected with the reflectance signature of these canopies. Once linked, we can develop PLSR algorithms conceptually similar to that of leaves resulting in canopy-scale spectra-trait models capable of mapping functional traits across the broader landscape

While the PLSR approach produces algorithms that "weight" wavelengths by their importance in the prediction (Wold et al. 2001) of the functional traits of interest (e.g., Serbin et al. 2014), some researchers have also explored modifications to the standard PLSR approach that provide additional reductions in data dimensionality. For example, Li et al. (2008) coupled PLSR with a genetic algorithm (GA) approach to select a smaller subset of wavelengths to use in the final PLSR model for predicting leaf water content, measured as equivalent water thickness (EWT). DuBois et al. (2018) combined the SVI and PLSR approach by using all two-band AVIRIS wavelength combinations to model the relationship between spectral reflectance and ecosystem carbon fluxes across a water-limited environment. To date, the spectra-trait PLSR modeling approach has shown the capacity to characterize the widest array of leaf functional traits using the optical properties of plants across a broad range of species and ecosystems (e.g., Dahlin et al. 2013; Asner et al. 2014; Asner et al. 2015; Serbin et al. 2015; Singh et al. 2015; Couture et al. 2016).

Similar to the PLSR approach, researchers have leveraged various machine-learning approaches to connect RS observations to functional traits (e.g., Féret et al. 2018). Schweiger (Chap. 15) describes two commonly used machine-learning approaches in RS; several other approaches have also been used to model trait variation as a function of spectral measurements. More recently, Gaussian processes regression (GPR) has been recommended as superior to other machine-learning approaches for trait mapping from imaging spectroscopy data (Verrelst et al. 2012; Verrelst et al. 2016). GPR is a nonlinear nonparametric probabilistic approach similar to kernel ridge regression that directly generates uncertainty (or confidence) levels for the prediction (Wang et al. 2019). This is in contrast to PLSR uncertainties, generally assessed through permutation (Singh et al. 2015; Serbin et al. 2015). PLSR and GPR yield very similar results, both in terms of absolute trait predictions and relative scaling of uncertainties (Wang et al. 2019). PLSR is much more computationally efficient, and results are readily interpretable in terms of wavelength quantitative contribution to prediction (see Fig. 3.1 in Schimel et al., Chap. 19), whereas GPR only identifies relatively important wavelengths.

The challenge with most machine-learning approaches is that some level of data reduction is required for optimal performance. Standard approaches, such as principle component analysis (PCA) or minimum noise fraction (MNF) transformations, may reduce data dimensionality. However, features important to trait estimation may be buried in lower principle components, as high contrast variation (albedo, greenness, water content) dominate scene properties. In contrast, PLSR rotates the data into latent vectors optimized to the empirical dependent variables, which generally yields strong models for calibration data but can lead to poor model performance when confronted with new data that differ considerably from the model-building data sets.

3.2.2.2 Radiative Transfer Models and Scaling Functional Traits

An alternative to statistical, field-based, and empirical approaches for connecting leaf and canopy optical properties with plant functional traits, RTMs can be used either at the leaf and canopy scales to directly retrieve leaf traits (e.g., Colombo et al. 2008; Darvishzadeh et al. 2008; Feret et al. 2011; Banskota et al. 2015; Shiklomanov et al. 2016) or in hybrid approaches where statistical algorithms are developed based on RTM simulations (e.g., Asner et al. 2011). RTMs encapsulate our best mechanistic understanding of the coordination among leaf properties, canopy structure, and resulting spectral signatures at the leaf and canopy scales, but abstracted to operate with different degrees of complexity and assumptions (Bacour et al. 2002; Nilson et al. 2003; Kobayashi and Iwabuchi 2008; See also Morsdorf et al., Chap. 4; Ustin and Jacquemoud, Chap. 14).

At the leaf scale, RTMs were generally spawned from earlier work that identified the relationships between fresh and dried leaf reflectance and a range of foliar traits, including pigments, water content, nitrogen, dry matter, cellulose, and lignin. The realization that leaf optical properties were fundamentally tied to the concentration and distribution of leaf traits led to the development of models that could closely mimic the spectral patterns across the shortwave spectral region (0.4–2.5 microns) based on select leaf properties, such as chlorophyll and water content, as well as structural variables. By far the most widely and commonly used leaf-level RTM is the PROSPECT model (Jacquemoud and Baret 1990; Feret et al. 2008), which simulates leaf directional-hemispherical reflectance (R) and transmittance (T), allowing for the calculation of leaf absorption (1-R+T) based on leaf biochemical and morphological properties, primary and accessory pigments, water content, LMA, or dry matter content, brown material, and an approximation of the thickness of the internal leaf mesophyll layer (Féret et al. 2008; Féret et al. 2017). PROSPECT then simulates leaf optical properties based on a generalized plate model describing leaves as a stack of N homogenous absorbing layers that are calculated based on the values of input leaf traits and their corresponding spectral absorption coefficient. Other prominent leaf models include the Leaf Incorporating Biochemistry Exhibiting Reflectance and Transmittance Yields (LIBERTY) model (Dawson et al. 1998) and LEAFMOD (Ganapol et al. 1998). In particular, LIBERTY is notable given its original application focusing on improving the modeling of needle-leaf evergreen conifer species and their leaf optical properties based on several leaf traits, similar to PROSPECT, but also including foliar lignin and nitrogen content.

Moving to the canopy scale, RTMs are far more numerous with a wide variety of complexities, assumptions, and requirements (Verhoef and Bach 2007; Widlowski et al. 2015; Kuusk 2018). Most canopy RTMs leverage leaf-scale models, such as PROSPECT, to provide the leaf optical properties (i.e., leaf single-scattering albedo) needed to simulate canopy directional-hemispherical reflectance across select wavelengths, simulated spectral bands, or specific SVIs. Generally, the soil boundary layer is either prescribed or simulated using a simple model of soil BRDF (e.g., Hapke model, Verhoef and Bach 2007), and stem or woody material reflectance and transmittance (when used) is prescribed. Canopy RTMs can be separated into two main classes, homogenous and heterogenous models. Homogenous models assume the canopy to be horizontally unlimited and treated as a turbid medium of sufficiently large number of phytoelements (leaves, stems, other materials). For example, the Ross–Nilson model of plate medium (Ross 1981) assumes these elements to be composed of small bi-Lambertian "plates" described by their reflectance and transmittance properties with a specific leaf angle distribution (LAD). Leaves are small compared to the full canopy medium, with no self-shading, and transmittance is a function of optical properties and leaf area index (LAI). Additional canopy parameters were added, including the hot-spot and canopy clumping to describe sun-sensor illumination effects and the inhomogeneity of the canopy elements (Kuusk 2018). Early SAIL models also fall into this classification (e.g., Verhoef 1984). On the other hand, heterogenous canopy RTM models, including 3-D models, address the fact that vegetation canopies are heterogenous (e.g., gaps between crowns, spatial structure, differing canopy architectures) but range widely in their complexity and implementations. These models provide enhanced detail in the modeling of vegetation canopies but are necessarily more complex. Often these models require additional information to model vegetation "scenes," which can include information on tree crown shape, stem location, and other properties (e.g., hot spot, clumping) in addition to leaf optical properties, sun-sensor geometry, and LAI. These models range from 3-D Monte Carlo ray-tracing models, such as FLIGHT (North 1996) and FLiES (Kobayashi and Iwabuchi 2008), to analytical and hybrid approaches using a variety of canopy structure schemes including geometric optical (GO) representation of individual plants where tree placement follows a statistical distribution and leaf and stem scattering elements are homogenously distributed (e.g., Kuusk and Nilson 2000; Nilson et al. 2003). For example, multiple stream, including four-stream, two-layer models often utilize simplifying assumptions, to model canopies as homogenous and continuous (i.e., "slab canopies"), but which are composed of a large number of small scattering elements (leaves, sometimes leaves and stems) with arbitrary inclination angles (e.g., 4SAIL2, Verhoef and Bach 2007). The scattering elements and the soil can be prescribed with specific optical properties using observed data or based on a leaf RTM, such as PROSPECT (Jacquemoud et al. 2009). In addition, some models can divide complex scenes into smaller cells to perform the radiative transfer calculations (e.g., DART, Gastellu-Etchegorry et al. 2015) where the level of simulation detail is based on the size of the cells and the degree of detail built into the model scene components. See the review by Kuusk (2018) for more details regarding canopy RTMs and their design, diversity, assumptions, and approaches.

The use of RTMs allows for the estimation of leaf and canopy traits using simulated canopy reflectance, without some of the limitations or challenges of empirical approaches (3.3.1), such as the requirement of field sampling, scaling leaf traits to the canopy, and other issues such as the timing of field and imagery collections. Furthermore, RTMs can provide a more mechanistic connection between traits and reflectance allowing for potentially broader application than empirical approaches in areas were ground sampling may be sparse (e.g., remote regions such as the Arctic or the tropics). In addition, RTMs provide the opportunity to prototype inversion approaches across a range of remote sensing platforms and evaluate the trade-offs between different sensor designs, spectral resolutions, and temporal coverage (Shiklomanov et al. 2016), enabling the development of cross-platform retrieval algorithms.

Depending on the application, and RTM complexity, inversion can be conducted at the pixel or larger patch scales (i.e., collections of relatively homogenous areas of vegetation) to characterize spatial and temporal patterns in plant functional (e.g., pigments) and structural (e.g., LAI) properties. In RTM inversion, the leaf-scale model is often the focus, where the goal is to invert the canopy and leaf models jointly to extract estimated foliar traits based on observed canopy reflectance (e.g., Colombo et al. 2008). Many other studies have focused on retrieving canopy-scale parameters, such as LAI (e.g., Darvishzadeh et al. 2008; Banskota et al. 2015). Early approaches leveraged RTM inversions that focused on numerical optimization techniques to minimize the difference between modeled and observed reflectance across similar wavelengths (e.g., Jacquemoud et al. 1995). Other methods have utilized look-up table (LUT) inversion (e.g., Weiss et al. 2000) where a range of simulated canopy reflectance patterns are generated in advanced by varying leaf and canopy inputs across predetermined values. These simulated spectra are then compared to observations where either a single or select number of closely matching modeled spectra, and their associated inputs, are selected as the solution to the inversion. Bayesian RTM inversion methods have also been utilized (e.g., Shiklomanov et al. 2016) as a means to retrieve leaf and canopy properties as joint posterior probability distributions through iterative sampling of the input parameter space. The use of RTMs ranges from retrieval of vegetation functional and structural traits to the characterization of landscape functional diversity (Kattenborn et al. 2017; Kattenborn et al. 2019).

3.3 Important Considerations, Caveats, and Future Opportunities

3.3.1 Field Sampling and Scaling Considerations

There are several important considerations and best practices when developing algorithms for the remote estimation of plant traits (see Schweiger, Chap. 15). We will only briefly touch on these here. A key first step is to consider the scope of the research and area of interest, focusing specifically on considerations such as local climate conditions, terrain, vegetation, and canopy access. Specifically, the spatial locations, site, and canopy access (e.g., is it possible to reach canopy foliage?); vegetation composition and canopy architecture; timing of collection; and methods for sample retrieval are key to identify prior to field campaigns in order to maximize the utility of the field samples for conversion of RS signatures to accurate trait maps. Furthermore, it may be important to consider what approach may be best to characterize the vegetation canopy architecture and/or composition to facilitate scaling of each trait to the pixel or plot scale (e.g., using basal area, LAI). This may strongly depend on the dominant vegetation types, where more open canopies may require a different approach to a closed canopy, or on the spatial resolution of the imagery. Observational data range is a primary consideration (see Schweiger, Chap. 15), and sample locations should be chosen to cover the range of canopy types and vegetation communities that will fall within the RS observations. The timing of the field sampling should be as close to the RS collection date as possible, as an optimal approach, but at least be selected to match the phenological stage of the vegetation during the imagery collection, if leveraging sample campaigns in following year(s).

A number of different methods have been used to collect plant functional traits to link with RS imagery (e.g., Wang et al. 2019). Common approaches for the collection of canopy leaf samples include the use of slingshot, pruning pole, and shotgun (Lausch et al., Chap. 13), but also include line-launcher and air cannon (e.g., Serbin et al. 2014); simpler tools and hand shears are often used for accessible, shorter canopies. Regardless of the sample collection approach, harvested leaves should be reasonably intact and minimally damaged in order to avoid any issues with changes in leaf chemistry from physical damage or stress. In addition, leaves should be immediately measured for leaf optical properties and fresh mass, if these are of interest, then stored in humidified and sealed bags and placed in a cool, dark place prior to transport for further processing. Processing should then be completed within 2–4 hours of sampling—though a much shorter time between sample and measurement or different sample storage and handling (e.g., flash freezing in liquid nitrogen) may be needed for specific biochemical traits. Typically top-of-canopy, sunlit samples have been the main focus; however, more recent work has also begun to focus on collection of canopy and subcanopy samples (e.g., Serbin et al. 2014; Singh et al. 2015). This provides the ability to evaluate the depth in the canopy needed to link traits with image, which may vary by vegetation type or LAI.

3.3.2 Evaluating Functional Trait Maps and the Need to Quantify Uncertainties

Maps of plant functional traits are useful for a wide variety of applications. From an ecological perspective, maps of plant traits across broad biotic and abiotic gradients can be used to explore the drivers of plant trait variation in relation to climate, soils, and vegetation types (e.g., McNeil et al. 2008). Modeling activities can leverage these trait maps as either inputs for model parameterization across space and time (Ollinger and Smith 2005) or to evaluate prognostic plant trait predictions. However, to maximize the utility of functional trait maps a detailed understanding of the their uncertainties across space and time is required.

In the earliest functional trait mapping work, predictive model uncertainties were limited to the "goodness of fit" and overall model root mean square error (RMSE) statistics provided by the modeling approach (e.g., Wessman et al. 1988; Martin and Aber 1997; Townsend et al. 2003). While this information is helpful to understand the accuracy of the model fit, that level of accuracy assessment is insufficient for characterizing the uncertainty of the trait maps themselves. Mapping efforts should instead provide an accounting of the trait measurement, scaling, and algorithm uncertainties and provide this information in the resulting trait map data products. However, detailed error propagation is not trivial, particularly with respect to empirical modeling approaches, and is an ongoing and active area of research in the RS sciences and not discussed in detail here. On the other hand, efforts to provide product uncertainties do exist. Serbin et al. (2014) and Singh et al. (2015) illustrate how to incorporate data and modeling uncertainties at the leaf and canopy scales in the mapping of plant functional traits. This approach captures the uncertainties stemming from the leaf-level estimation of traits (Serbin et al. 2014) and the modeling of plot-level spectra and trait values (Singh et al. 2015) using a similar PLSR and uncertainty analysis approach. The result is an ensemble of PLSR models to apply to new RS data providing mean and error metrics for every pixel in the image. However, even approaches such as these fail to incorporate and propagate the uncertainties stemming from the atmospheric correction workflow given the challenge of extract the information needed to enable this on a pixel-by-pixel or even a scene-by-scene basis. Future work will be required to focus on capturing this information and providing it to the end-user who conducts the trait mapping efforts.

Uncertainty in RTM approaches have generally been derived based on inversion approaches applied to imagery. For example, as described in Sect. 3.2.2.2, a commonly used approach to the inversion of RTM simulations for the RS of functional traits is the use of LUTs. Some LUT approaches provide results based on the "best fit" of the model inversion results to the RS observations. However, this only provides an assessment of error where field measurements can be used to evaluate the retrieved values. Given the challenge of equifinality in RTM approaches, later efforts have used an ensemble of best fit results to provide a mean and distribution of values that provide a good fit of modeled reflectance to observed (e.g., Weiss et al. 2000; Banskota et al. 2015). Using this approach allows for the description of pixel-level uncertainty based on the best fit ensembles; however, these need to be combined with an accuracy assessment to get a true uncertainty of the functional trait retrievals. More recent approaches have leveraged Bayesian inversion approaches that provide output that is not a point estimate for each parameter but rather the joint probability distribution that includes estimates of parameter uncertainties and covariance structure (Shiklomanov et al. 2016). Regardless of the approach, the key is that the derived products provide a reasonable assessment of trait uncertainty across the spatial and temporal domain (where appropriate).

3.3.3 Current and Future Opportunities in the Use of Remote Sensing to Characterize Functional Traits and Biodiversity

The ability to map foliar functional traits from imaging spectroscopy greatly expands the potential for understanding patterns of vegetation function and functional diversity both locally and broadly across biomes, especially in comparison to the challenges of fully characterizing spatial and temporal (across seasons and between years) variation using field data (e.g., the TRY database). With forthcoming spaceborne sensors (see Schimel et al., Chap. 19) and continental-scale experiments like the US National Ecological Observatory Network (NEON), we are able to test relationships among traits and characterize functional diversity at unprecedented scales. For example, NEON is collecting imaging spectroscopy data at 1 m resolution and waveform lidar data almost annually for 30 years at 81 10 km × 10 km sites covering 20 biomes defined for the USA. With the addition of lidar, which enables measuring traits such as plant area index, canopy height, canopy volume, and aboveground biomass (of forests), a broad suite of traits can be leveraged to test relationships that have been published in the literature (e.g., the leaf economics spectrum) and are generally tested now at global scales using extensive—but still not comprehensive—databases such as TRY. With spaceborne imaging, phenological variation in traits (e.g., Yang et al. 2016) can be further explored. For example, preliminary mapping of key functional traits across all NEON biomes in the USA shows the leaf economics spectrum relationship between LMA and nitrogen for forest and grassland ecosystems east of the US Rocky Mountains (Fig. 3.8.) in comparison to the data set used for the original LES studies, GLOPNET (Global Plant Trait Network, Wright et al. 2004; Reich et al. 2007). Importantly, the use of data from RS platforms, such as NEON, AVIRIS, and upcoming spaceborne sensors (see Schimel et al., Chap. 19), enables the filling of critical research gaps and global coverage in remote regions, as suggested by Jetz et al. (2016) and Schimel et al. (2015). The relationship does not differ significantly from published relationships but does suggest a breadth of the relationship as well as outliers for a number of observations many orders of magnitude higher than is possible from field databases. Field databases are still required for basic science studies, as well as inventory, calibration, and validation, but RS offers new possibilities for baseline characterization of Earth's functional diversity and thus testing new hypotheses about the drivers of such variation, using the range of traits detectable from RS (Tables 3.1 and 3.2).

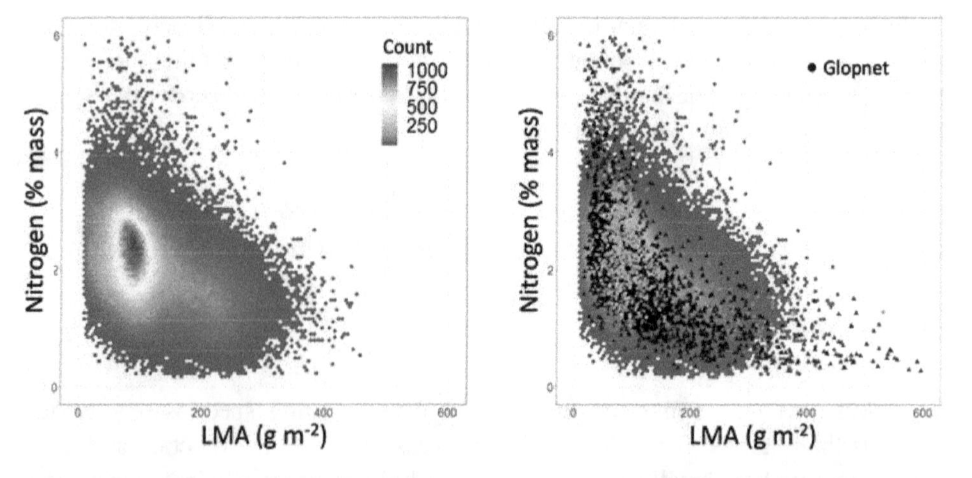

Fig. 3.8. LMA versus nitrogen for NEON for GLOPNET observations (black dots, truncated to observations with LMA <600) vs. pixel predictions derived for NEON sites east of the US Rocky Mountains (color gradient). Color gradient is density of pixel observations based on 333,500 pixel values randomly extracted from 447 flight NEON flight lines in 18 sites across 6 biomes

Furthermore, coupling of spectral and functional trait databases (e.g., ecosis.org) will facilitate more rapid development and testing of new functional algorithms or the expansion of the scope of inference of existing models. In addition, the inclusion of high spectral resolution sensors on unmanned aerial systems (UASs, Shiklomanov et al. 2019) provides the opportunity to leverage similar scaling approaches as presented in this chapter with UAS observations to provide unprecedented temporal coverage and targeted spatial sampling that can be used to understand ecosystem in new detail or aid in the scaling from the plant to grid cell. In all, functional trait maps from imaging spectroscopy will supplement data and approaches presented by Butler et al. (2017) or Moreno-Martínez et al. (2018) for broad-scale trait characterization.

Acknowledgments The authors would like to thank Anna Schweiger, Erin Hestir, and Jeannine Cavender-Bares for their careful reviews, input, and suggestions on earlier versions of this chapter as part of the and the National Institute of Mathematical Biology and Synthesis Working Group on Remotely Sensing Biodiversity. Special thanks to Tiffany Bowman and Yelena Belyavina for assistance with graphics. S.P.S was supported by the Next-Generation Ecosystem Experiments (NGEEs) in the Arctic and tropics that are supported by the Office of Biological and Environmental Research in the Department of Energy, Office of Science, and through the United States Department of Energy contract No. DE-SC0012704 to Brookhaven National Laboratory. P.T. acknowledges support from NSF Emerging Frontiers Macrosystems Biology and NEON-Enabled Science (MSB-NES) grant 1638720, USDA McIntire-Stennis WIS01809 and Hatch WIS01874, NASA Biodiversity Program grant 80NSSC17K0677, and AIST program grant 80NSSC17K0244.

References

Aber JD, Melillo JM (1982) Nitrogen immobilization in decaying hardwood leaf litter as a function of initial nitrogen and lignin content. Can J Bot 60:2261–2269

Ahl DE, Gower ST, Burrows SN, Shabanov NV, Myneni RB, Knyazikhin Y (2006) Monitoring spring canopy phenology of a deciduous broadleaf forest using modis. Remote Sens Environ 104:88–95

Ainsworth EA, Serbin SP, Skoneczka JA, Townsend PA (2014) Using leaf optical properties to detect ozone effects on foliar biochemistry. Photosynth Res 119:65–76

Asner GP (1998) Biophysical and biochemical sources of variability in canopy reflectance. Remote Sens Environ 64:234–253

Asner GP, Nepstad D, Cardinot G, Ray D (2004) Drought stress and carbon uptake in an Amazon forest measured with spaceborne imaging spectroscopy. Proc Natl Acad Sci U S A 101:6039–6044

Asner GP, Knapp DE, Boardman J, Green RO, Kennedy-Bowdoin T, Eastwood M, Martin RE, Anderson C, Field CB (2012) Carnegie Airborne Observatory-2: increasing science data dimensionality via high-fidelity multi-sensor fusion. Remote Sens Environ 124:454–465

Asner GP, Martin RE (2008) Spectral and chemical analysis of tropical forests: scaling from leaf to canopy levels. Remote Sens Environ 112:3958–3970

Asner PG, Martin ER (2015) Spectroscopic remote sensing of non-structural carbohydrates in forest canopies. Remote Sens 7(4)

Asner GP, Martin RE, Anderson CB, Knapp DE (2015) Quantifying forest canopy traits: imaging spectroscopy versus field survey. Remote Sens Environ 158:15–27

Asner GP, Brodrick PG, Anderson CB, Vaughn N, Knapp DE, Martin RE (2016) Progressive forest canopy water loss during the 2012-2015 California drought. Proc Natl Acad Sci U S A 113:E249–E255

Asner GP, Martin RE, Carranza-Jiménez L, Sinca F, Tupayachi R, Anderson CB, Martinez P (2014) Functional and biological diversity of foliar spectra in tree canopies throughout the Andes to Amazon region. New Phytol 204:127–139

Asner GP, Martin RE, Knapp DE, Tupayachi R, Anderson C, Carranza L, Martinez P, Houcheime M, Sinca F, Weiss P (2011) Spectroscopy of canopy chemicals in humid tropical forests. Remote Sens Environ 115:3587–3598

Asner GP, Wessman CA, Bateson CA, Privette JL (2000) Impact of tissue, canopy, and landscape factors on the hyperspectral reflectance variability of arid ecosystems. Remote Sens Environ 74:69–84

Bacour C, Jacquemoud S, Tourbier Y, Dechambre M, Frangi JP (2002) Design and analysis of numerical experiments to compare four canopy reflectance models. Remote Sens Environ 79:72–83

Banskota A, Serbin SP, Wynne RH, Thomas VA, Falkowski MJ, Kayastha N, Gastellu-Etchegorry JP, Townsend PA (2015) An LUT-based inversion of DART model to estimate forest LAI from hyperspectral data. IEEE J Sel Top Appl Earth Obs Remote Sens 8:3147–3160

Baret F, Vanderbilt VC, Steven MD, Jacquemoud S (1994) Use of spectral analogy to evaluate canopy reflectance sensitivity to leaf optical properties. Remote Sens Environ 48:253–260

Billings WD, Morris RJ (1951) Reflection of visible and infrared radiation from leaves of different ecological groups. Am J Bot 38(5):327–331

Blackburn GA (2007) Hyperspectral remote sensing of plant pigments. J Exp Bot 58:855–867

Blackburn GA, Ferwerda JG (2008) Retrieval of chlorophyll concentration from leaf reflectance spectra using wavelet analysis. Remote Sens Environ 112:1614–1632

Bolster KL, Martin ME, Aber JD (1996) Determination of carbon fraction and nitrogen concentration in tree foliage by near infrared reflectance: a comparison of statistical methods. Can J For Res 26:590–600

Bonan GB, Levis S, Kergoat L, Oleson KW (2002) Landscapes as patches of plant functional types: an integrating concept for climate and ecosystem models. Global Biogeochem Cy 16:5-1–5-23

Butler EE, Datta A, Flores-Moreno H, Chen M, Wythers KR, Fazayeli F, Banerjee A, Atkin OK, Kattge J, Amiaud B, Blonder B, Boenisch G, Bond-Lamberty B, Brown KA, Byun C, Campetella G, Cerabolini BEL, Cornelissen JHC, Craine JM, Craven D, de Vries FT, Díaz S, Domingues TF, Forey E, González-Melo A, Gross N, Han W, Hattingh WN, Hickler T, Jansen S, Kramer K, Kraft NJB, Kurokawa H, Laughlin DC, Meir P, Minden V, Niinemets Ü, Onoda Y, Peñuelas J, Read Q, Sack L, Schamp B, Soudzilovskaia NA, Spasojevic MJ, Sosinski E, Thornton PE, Valladares F, van Bodegom PM, Williams M, Wirth C, Reich PB (2017) Mapping local and global variability in plant trait distributions. Proc Natl Acad Sci 114:E10937–E10946

Card DH, Peterson DL, Matson PA, Aber JD (1988) Prediction of leaf chemistry by the use of visible and near infrared reflectance spectroscopy. Remote Sens Environ 26:123–147

Carrascal LM, Galván I, Gordo O (2009) Partial least squares regression as an alternative to current regression methods used in ecology. Oikos 118:681–690

Cavender-Bares J, Gamon JA, Hobbie SE, Madritch MD, Meireles JE, Schweiger AK, Townsend PA (2017) Harnessing plant spectra to integrate the biodiversity sciences across biological and spatial scales. Am J Bot 104:966–969

Chavana-Bryant C, Malhi Y, Wu J, Asner GP, Anastasiou A, Enquist BJ, Cosio Caravasi EG, Doughty CE, Saleska SR, Martin RE, Gerard FF (2017) Leaf aging of Amazonian canopy trees as revealed by spectral and physiochemical measurements. New Phytol 214:1049–1063

Chen JM, Cihlar J (1996) Retrieving leaf area index of boreal conifer forests using landsat TM images. Remote Sens Environ 55:153–162

Cheng T, Rivard B, Sanchez-Azofeifa A (2011) Spectroscopic determination of leaf water content using continuous wavelet analysis. Remote Sens Environ 115:659–670

Clark W (1946) Photography by infrared: its principles and applications: J. Wiley & sons, Incorporated

Cocks T, Jensen R, Stewart A, Wilson I, and Shields T (1998) The HyMap airborne hyperspectral sensor: the system, calibration and performance. In: Proceedings of 1st EARSeL Workshop on Imaging Spectroscopy, Zurich, Switzerland, pp 37–42

Colombo R, Merom M, Marchesi A, Busetto L, Rossini M, Giardino C, Panigada C (2008) Estimation of leaf and canopy water content in poplar plantations by means of hyperspectral indices and inverse modeling. Remote Sens Environ 112:1820–1834

Cotrozzi L, Townsend PA, Pellegrini E, Nali C, Couture JJ (2018) Reflectance spectroscopy: a novel approach to better understand and monitor the impact of air pollution on Mediterranean plants. Environ Sci Pollut Res 25:8249–8267

Couture J, Singh A, Rubert-Nason KF, Serbin SP, Lindroth RL, Townsend PA (2016) Spectroscopic determination of ecologically relevant plant secondary metabolites. Methods Ecol Evol (in press).

Couture JJ, Serbin SP, Townsend PA (2013) Spectroscopic sensitivity of real-time, rapidly induced phytochemical change in response to damage. New Phytol 198:311–319

Curran PJ (1989) Remote-sensing of foliar chemistry. Remote Sens Environ 30:271–278

Dahlin KM, Asner GP, Field CB (2013) Environmental and community controls on plant canopy chemistry in a Mediterranean-type ecosystem. Proc Natl Acad Sci 110:6895–6900

Darvishzadeh R, Skidmore A, Schlerf M, Atzberger C (2008) Inversion of a radiative transfer model for estimating vegetation LAI and chlorophyll in a heterogeneous grassland. Remote Sens Environ 112:2592–2604

Datt B (1998) Remote sensing of chlorophyll a, chlorophyll b, chlorophyll a+b, and total carotenoid content in eucalyptus leaves. Remote Sens Environ 66(2):111–121

Davies T (1998) The history of near infrared spectroscopic analysis: Past, present and future "From sleeping technique to the morning star of spectroscopy". Analusis 26(4):17–19

Dawson TP, Curran PJ, Plummer SE (1998) LIBERTY – Modeling the Effects of Leaf Biochemical Concentration on Reflectance Spectra. Remote Sens Environ 65:50–60

Díaz S, Kattge J, Cornelissen JHC, Wright IJ, Lavorel S, Dray S, Reu B, Kleyer M, Wirth C, Colin Prentice I, Garnier E, Bönisch G, Westoby M, Poorter H, Reich PB, Moles AT, Dickie J, Gillison AN, Zanne AE, Chave J, Joseph Wright S, Sheremet'ev SN, Jactel H, Baraloto C, Cerabolini B, Pierce S, Shipley B, Kirkup D, Casanoves F, Joswig JS, Günther A, Falczuk V, Rüger N, Mahecha MD, Gorné LD (2015) The global spectrum of plant form and function. Nature 529:167

Drolet GG, Middleton EM, Huemmrich KF, Hall FG, Amiro BD, Barr AG, Black TA, McCaughey JH, Margolis HA (2008) Regional mapping of gross light-use efficiency using MODIS spectral indices. Remote Sens Environ 112:3064–3078

DuBois S, Desai AR, Singh A, Serbin SP, Goulden ML, Baldocchi DD, Ma S, Oechel WC, Wharton S, Kruger EL, Townsend PA (2018) Using imaging spectroscopy to detect variation in terrestrial ecosystem productivity across a water-stressed landscape. Ecol Appl 28:1313–1324

Elvidge CD (1990) Visible and near-infrared reflectance characteristics of dry plant materials. Int J Remote Sens 11:1775–1795

Ely KS, Burnett AC, Lieberman-Cribbin W, Serbin S, and Rogers A (2019) Spectroscopy can predict key leaf traits associated with source–sink balance and carbon–nitrogen status. J Exp Bot. 70:1789–1799

Enquist BJ, Bentley LP, Shenkin A, Maitner B, Savage V, Michaletz S, Blonder B, Buzzard V, Espinoza TEB, Farfan-Rios W, Doughty CE, Goldsmith GR, Martin RE, Salinas N, Silman M, Díaz S, Asner GP, Malhi Y (2017) Assessing trait-based scaling theory in tropical forests spanning a broad temperature gradient. Global Ecol Biogeogr 26:1357–1373

Evans JR, Clarke VC (2018) The nitrogen cost of photosynthesis. J Exp Bot 70:7–15

Falster DS, Westoby M (2003) Leaf size and angle vary widely across species: what consequences for light interception? New Phytol 158:509–525

Feilhauer H, Asner GP, Martin RE, Schmidtlein S (2010) Brightness-normalized partial least squares regression for hyperspectral data. J Quant Spectrosc Radiat Transf 111:1947–1957

Feret J-B, Francois C, Gitelson A, Asner GP, Barry KM, Panigada C, Richardson AD, Jacquemoud S (2011) Optimizing spectral indices and chemometric analysis of leaf chemical properties using radiative transfer modeling. Remote Sens Environ 115:2742–2750

Féret JB, Francois C, Asner GP, Gitelson AA, Martin RE, Bidel LPR, Ustin SL, le Maire G, Jacquemoud S (2008) PROSPECT-4 and 5: Advances in the leaf optical properties model separating photosynthetic pigments. Remote Sens Environ 112(6):3030–3043

Féret JB, Gitelson AA, Noble SD, Jacquemoud S (2017) PROSPECT-D: towards modeling leaf optical properties through a complete lifecycle. Remote Sens Environ 193:204–215

Féret JB, le Maire G, Jay S, Berveiller D, Bendoula R, Hmimina G, Cheraiet A, Oliveira JC, Ponzoni FJ, Solanki T, de Boissieu F, Chave J, Nouvellon Y, Porcar-Castell A, Proisy C, Soudani K, Gastellu-Etchegorry JP, Lefèvre-Fonollosa MJ (2018) Estimating leaf mass per area and equivalent water thickness based on leaf optical properties: Potential and limitations of physical modeling and machine learning. In: Remote Sens Environ

Field C, Mooney HA (1986) The photosynthesis-nitrogen relationship in wild plants. In: Givnish T (ed) On the economy of plant form and function. Cambridge University Press, Cambridge, pp 22–55

Foster JR, Townsend PA, Zganjar CE (2008) Spatial and temporal patterns of gap dominance by low-canopy lianas detected using EO-1 Hyperion and Landsat Thematic Mapper. Remote Sens Environ 112:2104–2117

Friedlingstein P, Meinshausen M, Arora VK, Jones CD, Anav A, Liddicoat SK, Knutti R (2014) Uncertainties in CMIP5 climate projections due to carbon cycle feedbacks. J Clim 27:511–526

Funk JL, Larson JE, Ames GM, Butterfield BJ, Cavender-Bares J, Firn J, Laughlin DC, Sutton-Grier AE, Williams L, Wright J (2017) Revisiting the Holy Grail: using plant functional traits to understand ecological processes. Biol Rev 92:1156–1173

Gamon JA, Field CB, Goulden ML, Griffin KL, Hartley AE, Joel G, Penuelas J, Valentini R (1995) Relationships between NDVI, canopy structure, and photosynthesis in 3 Californian vegetation types. Ecol Appl 5:28–41

Gamon JA, Penuelas J, Field CB (1992) A narrow-waveband spectral index that tracks diurnal changes in photosynthetic efficiency. Remote Sens Environ 41:35–44

Gamon JA, Serrano L, Surfus JS (1997) The photochemical reflectance index: an optical indicator of photosynthetic radiation use efficiency across species, functional types, and nutrient levels. Oecologia 112:492–501

Ganapol BD, Johnson LF, Hammer PD, Hlavka CA, Peterson DL (1998) LEAFMOD: a new within-leaf radiative transfer model. Remote Sens Environ 63:182–193

Gao BC (1996) NDWI—A normalized difference water index for remote sensing of vegetation liquid water from space. Remote Sens Environ 58(3):257–266

Gao BC, Goetz AFH (1995) Retrieval of equivalent water thickness and information related to biochemical components of vegetation canopies from AVIRIS data. Remote Sens Environ 52(3):155–162

Gao BC, Heidebrecht KB, Goetz AFH (1993) Derivation of scaled surface reflectances from AVIRIS data. Remote Sens Environ, 44:165–178

Gastellu-Etchegorry JP, Zagolski F, Mougtn E, Marty G, Giordano G (1995). An assessment of canopy chemistry with AVIRIS—a case study in the Landes Forest, South-west France. Int J Remote Sens 16(3):487–501

Gastellu-Etchegorry J-P, Yin T, Lauret N, Cajgfinger T, Gregoire T, Grau E, Feret J-B, Lopes M, Guilleux J, Dedieu G, Malenovský Z, Cook BD, Morton D, Rubio J, Durrieu S, Cazanave G, Martin E, Ristorcelli T (2015) Discrete anisotropic radiative transfer (DART 5) for modeling airborne and satellite spectroradiometer and LIDAR acquisitions of natural and urban landscapes. Remote Sens 7:1667–1701

Gates DM, Keegan HJ, Schleter JC, Weidner VR (1965) Spectral properties of plants. Appl Opt 4(1):11–20

Geladi P, Kowalski BR (1986) Partial least-squares regression - A tutorial. Anal Chim Acta 185:1–17

Gil-Pérez B, Zarco-Tejada PJ, Correa-Guimaraes A, Relea-Gangas E, Navas-Gracia LM, Hernández-Navarro S, Sanz-Requena JF, Berjón A, Martín-Gil J (2010). Vitis-Journal of Grapevine Research 49(4):167–173

Gitelson AA (2004) Wide dynamic range vegetation index for remote quantification of biophysical characteristics of vegetation. J Plant Physiol 161:165–173
Gitelson AA, Vina A, Verma SB, Rundquist DC, Arkebauer TJ, Keydan G, Leavitt B, Ciganda V, Burba GG, Suyker AE (2006) Relationship between gross primary production and chlorophyll content in crops: Implications for the synoptic monitoring of vegetation productivity. J Geophys Res-Atmos 111:13
Glenn EP, Huete AR, Nagler PL, Nelson SG (2008) Relationship Between Remotely-sensed Vegetation Indices, Canopy Attributes and Plant Physiological Processes: What Vegetation Indices Can and Cannot Tell Us About the Landscape. Sensors (Basel) 8:2136–2160
Goetz SJ, Bunn AG, Fiske GJ, Houghton RA (2005) Satellite-observed photosynthetic trends across boreal north america associated with climate and fire disturbance. Proc Natl Acad Sci U S A 102:13521–13525
Goetz SJ, Fiske GJ, Bunn AG (2006) Using satellite time-series data sets to analyze fire disturbance and forest recovery across Canada. Remote Sens Environ 101:352–365
Gökkaya K, Thomas V, Noland TL, McCaughey H, Morrison I, Treitz P (2015) Prediction of macronutrients at the canopy level using spaceborne imaging spectroscopy and LiDAR data in a mixedwood boreal forest. Remote Sens 7:9045–9069
Goward SN, Huemmrich KF (1992) Vegetation canopy PAR absorptance and the normalized difference vegetation index – An assessment using the SAIL model. Remote Sens Environ 39:119–140
Green DS, Erickson JE, Kruger EL (2003) Foliar morphology and canopy nitrogen as predictors of light-use efficiency in terrestrial vegetation. Agric For Meteorol 115:165–173
Green RO, Eastwood ML, Sarture CM, Chrien TG, Aronsson M, Chippendale BJ, Faust JA, Pavri BE, Chovit CJ, Solis MS, Olah MR, Williams O (1998) Imaging spectroscopy and the Airborne Visible Infrared Imaging Spectrometer (AVIRIS). Remote Sens Environ 65:227–248
Grossman YL, Ustin SL, Jacquemoud S, Sanderson EW, Schmuck G, Verdebout J (1996) Critique of stepwise multiple linear regression for the extraction of leaf biochemistry information from leaf reflectance data. Remote Sens Environ 56:182–193
Hilker T, Coops NC, Wulder MA, Black TA, Guy RD (2008) The use of remote sensing in light use efficiency based models of gross primary production: A review of current status and future requirements. Sci Total Environ 404:411–423
Huemmrich KF (2013) Simulations of seasonal and latitudinal variations in leaf inclination angle distribution: implications for remote sensing. J Adv Remote Sens 02:9
Hunt ER, Rock BN (1989) Detection of changes in leaf water content using near-infrared and middle-infrared reflectances. Remote Sens Environ 30:43–54
Inoue Y, Penuelas J, Miyata A, Mano M (2008) Normalized difference spectral indices for estimating photosynthetic efficiency and capacity at a canopy scale derived from hyperspectral and CO2 flux measurements in rice. Remote Sens Environ 112:156–172
IPCC (2018) Summary for Policymakers. In: Global Warming of 1.5°C. An IPCC Special Report on the impacts of global warming of 1.5°C above pre-industrial levels and related global greenhouse gas emission pathways, in the context of strengthening the global response to the threat of climate change, sustainable development, and efforts to eradicate poverty [Masson-Delmotte, V., P. Zhai, H.-O. Pörtner, D. Roberts, J. Skea, P.R. Shukla, A. Pirani, W. Moufouma-Okia, C. Péan, R. Pidcock, S. Connors, J.B.R. Matthews, Y. Chen, X. Zhou, M.I. Gomis, E. Lonnoy, T. Maycock, M. Tignor, and T. Waterfield (eds.)]. World Meteorological Organization, Geneva, Switzerland, 32 pp.
IPBES (2018) Summary for policymakers of the regional assessment report on biodiversity and ecosystem services for the Americas of the Intergovernmental Science-Policy Platform on Biodiversity and Ecosystem Services. IPBES secretariat, Bonn, Germany
Jacquemoud S, Baret F (1990) PROSPECT – a model of leaf optical-properties of spectra. Remote Sens Environ 34:75–91
Jacquemoud S, Baret F, Andrieu B, Danson FM, Jaggard K (1995) Extraction of vegetation biophysical parameters by inversion of the PROSPECT + SAIL models on sugar beet canopy reflectance data. Application to TM and AVIRIS sensors. Remote Sens Environ 52:163–172

Jacquemoud S, Ustin SL, Verdebout J, Schmuck G, Andreoli G, Hosgood B (1996) Estimating leaf biochemistry using the PROSPECT leaf optical properties model. Remote Sens Environ 56:194–202

Jacquemoud S, Verhoef W, Baret F, Bacour C, Zarco-Tejada PJ, Asner GP, Francois C, Ustin SL (2009) PROSPECT + SAIL models: a review of use for vegetation characterization. Remote Sens Environ 113:S56–S66

Jensen DJ, Simard M, Cavanaugh KC, Thompson DR (2018) Imaging spectroscopy BRDF correction for mapping Louisiana's coastal ecosystems. IEEE Trans Geosci Remote Sens 56:1739–1748

Jetz W, Cavender-Bares J, Pavlick R, Schimel D, Davis FW, Asner GP, Guralnick R, Kattge J, Latimer AM, Moorcroft P, Schaepman ME, Schildhauer MP, Schneider FD, Schrodt F, Stahl U, Ustin SL (2016) Monitoring plant functional diversity from space. Nature Plants 2:16024

Jetz W, Wilcove DS, Dobson AP (2007) Projected impacts of climate and land-use change on the global diversity of birds. PLoS Biol 5:e157

Johnson LF, Hlavka CA, Peterson DL (1994) Multivariate analysis of AVIRIS data for canopy biochemical estimation along the oregon transect. Remote Sens Environ 47(2):216–230

Jones LA, Kimball JS, McDonald KC, Chan STK, Njoku EG, Oechel WC (2007) Satellite microwave remote sensing of boreal and arctic soil temperatures from AMSR-E. IEEE Trans Geosci Remote Sens 45:2004–2018

Kalacska M, Lalonde M, Moore TR (2015) Estimation of foliar chlorophyll and nitrogen content in an ombrotrophic bog from hyperspectral data: Scaling from leaf to image. Remote Sens Environ 169:270–279

Kampe TU, Johnson BR, Kuester M, Keller M (2010) NEON: the first continental-scale ecological observatory with airborne remote sensing of vegetation canopy biochemistry and structure. J Appl Remote Sens 4:043510

Kasischke ES, French NHF, Harrell P, Christensen NL, Ustin SL, Barry D (1993) Monitoring of wildfires in boreal forests using large-area AVHRR NDVI composite image data. Remote Sens Environ 45:61–71

Kattenborn T, Fassnacht FE, Pierce S, Lopatin J, Grime JP, Schmidtlein S (2017) Linking plant strategies and plant traits derived by radiative transfer modelling. J Veg Sci 28:717–727

Kattenborn T, Fassnacht FE, Schmidtlein S (2019) Differentiating plant functional types using reflectance: which traits make the difference? Remote Sens Ecol Conserv 5(1):5–19

Kattge J, DÍaz S, Lavorel S, Prentice IC, Leadley P, Bönisch G, Garnier E, Westoby M, Reich PB, Wright IJ, Cornelissen JHC, Violle C, Harrison SP, Van Bodegom PM, Reichstein M, Enquist BJ, Soudzilovskaia NA, Ackerly DD, Anand M, Atkin O, Bahn M, Baker TR, Baldocchi D, Bekker R, Blanco CC, Blonder B, Bond WJ, Bradstock R, Bunker DE, Casanoves F, Cavender-Bares J, Chambers JQ, Chapin Iii FS, Chave J, Coomes D, Cornwell WK, Craine JM, Dobrin BH, Duarte L, Durka W, Elser J, Esser G, Estiarte M, Fagan WF, Fang J, Fernández-Méndez F, Fidelis A, Finegan B, Flores O, Ford H, Frank D, Freschet GT, Fyllas NM, Gallagher RV, Green WA, Gutierrez AG, Hickler T, Higgins SI, Hodgson JG, Jalili A, Jansen S, Joly CA, Kerkhoff AJ, Kirkup D, Kitajima K, Kleyer M, Klotz S, Knops JMH, Kramer K, Kühn I, Kurokawa H, Laughlin D, Lee TD, Leishman M, Lens F, Lenz T, Lewis SL, Lloyd J, Llusià J, Louault F, Ma S, Mahecha MD, Manning P, Massad T, Medlyn BE, Messier J, Moles AT, Mülier SC, Nadrowski K, Naeem S, Niinemets Ü, Nöliert S, Nüske A, Ogaya R, Oleksyn J, Onipchenko VG, Onoda Y, Ordoñez J, Overbeck G, Ozinga WA, Patiño S, Paula S, Pausas JG, Peñuelas J, Phillips OL, Pillar V, Poorter H, Poorter L, Poschlod P, Prinzing A, Proulx R, Rammig A, Reinsch S, Reu B, Sack L, Salgado-Negret B, Sardans J, Shiodera S, Shipley B, Siefert A, Sosinski E, Soussana JF, Swaine E, Swenson N, Thompson K, Thornton P, Waldram M, Weiher E, White M, White S, Wright SJ, Yguel B, Zaehle S, Zanne AE, Wirth C (2011) TRY – a global database of plant traits. Glob Chang Biol 17:2905–2935

Knyazikhin Y, Schull MA, Stenberg P, Mõttus M, Rautiainen M, Yang Y, Marshak A, Latorre Carmona P, Kaufmann RK, Lewis P, Disney MI, Vanderbilt V, Davis AB, Baret F, Jacquemoud S, Lyapustin A, Myneni RB (2013) Hyperspectral remote sensing of foliar nitrogen content. Proc Natl Acad Sci 110:E185–E192

Kobayashi H, Iwabuchi H (2008) A coupled 1-D atmosphere and 3-D canopy radiative transfer model for canopy reflectance, light environment, and photosynthesis simulation in a heterogeneous landscape. Remote Sens Environ 112:173–185

Kokaly RF, Asner GP, Ollinger SV, Martin ME, Wessman CA (2009) Characterizing canopy biochemistry from imaging spectroscopy and its application to ecosystem studies. Remote Sens Environ 113:S78–S91

Kokaly RF, Clark RN (1999) Spectroscopic determination of leaf biochemistry using band-depth analysis of absorption features and stepwise multiple linear regression. Remote Sens Environ 67:267–287

Kokaly RF, Skidmore AK (2015) Plant phenolics and absorption features in vegetation reflectance spectra near 1.66μm. Int J Appl Earth Obs Geoinf 43:55–83

Krinov EL (1953) Spectral reflectance properties of natural formations. National Research Council of Canada (Ottawa) Technical Translations TT-439

Kuusk A (2018) 3.03 - Canopy radiative transfer modeling. In: Liang S (ed). Comprehensive Remote Sensing. Oxford: Elsevier, 9–22, https://doi.org/10.1016/B978-0-12-409548-9.10534-2

Kuusk A, Nilson T (2000) A directional multispectral forest reflectance model. Remote Sens Environ 72:244–252

Lavorel S, Garnier E (2002) Predicting changes in community composition and ecosystem functioning from plant traits: revisiting the Holy Grail. Funct Ecol 16:545–556

LeBauer D, Kooper R, Mulrooney P, Rohde S, Wang D, Long SP, Dietze MC (2018) BETYdb: a yield, trait, and ecosystem service database applied to second-generation bioenergy feedstock production. GCB Bioenergy 10(1):61–71

Li D, Wang X, Zheng H, Zhou K, Yao X, Tian Y, Zhu Y, Cao W, Cheng T (2018) Estimation of area- and mass-based leaf nitrogen contents of wheat and rice crops from water-removed spectra using continuous wavelet analysis. Plant Methods 14:76

Li L, Cheng YB, Ustin S, Hu XT, Riaño D (2008) Retrieval of vegetation equivalent water thickness from reflectance using genetic algorithm (GA)-partial least squares (PLS) regression. Adv Space Res 41:1755–1763

Mand P, Hallik L, Penuelas J, Nilson T, Duce P, Emmett BA, Beier C, Estiarte M, Garadnai J, Kalapos T, Schmidt IK, Kovacs-Lang E, Prieto P, Tietema A, Westerveld JW, Kull O (2010) Responses of the reflectance indices PRI and NDVI to experimental warming and drought in European shrublands along a north-south climatic gradient. Remote Sens Environ 114:626–636

Martin ME, Aber JD (1997) High spectral resolution remote sensing of forest canopy lignin, nitrogen, and ecosystem processes. Ecol Appl 7:431–443

Martin ME, Plourde LC, Ollinger SV, Smith ML, McNeil BE (2008) A generalizable method for remote sensing of canopy nitrogen across a wide range of forest ecosystems. Remote Sens Environ 112:3511–3519

Matson P, Johnson L, Billow C, Miller J, Pu RL (1994) Seasonal patterns and remote spectral estimation of canopy chemistry across the Oregon transect. Ecol Appl 4:280–298

McNeil BE, Read JM, Sullivan TJ, McDonnell TC, Fernandez IJ, Driscoll CT (2008) The spatial pattern of nitrogen cycling in the Adirondack Park, New York. Ecol Appl 18:438–452

McNicholas HJ (1931) The visible and ultraviolet absorption spectra of carotin and xanthophyll and the changes accompanying oxidation. Bureau of Standards Journal of Research 7(1):171. Research Paper 337 (RP337)

Middleton EM, Ungar SG, Mandl DJ, Ong L, Frye SW, Campbell PE, Landis DR, Young JP, Pollack NH (2013) The Earth observing one (EO-1) satellite mission: over a decade in space. IEEE J Sel Top Appl Earth Obs Remote Sens 6:243–256

Miller CE, Green RO, Thompson DR, Thorpe AK, Eastwood M, Mccubbin IB, Olson-Duvall W, Bernas M, Sarture CM, Nolte S, Rios LM, Hernandez MA, Bue BD, Lundeen SR (2019) ABoVE: Hyperspectral Imagery from AVIRIS-NG, Alaskan and Canadian Arctic, 2017-2018. ORNL DAAC, Oak Ridge, Tennessee, USA. https://doi.org/10.3334/ORNLDAAC/1569

Mirik M, Norland JE, Crabtree RL, Biondini ME (2005) Hyperspectral one-meter-resolution remote sensing in yellowstone National Park, Wyoming: I. Forage nutritional values. Rangel Ecol Manag 58:452–458

Moorthy I, Miller JR, Noland TL (2008) Estimating chlorophyll concentration in conifer needles with hyperspectral data: An assessment at the needle and canopy level. Remote Sens Environ 112(6):2824–2838

Moreno-Martínez Á, Camps-Valls G, Kattge J, Robinson N, Reichstein M, van Bodegom P, Kramer K, Cornelissen JHC, Reich P, Bahn M, Niinemets Ü, Peñuelas J, Craine JM, Cerabolini BEL, Minden V, Laughlin DC, Sack L, Allred B, Baraloto C, Byun C, Soudzilovskaia NA, Running SW (2018) A methodology to derive global maps of leaf traits using remote sensing and climate data. Remote Sens Environ 218:69–88

Mutanga O, Kumar L (2007) Estimating and mapping grass phosphorus concentration in an African savanna using hyperspectral image data. Int J Remote Sens 28(21):4897–4911

Myneni RB, Williams DL (1994) On the relationship between FAPAR and NDVI. Remote Sens Environ 49:200–211

Neyret M, Bentley LP, Oliveras I, Marimon BS, Marimon-Junior BH, Almeida de Oliveira E, Barbosa Passos F, Castro Ccoscco R, dos Santos J, Matias Reis S, Morandi PS, Rayme Paucar G, Robles Cáceres A, Valdez Tejeira Y, Yllanes Choque Y, Salinas N, Shenkin A, Asner GP, Díaz S, Enquist BJ, Malhi Y (2016) Examining variation in the leaf mass per area of dominant species across two contrasting tropical gradients in light of community assembly. Ecology and Evolution 6:5674–5689

Niinemets U (2007) Photosynthesis and resource distribution through plant canopies. Plant Cell Environ 30:1052–1071

Niinemets Ü (2016) Leaf age dependent changes in within-canopy variation in leaf functional traits: a meta-analysis. J Plant Res 129:313–338

Nilson T, Kuusk A, Lang M, Lükk T (2003) Forest reflectance modeling: theoretical aspects and applications. Ambio 32:535–541

North PRJ (1996) Three-dimensional forest light interaction model using a Monte Carlo method. IEEE Trans Geosci Remote Sens 34:946–956

Norris KH, Hart JR (1965) Direct spectrophotometric determination of moisture content of grain and seeds. Proceedings of the 1963 International Symposium on Humidity and Moisture, Reinhold, New York, vol. 4, pp 19–25

Norris KH, Barnes RF, Moore JE, Shenk JS (1976) Predicting forage quality by infrared replectance spectroscopy. J Anim Sci 43(4):889–897

Ollinger SV (2011) Sources of variability in canopy reflectance and the convergent properties of plants. New Phytol 189:375–394

Ollinger SV, Smith ML (2005) Net primary production and canopy nitrogen in a temperate forest landscape: An analysis using imaging spectroscopy, modeling and field data. Ecosystems 8:760–778

Ollinger SV, Smith ML, Martin ME, Hallett RA, Goodale CL, Aber JD (2002) Regional variation in foliar chemistry and N cycling among forests of diverse history and composition. Ecology 83:339–355

Osnas JLD, Katabuchi M, Kitajima K, Wright SJ, Reich PB, Van Bael SA, Kraft NJB, Samaniego MJ, Pacala SW, Lichstein JW (2018) Divergent drivers of leaf trait variation within species, among species, and among functional groups. Proc Natl Acad Sci 115:5480–5485

Peckham SD, Ahl DE, Serbin SP, Gower ST (2008) Fire-induced changes in green-up and leaf maturity of the Canadian boreal forest. Remote Sens Environ 112:3594–3603

Penuelas J, Filella I, Gamon JA (1995) Assessment of photosynthetic radiation-use efficiency with spectral reflectance. New Phytol 131:291–296

Peterson DL, Aber JD, Matson PA, Card DH, Swanberg N, Wessman C, Spanner M (1988) Remote-sensing of forest canopy and leaf biochemical contents. Remote Sens Environ 24:85–108

Pettorelli N, Wegmann M, Skidmore A, Mücher S, Dawson TP, Fernandez M, Lucas R, Schaepman ME, Wang T, O'Connor B, Jongman RHG, Kempeneers P, Sonnenschein R, Leidner AK, Böhm M, He KS, Nagendra H, Dubois G, Fatoyinbo T, Hansen MC, Paganini M, de Klerk HM, Asner GP, Kerr JT, Estes AB, Schmeller DS, Heiden U, Rocchini D, Pereira HM, Turak E, Fernandez N, Lausch A, Cho MA, Alcaraz-Segura D, McGeoch MA, Turner W, Mueller A, St-Louis V,

Penner J, Vihervaara P, Belward A, Reyers B, Geller GN (2016) Framing the concept of satellite remote sensing essential biodiversity variables: challenges and future directions. Remote Sens Ecol Conser 2:122–131
Poorter H, Niinemets U, Poorter L, Wright IJ, Villar R (2009) Causes and consequences of variation in leaf mass per area (LMA): a meta-analysis. New Phytol 182:565–588
Raabe K, Pisek J, Sonnentag O, Annuk K (2015) Variations of leaf inclination angle distribution with height over the growing season and light exposure for eight broadleaf tree species. Agric For Meteorol 214-215:2–11
Rabideau GS, French CS, Holt AS (1946) The absorption and reflection spectra of leaves, chloroplast suspensions, and chloroplast fragments as measured in an Ulbricht sphere. Am J Bot 33(10):769–777
Reich PB (2014) The world-wide 'fast–slow' plant economics spectrum: a traits manifesto. J Ecol 102:275–301
Reich PB, Walters MB, Ellsworth DS (1997) From tropics to tundra: Global convergence in plant functioning. Proc Natl Acad Sci U S A 94:13730–13734
Reich PB, Wright IJ, Cavender-Bares J, Craine JM, Oleksyn J, Westoby M, Walters MB (2003) The evolution of plant functional variation: traits, spectra, and strategies. Int J Plant Sci 164:S143–S164
Reich PB, Wright IJ, Lusk CH (2007) Predicting leaf physiology from simple plant and climate attributes: a global GLOPNET analysis. Ecol Appl 17:1982–1988
Rice J, Seixas CS, Zaccagnini ME, BedoyaGaitán M, Valderrama N, Anderson CB, Arroyo MTK, Bustamante M, Cavender-Bares J, Diaz-de-Leon A, Fennessy S, Márquez JRG, Garcia K, Helmer EH, Herrera B, Klatt B, Ometo JP, Osuna VR, Scarano FR, Schill S, and Farinaci JS (2018) IPBES, The regional assessment report on biodiversity and ecosystem services for the Americas. Bonn, Germany
Richardson AD, Duigan SP, Berlyn GP (2002) An evaluation of noninvasive methods to estimate foliar chlorophyll content. New Phytol 153:185–194
Roberts DA, Ustin SL, Ogunjemiyo S, Greenberg J, Dobrowski SZ, Chen JQ, Hinckley TM (2004) Spectral and structural measures of northwest forest vegetation at leaf to landscape scales. Ecosystems 7:545–562
Ross J (1981) Optical properties of phytoelements. In: Ross J (ed) The radiation regime and architecture of plant stands. Springer Netherlands, Dordrecht, pp 175–187
Schaepman ME, Jehle M, Hueni A, D'Odorico P, Damm A, Weyermann J, Schneider FD, Laurent V, Popp C, Seidel FC, Lenhard K, Gege P, Küchler C, Brazile J, Kohler P, De Vos L, Meuleman K, Meynart R, Schläpfer D, Kneubühler M, Itten KI (2015) Advanced radiometry measurements and Earth science applications with the Airborne Prism Experiment (APEX). Remote Sens Environ 158:207–219
Schimel D, Pavlick R, Fisher JB, Asner GP, Saatchi S, Townsend P, Miller C, Frankenberg C, Hibbard K, Cox P (2015) Observing terrestrial ecosystems and the carbon cycle from space. Glob Chang Biol 21:1762–1776
Schläpfer D, Richter R, Feingersh T (2015) Operational BRDF effects correction for wide-field-of-view optical scanners (BREFCOR). IEEE Trans Geosci Remote Sens 53:1855–1864
Serbin SP, Singh A, Desai AR, Dubois SG, Jablonski AD, Kingdon CC, Kruger EL, Townsend PA (2015) Remotely estimating photosynthetic capacity, and its response to temperature, in vegetation canopies using imaging spectroscopy. Remote Sens Environ 167:78–87
Serbin SP, Singh A, McNeil BE, Kingdon CC, Townsend PA (2014) Spectroscopic determination of leaf morphological and biochemical traits for northern temperate and boreal tree species. Ecol Appl 24:1651–1669
Serrano L, Ustin SL, Roberts DA, Gamon JA, Penuelas J (2000) Deriving water content of chaparral vegetation from AVIRIS data. Remote Sens Environ 74(3):570–581
Shabanov NV, Huang D, Yang WZ, Tan B, Knyazikhin Y, Myneni RB, Ahl DE, Gower ST, Huete AR, Aragao L, Shimabukuro YE (2005) Analysis and optimization of the MODIS leaf area index algorithm retrievals over broadleaf forests. IEEE Trans Geosci Remote Sens 43:1855–1865

Shenk JS, Landa I, Hoover MR, Westerhaus MO (1981) Description and evaluation of a near infrared reflectance spectro-computer for forage and grain analysis1. Crop Sci 21:355–358

Shiklomanov AN, Dietze MC, Viskari T, Townsend PA, Serbin SP (2016) Quantifying the influences of spectral resolution on uncertainty in leaf trait estimates through a Bayesian approach to RTM inversion. Remote Sens Environ 183:226–238

Shiklomanov A, Bradley BA, Dahlin K, Fox A, Gough C, Hoffman FM, Middleton E, Serbin S, Smallman L, Smith WK (2019) Enhancing global change experiments through integration of remote-sensing techniques. Front Ecol Environ 17(4):215–224

Shull CA (1929) A Spectrophotometric Study of Reflection of Light from Leaf Surfaces. Bot Gaz 87(5):583–607

Sims DA, Gamon JA (2002) Relationships between leaf pigment content and spectral reflectance across a wide range of species, leaf structures and developmental stages. Remote Sens Environ 81:337–354

Sims DA, Gamon JA (2003) Estimation of vegetation water content and photosynthetic tissue area from spectral reflectance: a comparison of indices based on liquid water and chlorophyll absorption features. Remote Sens Environ 84:526–537

Singh A, Serbin SP, McNeil BE, Kingdon CC, Townsend PA (2015) Imaging spectroscopy algorithms for mapping canopy foliar chemical and morphological traits and their uncertainties. Ecol Appl 25:2180–2197

Sonnentag O, Chen JM, Roberts DA, Talbot J, Halligan KQ, Govind A (2007) Mapping tree and shrub leaf area indices in an ombrotrophic peatland through multiple endmember spectral unmixing. Remote Sens Environ 109:342–360

Stimson HC, Breshears DD, Ustin SL, Kefauver SC (2005) Spectral sensing of foliar water conditions in two co-occurring conifer species: Pinus edulis and Juniperus monosperma. Remote Sens Environ 96:108–118

Stylinski CD, Gamon JA, Oechel WC (2002) Seasonal patterns of reflectance indices, carotenoid pigments and photosynthesis of evergreen chaparral species. Oecologia 131:366–374

Thompson DR, Gao B-C, Green RO, Roberts DA, Dennison PE, Lundeen SR (2015) Atmospheric correction for global mapping spectroscopy: ATREM advances for the HyspIRI preparatory campaign. Remote Sens Environ 167:64–77

Thompson DR, Natraj V, Green RO, Helmlinger MC, Gao B-C, Eastwood ML (2018) Optimal estimation for imaging spectrometer atmospheric correction. Remote Sens Environ 216:355–373

Thulin S, Hill MJ, Held A, Jones S, Woodgate P (2014) Predicting levels of crude protein, digestibility, lignin and cellulose in temperate pastures using hyperspectral image data. Am J Plant Sci 5:997–1019

Townsend PA (2002) Estimating forest structure in wetlands using multitemporal SAR. Remote Sens Environ 79:288–304

Townsend PA, Foster JR, Chastain RA, Currie WS (2003) Application of imaging spectroscopy to mapping canopy nitrogen in the forests of the central Appalachian Mountains using Hyperion and AVIRIS. IEEE Trans Geosci Remote Sens 41:1347–1354

Tucker CJ, Slayback DA, Pinzon JE, Los SO, Myneni RB, Taylor MG (2001) Higher northern latitude normalized difference vegetation index and growing season trends from 1982 to 1999. Int J Biometeorol 45:184–190

Turner DP, Cohen WB, Kennedy RE, Fassnacht KS, Briggs JM (1999) Relationships between leaf area index and landsat Tm spectral vegetation indices across three temperate zone sites. Remote Sens Environ 70:52–68

Ustin SL, Gamon JA (2010) Remote sensing of plant functional types. New Phytol 186:795–816

Ustin SL, Gitelson AA, Jacquemoud S, Schaepman M, Asner GP, Gamon JA, Zarco-Tejada P (2009) Retrieval of foliar information about plant pigment systems from high resolution spectroscopy. Remote Sens Environ 113:S67–S77

Ustin SL, Roberts DA, Gamon JA, Asner GP, Green RO (2004) Using imaging spectroscopy to study ecosystem processes and properties. Bioscience 54:523–534

Vane G (1987) First results from the airborne visible/infrared imaging spectrometer (AVIRIS): SPIE.

Vane G, Goetz AFH (1988) Terrestrial imaging spectroscopy. Remote Sens Environ 24(1):1–29
van den Berg AK, Perkins TD (2005) Nondestructive estimation of anthocyanin content in autumn sugar maple leaves. HortScience HortSci 40(3): 685-686.
Verhoef W (1984) Light scattering by leaf layers with application to canopy reflectance modeling: the SAIL model. Remote Sens Environ 16:125–141
Verhoef W, Bach H (2007) Coupled soil-leaf-canopy and atmosphere radiative transfer modeling to simulate hyperspectral multi-angular surface reflectance and TOA radiance data. Remote Sens Environ 109:166–182
Verrelst J, Alonso L, Camps-Valls G, Delegido J, Moreno J (2012) Retrieval of vegetation biophysical parameters using gaussian process techniques. IEEE Trans Geosci Remote Sens 50:1832–1843
Verrelst J, Rivera JP, Gitelson A, Delegido J, Moreno J, Camps-Valls G (2016) Spectral band selection for vegetation properties retrieval using Gaussian processes regression. Int J Appl Earth Obs Geoinf 52:554–567
Wang J, Chen Y, Chen F, Shi T, Wu G (2018) Wavelet-based coupling of leaf and canopy reflectance spectra to improve the estimation accuracy of foliar nitrogen concentration. Agric For Meteorol 248:306–315
Wang Z, Townsend PA, Schweiger AK, Couture JJ, Singh A, Hobbie SE, Cavender-Bares J (2019) Mapping foliar functional traits and their uncertainties across three years in a grassland experiment. Remote Sens Environ 221:405–416
Weiss M, Baret F, Myneni RB, Pragnere A, Knyazikhin Y (2000) Investigation of a model inversion technique to estimate canopy biophysical variables from spectral and directional reflectance data. Agronomie 20:3–22
Wessman CA, Aber JD, Peterson DL (1989) An evaluation of imaging spectrometry for estimating forest canopy chemistry. Int J Remote Sens 10:1293–1316
Wessman CA, Aber JD, Peterson DL, Melillo JM (1988) Remote-sensing of canopy chemistry and nitrogen cycling in temperate forest ecosystems. Nature 335:154–156
Weyermann J, Kneubühler M, Schläpfer D, Schaepman ME (2015) Minimizing reflectance anisotropy effects in airborne spectroscopy data using Ross–Li model inversion with continuous field land cover stratification. IEEE Trans Geosci Remote Sens 53:5814–5823
Widlowski J-L, Mio C, Disney M, Adams J, Andredakis I, Atzberger C, Brennan J, Busetto L, Chelle M, Ceccherini G, Colombo R, Côté J-F, Eenmäe A, Essery R, Gastellu-Etchegorry J-P, Gobron N, Grau E, Haverd V, Homolová L, Huang H, Hunt L, Kobayashi H, Koetz B, Kuusk A, Kuusk J, Lang M, Lewis PE, Lovell JL, Malenovský Z, Meroni M, Morsdorf F, Mõttus M, Ni-Meister W, Pinty B, Rautiainen M, Schlerf M, Somers B, Stuckens J, Verstraete MM, Yang W, Zhao F, Zenone T (2015) The fourth phase of the radiative transfer model intercomparison (RAMI) exercise: actual canopy scenarios and conformity testing. Remote Sens Environ 169:418–437
Wilson KB, Baldocchi DD, Hanson PJ (2001) Leaf age affects the seasonal pattern of photosynthetic capacity and net ecosystem exchange of carbon in a deciduous forest. Plant Cell Environ 24:571–583
Wold S, Ruhe A, Wold H, Dunn WJ (1984) The collinearity problem in linear-regression – The partial least-squares (PLS) regression approach to generalized inverses. SIAM J Sci Stat Comput 5:735–743
Wold S, Sjostrom M, Eriksson L (2001) PLS-regression: a basic tool of chemometrics. Chemom Intell Lab Syst 58:109–130
Wolter PT, Townsend PA, Sturtevant BR, Kingdon CC (2008) Remote sensing of the distribution and abundance of host species for spruce budworm in Northern Minnesota and Ontario. Remote Sens Environ 112:3971–3982
Workman J, Weyer L (2012) Practical guide and spectral atlas for interpretive near-infrared spectroscopy. CRC Press, Boca Raton, 326. https://doi.org/10.1201/b11894
Wong CYS, Gamon JA (2015) The photochemical reflectance index provides an optical indicator of spring photosynthetic activation in evergreen conifers. New Phytol 206:196–208

Wright IJ, Reich PB, Westoby M, Ackerly DD, Baruch Z, Bongers F, Cavender-Bares J, Chapin T, Cornelissen JHC, Diemer M, et al. (2004) The worldwide leaf economics spectrum. Nature 428(6985):821–827

Wu J, Albert LP, Lopes AP, Restrepo-Coupe N, Hayek M, Wiedemann KT, Guan K, Stark SC, Christoffersen B, Prohaska N, Tavares JV, Marostica S, Kobayashi H, Ferreira ML, Campos KS, da Silva R, Brando PM, Dye DG, Huxman TE, Huete AR, Nelson BW, Saleska SR (2016) Leaf development and demography explain photosynthetic seasonality in Amazon evergreen forests. Science 351:972–976

Wu J, Chavana-Bryant C, Prohaska N, Serbin SP, Guan K, Albert LP, Yang X, Leeuwen WJD, Garnello AJ, Martins G, Malhi Y, Gerard F, Oliviera RC, Saleska SR (2017) Convergence in relationships between leaf traits, spectra and age across diverse canopy environments and two contrasting tropical forests. New Phytol 214:1033–1048

Wullschleger SD, Epstein HE, Box EO, Euskirchen ES, Goswami S, Iversen CM, Kattge J, Norby RJ, van Bodegom PM, Xu X (2014) Plant functional types in Earth system models: past experiences and future directions for application of dynamic vegetation models in high-latitude ecosystems. Ann Bot 114:1–16

Yang X, Shi H, Stovall A, Guan K, Miao G, Zhang Y, Zhang Y, Xiao X, Ryu Y, Lee J-E (2018) FluoSpec 2-an automated field spectroscopy system to monitor canopy solar-induced fluorescence. Sensors (Basel) 18:2063

Yang X, Tang J, Mustard JF, Wu J, Zhao K, Serbin S, Lee J-E (2016) Seasonal variability of multiple leaf traits captured by leaf spectroscopy at two temperate deciduous forests. Remote Sens Environ 179:1–12

Yilmaz MT, Hunt ER, Jackson TJ (2008) Remote sensing of vegetation water content from equivalent water thickness using satellite imagery. Remote Sens Environ 112:2514–2522

Zarco-Tejada, Pablo J (2000a) Hyperspectral remote sensing of closed forest canopies: estimation of chlorophyll fluorescence and pigment content. Ph.D. Thesis, York University, Toronto Ontario, Canada

Zarco-Tejada PJ, Miller JR, Mohammed GH, Noland TL, Sampson PH (2000b) Chlorophyll fluorescence effects on vegetation apparent reflectance: II. laboratory and airborne canopy-level measurements with hyperspectral Data. Remote Sens Environ 74(3):596–608

Zarco-Tejada PJ, Miller JR, Mohammed GH, Noland TL, Sampson PH (2000c) Optical indices as bioindicators of forest condition from hyperspectral CASI data. Remote Sensing in the 21st Century: Economic and Environmental Applications, 517–522

Zhang Y, Chen JM, Miller JR, Noland TL (2008) Leaf chlorophyll content retrieval from airborne hyperspectral remote sensing imagery. Remote Sens Environ 112(7):3234–3247

Zhou LM, Tucker CJ, Kaufmann RK, Slayback D, Shabanov NV, Myneni RB (2001) Variations in northern vegetation activity inferred from satellite data of vegetation index during 1981 to 1999. J Geophys Res-Atmos 106:20069–20083

4

Remote Sensing of Plant Functional Diversity

Felix Morsdorf, Fabian D. Schneider, Carla Gullien, Daniel Kükenbrink, Reik Leiterer, and Michael E. Schaepman

4.1 Introduction

Global change is altering biodiversity in an unprecedented manner (Parmesan and Yohe 2003), and its impact on humankind may be large (Chapin III et al. 2000; Isbell et al. 2017). Forests are of special relevance because they hold most of the terrestrial biomass (Bar-On et al. 2018), are a hot spot of biodiversity (Wilson et al. 2012), and are subject to climate- and human-induced changes (Gardner 2010; Hansen et al. 2013). To monitor and potentially mitigate changes in biodiversity, Pereira et al. (2013) defined a set of essential biodiversity variables (EBVs), which should be comprehensive, concise, and standardized. Originally, most of these EBVs were to be measured in situ within ecosystems, but because forest plots are particularly scarce in the regions where change is happening the fastest (Chave et al. 2014), remote sensing (RS) has been acknowledged as a vital component to contribute to the aims of EBVs in the form of RS-enabled EBVs (RS-EBVs; Pettorelli et al. 2016; O'Connor et al. 2015). More specifically, RS technologies such as imaging spectroscopy and laser scanning have been attributed with the potential to play an important role in providing the necessary information for RS-EBVs, be it at regional, national, or global scale (Skidmore et al. 2015; Jetz et al. 2016).

Still in its early stage is the design and use of the EBV framework to include and combine RS-EBVs with in-situ measurements. In-situ measurements are often based on point measurements of individual species, whereas RS-EBVs are area-

F. Morsdorf (✉) · F. D. Schneider · C. Gullien · D. Kükenbrink · R. Leiterer · M. E. Schaepman
Remote Sensing Laboratories, Department of Geography, University of Zurich, Zurich, Switzerland
e-mail: felix.morsdorf@geo.uzh.ch

based, with spatial characteristics depending on sensor resolution and coverage, similar to the concept of grain and extent in ecology (Turner 1989).

For large-scale assessments (i.e., regional, continental, global), cost and effort of fieldwork is a limiting factor with respect to in-situ observations. Data from the newest generation of optical satellites (e.g., Landsat 8 and Sentinel-2) have high potential for a global biodiversity assessment due to their high spatial resolution (10–30 m), multispectral information, and temporal coverage, with repeat passes within 5–6 days, depending on the area of interest. Nevertheless, due to their recent launch, these sensors do not provide a long time series, and the complementarity of lower resolution satellite data or airborne or terrestrial RS data in combination with in-situ observation is beneficial to map changes at decadal or longer timescales.

All optical RS approaches use reflected light of the vegetation canopy to infer information about its state (Schaepman et al. 2009; Homolova et al. 2013). Leaf-level biochemistry (e.g., traits such as chlorophyll and water content) has strong links with leaf reflectance and transmittance (Jacquemoud and Baret 1990). However, when light interacts with the canopy, a multitude of scattering and absorption processes have to be considered (North 1996), taking place at different levels (e.g., leaf, tree, canopy; Niinemets et al. 1998) of the canopy. Thus, passive optical observational approaches of forested ecosystems are susceptible to the effects of forest structure because directional effects associated with illumination and observation geometry may interact with signals related to leaf-level biochemistry (Hilker et al. 2008; Knyazikhin et al. 2013). Consequently, the reflectance signal at the canopy level is influenced by both vegetation structure and leaf-level physiology, and disentangling those based on passive optical data alone remains a difficult problem (Kotz et al. 2004). The effect of vegetation structure on RS indices and products (e.g., RS-EBVs) is difficult to assess, and its impact on current observations and predictions may be large. The validation of advanced wall-to-wall RS products becomes increasingly difficult because of spatiotemporal mismatches of in-situ observations with RS data. Hence, we need a framework to be able to upscale and validate leaf-level physiological traits to the level of RS data to test potential observables for RS-informed EBVs.

Radiative transfer (RT) modeling has been used for several decades to simulate and understand the signals in passive optical data (Myneni et al. 1995, 1997; Meroni et al. 2004; Lewis and Disney 2007; Gastellu-Etchegorry et al. 1996). In addition, RT models (RTMs) have been used with existing medium- to low-resolution spaceborne missions for the retrieval of products such as leaf area index (LAI) or fraction of absorbed photosynthetic radiation (fAPAR) through inversion (Myneni et al. 1997; Running et al. 2004). One particular issue with RTMs of vegetation is their parameterization. While modeling approaches simulating low-resolution data [such as Moderate Resolution Imaging Spectroradiometer (MODIS) or MEdium Resolution Imaging Spectrometer (MERIS)] mainly used one-dimensional parameterizations of the vegetation (Jacquemoud 1993; Huemmrich 2001; Verhoef and Bach 2007), higher-resolution sensors will need 3-D parameterization to account for effects like shadowing and multiple scattering (Asner and Warner 2003; Disney et al. 2006; Widlowski et al. 2015). The first RTMs incorporating 3-D forest structure were called geometric-optical radiative transfer (GORT)-type models (Ni et al. 1999).

While they were better at modeling directional effects than 1-D models, they still lacked multiple scattering and did not have full energy balance closure of incoming and outgoing radiation across all spectral domains. More advanced models use Monte Carlo ray tracing (MCRT) to add multiple scattering and provide a sound physical representation of the photon's interaction with vegetation canopies (Disney et al. 2006). While the inclusion of more physical processes (e.g., multiple scattering) certainly improves MCRT-type models over simpler approaches, their parameterization and benchmarking remains an issue. A large effort in testing RT models was undertaken in the course of the radiation transfer modeling intercomparison (RAMI) exercise, where different models were tested using a set of artificial scenes of different complexity, including 3-D scenes, to see if the models produced comparable results (Widlowski et al. 2008, 2015). However, this benchmarking remained relative (i.e., representing actual forest patches that could be validated with real-world Earth observation (EO) data acquired over the same area was not an aim of the RAMI exercise). One reason, among others, for this was the lack of suitable technologies and methods to capture and represent the 3-D vegetation structure at small scales (e.g., branches, leaves, and/or shoots).

Today, laser scanning is an established tool for retrieving quantitative measures of canopy structure (Nelson 1997; Lefsky et al. 1999; Næsset 2002; Morsdorf et al. 2004; Popescu et al. 2002; Morsdorf et al. 2006, 2010; Nelson 2013; Wulder et al. 2012). Airborne (ALS)-, terrestrial (TLS)-, and unmanned aerial vehicle (UAV)-based laser scanning (Morsdorf et al. 2017) provide a direct means to assess vegetation structure by combining the known position and orientation of the sensor with the time of flight of a laser pulse to produce a point cloud of exact 3-D coordinates. Measurements can be made across scales (e.g., stand, tree, branch, and leaf level) with finer scales often captured by close-range laser scanning (Morsdorf et al. 2018). The amount of structural detail contained in the point cloud can be overwhelming, and the extraction of meaningful information remains a challenge (Wulder et al. 2013; Morsdorf et al. 2018). Due to large data sets, automated methods for the extraction of either semantic information, such as single-tree detection based on ALS (Hyyppa et al. 2001; Morsdorf et al. 2004; Kaartinen et al. 2012; Wang et al. 2016) or tree geometry reconstruction from TLS (Cote et al. 2009; Raumonen et al. 2013) or the derivation of biophysical variables such as LAI (Morsdorf et al. 2006), are preferable over manual and/or empirical approaches.

Figure 4.1 shows an example of a single-tree-based 3-D reconstruction using ALS- and TLS-derived information. Using the 3-D information derived by ALS and TLS, one can reconstruct a virtual representation of the forest that will be used by the RT model to simulate the radiative regime of the canopy. Such an approach can be utilized to upscale measurements of leaf biochemistry to the canopy scale and to validate imaging spectroscopy-derived RS-EBVs across larger regions. In addition, this approach uses a set of three physiological and tree morphological functional traits, derived from imaging spectroscopy and laser scanning, respectively, to showcase the potential of these technologies to map the functional diversity of forests and to provide relevant information for RS-enabled EBVs.

Here we describe how we (i) designed and implemented an observational scheme to gather in-situ and structural data across several scales to simulate the 3-D radiative

Fig. 4.1 Workflow of the 3-D reconstruction using ALS and TLS measurements

regime of the forest, (ii) tested the simulation by comparing simulated and actual RS data in their spectral and spatial information dimension, and (iii) use the approach to demonstrate how remotely sensed functional traits can be used to compute regional-scale functional richness, showcasing the information content of RS-EBVs.

4.2 The Laegeren Site: Description and History

The Laegeren site is located at N 47° 28′, 49″ and E 8° 21′, 05″ at 680 m a.s.l. on the southern slope of the Laegeren mountain, approximately 15 km northwest of Zürich, Switzerland (Fig. 4.2). The southern slope of the Laegeren marks the boundary of the Swiss Plateau, which is bordered by the Jura and the Alps. Since 1986, a 45-m-tall flux tower has provided micrometeorological data at high temporal resolution. Since April 2004, CO_2 and H_2O flux measurements are a routinely contribution to the FLUXNET/CarboEurope-IP network (Eugster et al. 2007). The mean annual temperature is 8°C. The mean annual precipitation is 1200 mm, and the growing season lasts 170–190 days. The natural vegetation cover around the tower is a mixed beech forest. The western part is dominated by broad-leaved trees, mainly beech (*Fagus sylvatica L.*) and ash (*Fraxinus excelsior L.*). In the eastern part, beech and

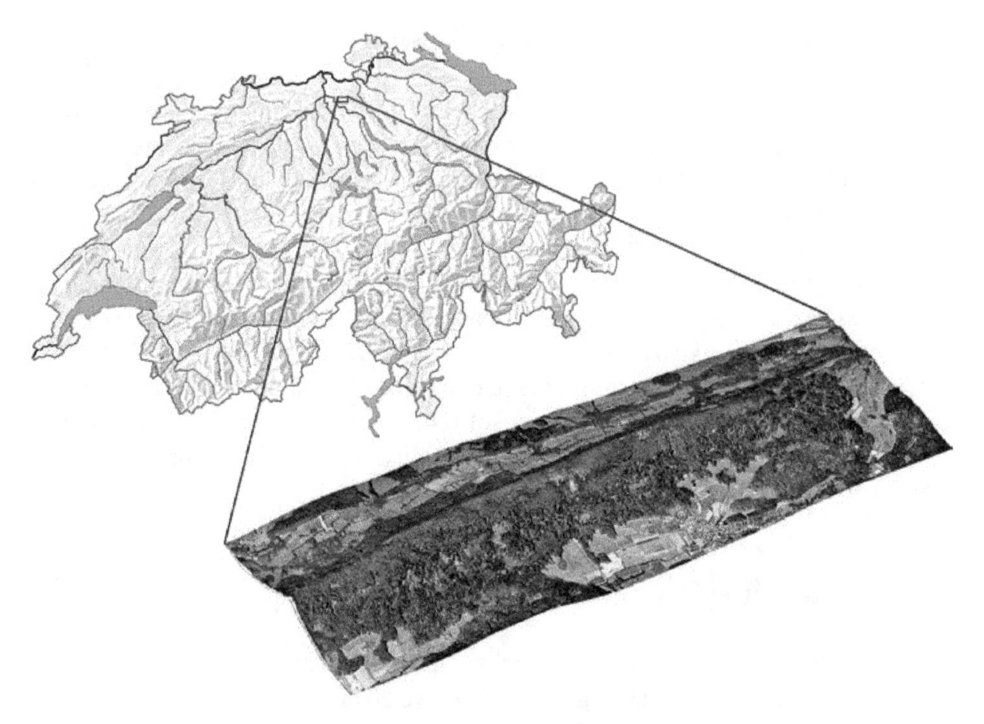

Fig. 4.2 Location of the Laegeren site within Switzerland

Norway spruce (*Picea abies (L.) Karst.*) are dominant. The forest stand has a relatively high diversity of species, ages, and diameters (Eugster et al. 2007). The ground cover mainly consists of bare soil, boulders, and litter, while the sparse understory vegetation is dominated by herbs and shrubs. Average canopy height (CH) is 24.9 m, with a maximum of 49 m, and the stem density is 270 stems per ha.

4.3 Data

4.3.1 In-Situ Data

Ground data with varying spatial, spectral, and temporal resolution allow for the 3-D reconstruction of the Laegeren, its attribution with leaf optical properties (LOPs), and generation of a reference database for parameterization and validation purposes. We used multitemporal TLS on a 60 m × 60 m plot (Sect. 4.3.2.2) and an extensive forest inventory for an area of 300 m × 300 m, which is extended to 300 m × 900 m for the simulation of EO data. In the inventory data, the type and accurate position of the trees, as well as their crown dimension and offset due to leaning stems, social position, and vertical stratification of the crown, were recorded (Sect. 4.3.1.2). In addition, the occurrence and characterization of the understory was mapped in the field and interpolated to a 2 m × 2 m grid using an ALS-based classification (Leiterer et al. 2013).

4.3.1.1 Measurements of Leaf Optical Properties

To obtain the optical properties of tree foliage, we used an integrating sphere coupled with an Analytical Spectral Devices (ASD) FieldSpec-3. Measurements of hemispherical and directional reflectance and transmittance, both from the abaxial and adaxial side of the leaves, were taken. To take into account the vertical variability of LOPs, we sampled in three different crown parts (top, middle, bottom), representing different lighting conditions in the canopy (e.g., sunlit, transitional, shaded). Deciduous leaves were collected from ten individual trees of five species (*Acer pseudoplatanus* spp., *Fagus excelsior*, *F. sylvatica*, *Ulmus glabra*, and *Tilia platyphyllos*). Measurements of aerosol optical depth (AOD) and precipitable amount of water (PAW) were provided by the aerosol robotic network (AERONET) as level 2.0 quality-assured data. For details of the sampling and measurement scheme, see Schneider et al. (2014).

4.3.1.2 Forest Inventory

An exhaustive forest inventory was carried out, individually addressing all single trees with a diameter at breast height (DBH) above 20 cm on the 300 m × 300 m site. Variables recorded for each tree included DBH, species, social status, and crown shift (i.e., an estimation of the magnitude and horizontal direction of the crown center in respect to the foot of the stem). The latter is of particular relevance on the Laegeren site because many trees have leaning stems due to topography and shallow soils. A geodetic tachymeter was used for the surveying, enabling fast and accurate electronic tree location measurements. Using all measured points, a polygonal traverse was calculated resulting in the x, y, z coordinates for each measurement position with an error range of millimeters for the TLS measurements and a maximum of 10 cm in x, y, and z for all other field measurements. The relative locations were transformed to absolute Swiss national coordinates using three differentially corrected global positioning system (GPS) base points, which were placed in canopy gaps. See Fig. 4.3 for a visualization of the tree inventory.

4.3.2 RS Data

4.3.2.1 Airborne Laser Scanning

To provide 3-D structure information across the whole study area, we relied on two airborne laser scanning campaigns, using a RIEGL LMS-Q680i scanner under leaf-on conditions and a RIEGL LMS-Q560 scanner under leaf-off conditions. Flight strips have an overlap of approximately 50%. Full-waveform features, namely, echo width and intensity, were extracted from the data using the software RiANALYZE and were assigned to the individual returns in the multiple-echo point cloud.

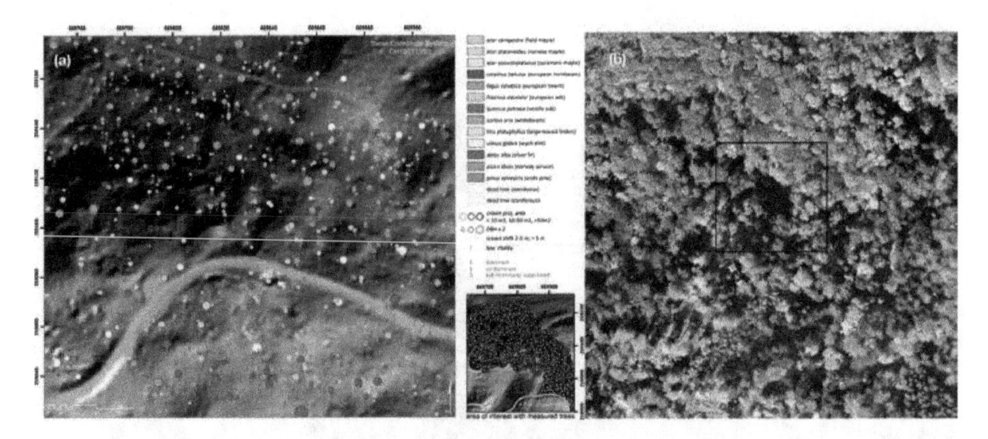

Fig. 4.3 Subset of single-tree ground inventory (**a**) and UAV-based RGB imagery acquired in fall (**b**). The black box in (**b**) denotes the subset presented in (**a**). The gray structure southwest of the bounding box is the flux tower, and the small inset shows the total extent of the single-tree ground inventory

The point cloud was filtered to classify ground and vegetation points, and the ground points were subsequently interpolated to a raster of 1 m resolution. For a detailed description of the digital terrain model (DTM) generation, see Leiterer et al. (2013). DTM accuracy was assessed using more than 500 TLS-measured road surface and bare soil points (see Sect. 4.3.2.2), which were related to the national land survey and resulted in a mean height uncertainty of about ±0.25 m. For each point of the full point cloud, the height above ground was calculated by subtracting the interpolated DTM value from the corresponding echo height above sea level, providing the vertical distance of the vegetation echoes to the terrain underneath.

4.3.2.2 Terrestrial Laser Scanning

On a subset of about 60 m × 60 m, a ground-based TLS survey was carried out using a Riegl VZ1000 instrument. A total of 40 scans on 20 scan locations were taken because each location had to be covered by two scans due to the VZ1000's camera scanning pattern (Morsdorf et al. 2018). About 50 reflective targets were placed within the scene and later used for co-registration of the scans. For co-registering RiSCAN Pro was used, and we used the ALS data to subsequently globally adjust (rotate and translate) the unified TLS point cloud. Due to the high and dense canopy, TLS needs to be complemented by laser data from above the canopy, providing more information in the upper part, either by ALS or UAV-based laser scanners. For biomass retrievals, the occlusion of upper canopy material in TLS data might be less of a problem because stems generally taper off toward the top. However, if simulation of the radiative regime and subsequent comparison with EO data gathered with a top-of-canopy perspective is the aim, TLS in denser forests needs to be complemented with laser scanning data from above the canopy (Morsdorf et al. 2017, 2018) (Fig. 4.4).

Fig. 4.4 Terrestrial laser scan of a beech-dominated part of the study area. Transect measures about 30 m (width) × 4 m (depth). One can observe a general thinning of the point cloud toward the top due to occlusion

4.3.3 Multispectral and Imaging Spectroscopy Data

Imaging spectroscopy data were acquired under clear sky conditions using the APEX imaging spectrometer (Schaepman et al. 2015). The average flight altitude was 4500 m a.s.l. resulting in an average ground pixel size of 2 m. APEX measured at-sensor radiances in 316 spectral bands ranging from 372 nm to 2540 nm. APEX data were processed to hemispherical-conical reflectance factors in the APEX processing and archiving facility (Hueni et al. 2009). Level 1 (L1) calibrated radiances were obtained by inverting the instrument model, applying coefficients established during calibration, and characterization at the APEX Calibration Home Base (CHB) in Oberpfaffenhofen, Germany. The position and orientation of each pixel in 3-D space was based on automatic geocoding in PARGE v3.269, using the swissALTI3D DTM. L1 data were then converted to hemispherical conical reflectance factors (HCRFs, Schaepman-Strub et al. 2006) by employing ATCOR4 v7.0 in the smile aware mode. The APEX data were complemented with other passive optical data of varying spatial and spectral resolution to build up an EO data set (Fig. 4.5). This EO data set enables cross-comparisons between the 3-D RT modeled and the actual, measured top-of-atmosphere (TOA) reflectance values at different spectral and spatial resolutions and thus an absolute evaluation of the 3-D reconstructed forest scenes and the RTM parameterization. The EO data acquired during the 2010–2014 growing seasons covers a variety of spectral and spatial resolutions:

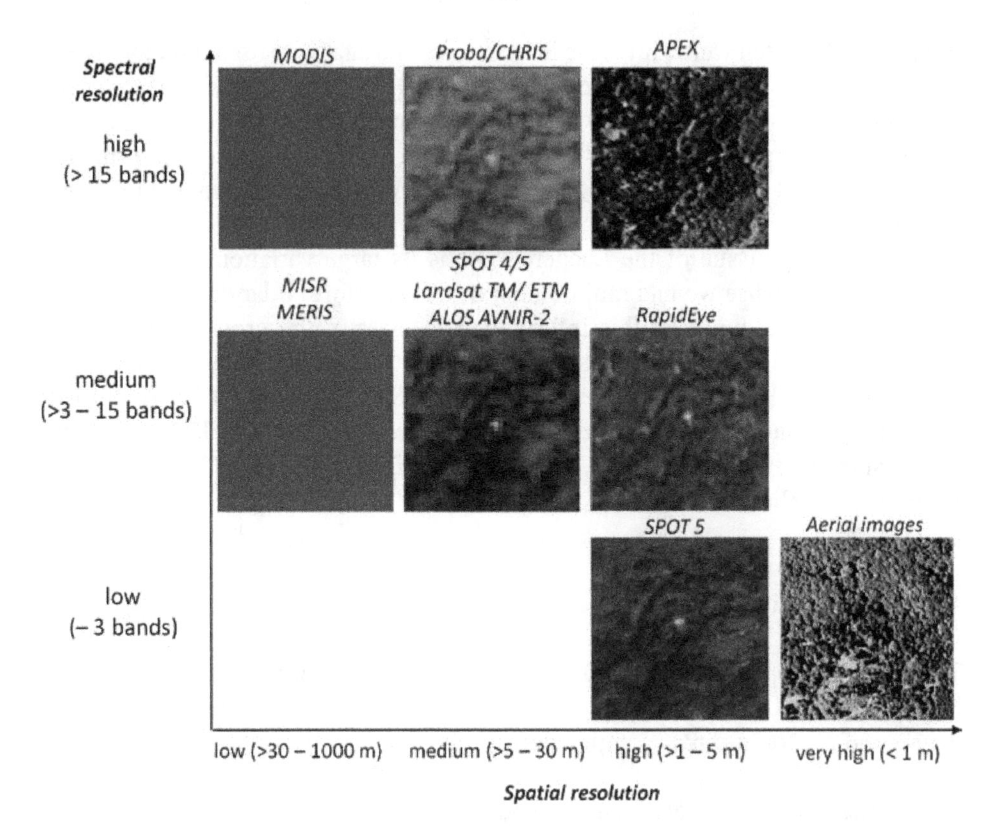

Fig. 4.5 The spatial and spectral scales covered by Earth observation (EO) data gathered for validation and up- and downscaling purposes

imaging spectrometer data from APEX (2 m × 2 m, see above for details) as well as multispectral data from RapidEye (5 m × 5 m, 4 scenes), SPOT HRG (10 m × 10 m, 5 scenes), PROBA CHRIS (17 m × 17 m, 1 scene), Landsat TM/ETM+/OLI (30 m × 30 m, 37 scenes), ENVISAT MERIS (300 m × 300 m, 8 scenes), and Aqua/Terra MODIS (250 m × 250 m, monthly). We use APEX data for the spectral validation and a RapidEye scene for the spatial validation of our 3-D RTM approach.

4.4 Methods

4.4.1 In-Situ Data Processing

4.4.1.1 Optical Properties

LOPs were calculated separately for deciduous and coniferous trees. A linear spectral forward mixing was applied to calculate the reflectance and transmittance spectra of sunlit, transitional, and shaded leaves and needles. Because the spectra were found to match well with those in literature, the data were used directly instead of a forward simulation of a LOP model (Feret et al. 2008). This was done to reduce the number of parameters and associated uncertainties. The broadleaf species composition used for spectral mixing was derived from the forest inventory information and is dominated by beech (about 50%), with lesser contributions from maple, elm, linden, and ash.

One particular issue of the Laegeren site is its large variation in the spectral background. Because we had multitemporal full-waveform lidar data available for the Laegeren site, we used this information to classify the ground into distinct classes (gravel, litter, soil) and assigned matching spectra from our field measurements to these classes (Leiterer et al. 2013). As Schneider et al. (2014) showed, using several understory classes instead of a homogenous (black) background makes simulated top-of-canopy (TOC) and top-of-atmosphere (TOA) reflectance values more realistic.

4.4.1.2 3-D Reconstruction

Two different approaches for 3-D reconstruction of the vegetation structure were implemented and tested. The first approach relied on a single-tree identification and the second one on a direct computation of plant area index (PAI) values inside a voxel cell. Voxels are basically 3-D pixels, dividing the 3-D space into equal-sized cubes. The single-tree detection (individual tree crown, ITC) method used was based on Morsdorf et al. (2004), which derives tree location, height, and crown diameter to reconstruct the forest in 3-D based on simple geometric primitives like rotational paraboloids. However, as with most local maxima detection-based ITC methods, its performance within the mixed forest stands of the Laegeren site was suboptimal, with tree detection rates of only 50–70%. This is much lower than what can be expected for conifer forests, where rates of up to 90% can be achieved (Kaartinen et al. 2012; Wang et al. 2016). Conifers generally have conical crowns with one distinct peak (treetop), greatly facilitating their detection as local maxima in a digital surface model (DSM). The main difference between the voxel-grid and ITC approaches is the added level of semantics (Morsdorf et al. 2018) in the single-tree case, which might be relevant for some species- and individual-focused experiments (i.e., when trying to link EO-based traits with genetic information of the individual tree). If the aim of the 3-D reconstruction is an accurate simulation of the radiative regime, single-tree identification adds a layer of unnecessary complexity, so the voxel-grid approach led to better results (Schneider et al. 2014) and was subsequently used for upscaling of the trait information (Schneider et al. 2017).

4.4.1.3 Linking Field and RS Data

The perspective of forest inventory is from within or beneath the canopy and the main sampling unit is the tree, quantified as diameter at breast height (DBH). RS, on the other hand, has a top-down perspective on the canopy, and the sampling unit

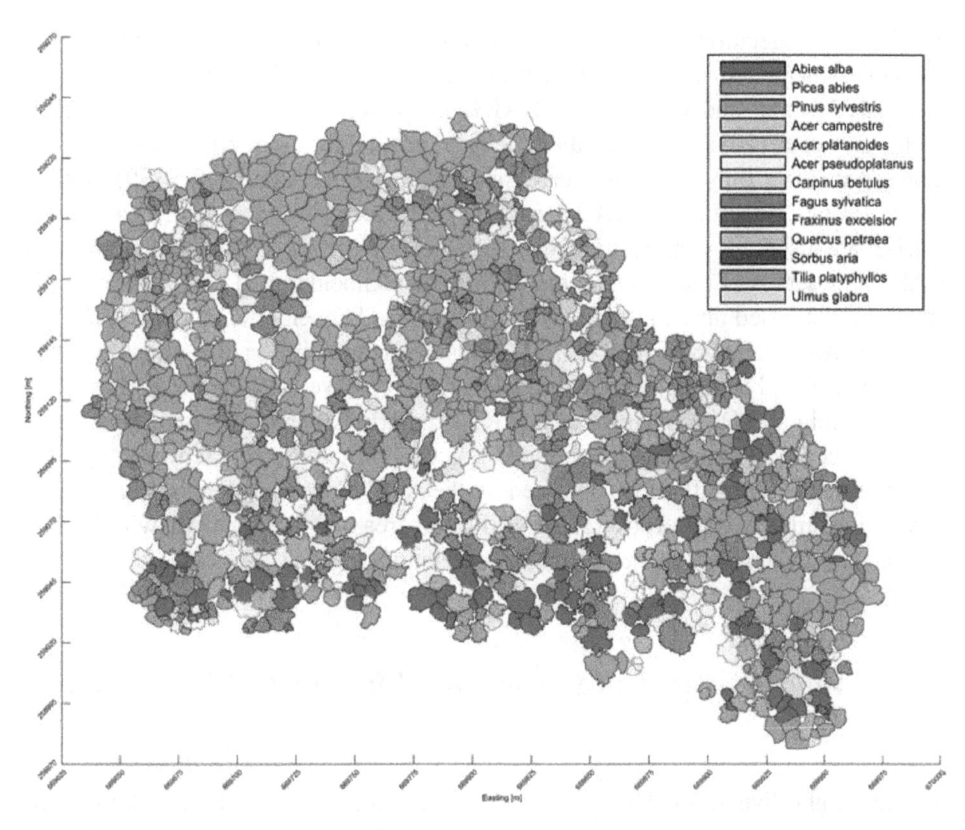

Fig. 4.6 Map of crown polygons determined from a combination of ALS and maximum leaf senescence (Fall) UAV data, linked with species information derived from the stem-referenced field inventory. Only the combination of these crown outlines and the stem map (see Fig. 4.3) allowed for individual specific computation of physiological and morphological traits

is normally a pixel. Linking these different perspectives can be difficult under any circumstances; the Laegeren site is situated on a steep slope, and trees have irregular crown shapes and different growing directions, which further complicates information matching. The first necessary step is to convert the pixel-based EO data to tree-based data by using 2-D polygons of crown boundaries. Considering the low success rates of ALS-based ITC, we manually delineated tree crowns based on UAV imagery acquired in the fall (Fig. 4.3) and matched each crown with the forest inventory data. The field inventory provided valuable additional information, such as magnitude and direction of the crown shift for trees with leaning stems, which hinders a direct stem and crown location matching based on location (Fig. 4.6).

4.4.2 Radiative Transfer Modeling

The RTM used to upscale and validate leaf-level traits such as chlorophyll and leaf water content is Discrete Anisotropic Radiative Transfer (DART; Gastellu-Etchegorry et al. 2015). Generally, a DART scene is built out of voxels with a predefined size.

To simulate vegetation such as grass or tree crowns, voxels can be filled by turbid media parameterized with PAI and leaf optical properties (LOPs). Further details of the DART model and examples of DART simulations can be found in Gastellu-Etchegorry et al. (2015). We use flux tracking in reflectance mode with the sun and the atmosphere as the only radiation sources and used DART version 5.6.0 (v739). Optical properties described in Sect. 4.4.1.1 and the forest reconstruction described in Sect. 4.4.1.2 are used to parameterize the forest canopy, background, and terrain in DART. For details of model parameterization, see Schneider et al. (2014); for details on the model-based upscaling of leaf-level traits, see Schneider et al. (2017). For the modeling results shown in Sect. 4.5.1.1, we used sun and observation angles as in the actual APEX and RapidEye acquisitions, respectively. We evaluate the performance of the combined 3-D reconstruction and RT simulation approach in two ways: spectrally, by comparing averaged simulated spectra on the core (i.e., covered by TLS measurements) site with those obtained by the APEX instrument, and, spatially, by comparing simulated bands of RapidEye over an area of 900 m × 300 m with DART-simulated reflectance at those particular wavelength regions.

4.4.3 Validation of Trait Predictions Using the RTM Approach

Three functional traits were derived from ALS data, canopy height (CH), PAI, and foliage height diversity (FHD), forming a set of morphological traits. These three were chosen because they are ecologically relevant and can be easily derived from airborne laser scanning data. Three additional functional traits—chlorophylls (CHL), carotenoids (CAR), and equivalent water thickness (EWT)—were chosen and computed using specific band ratios from the IS data (Schneider et al. 2017), forming a set of physiological traits. Both CH and CHL have been identified as primary observables for RS-EBVs, so their validation and scaling is particularly relevant. The traits were computed at a spatial aggregation unit of 6 m; for the ALS data, all echo values within a 6 m × 6 m grid cell were used for the computation, while for the IS data only sunlit pixels within the grid cell were retained for subsequent index computation. The shadow mask used for extracting sunlit pixels was derived from a DSM based on the ALS data and the solar illumination angle at the time of the IS overflight. For more details on the selected traits and their computation, please refer to Schneider et al. (2017).

The physiological traits used in this study are by definition leaf-level parameters, which need to be upscaled or averaged to be representative for the tree or canopy level. On the other hand, the morphological traits can be directly estimated from ALS data for any spatial unit. However, the chosen spatial scale and context might change how the data are interpreted (e.g., tree height needs to be estimated using single-tree information, whereas vegetation height can be derived at all different scales at the stand or plot level. See Fig. 4.7 for a map of the computed physiological and morphological traits.

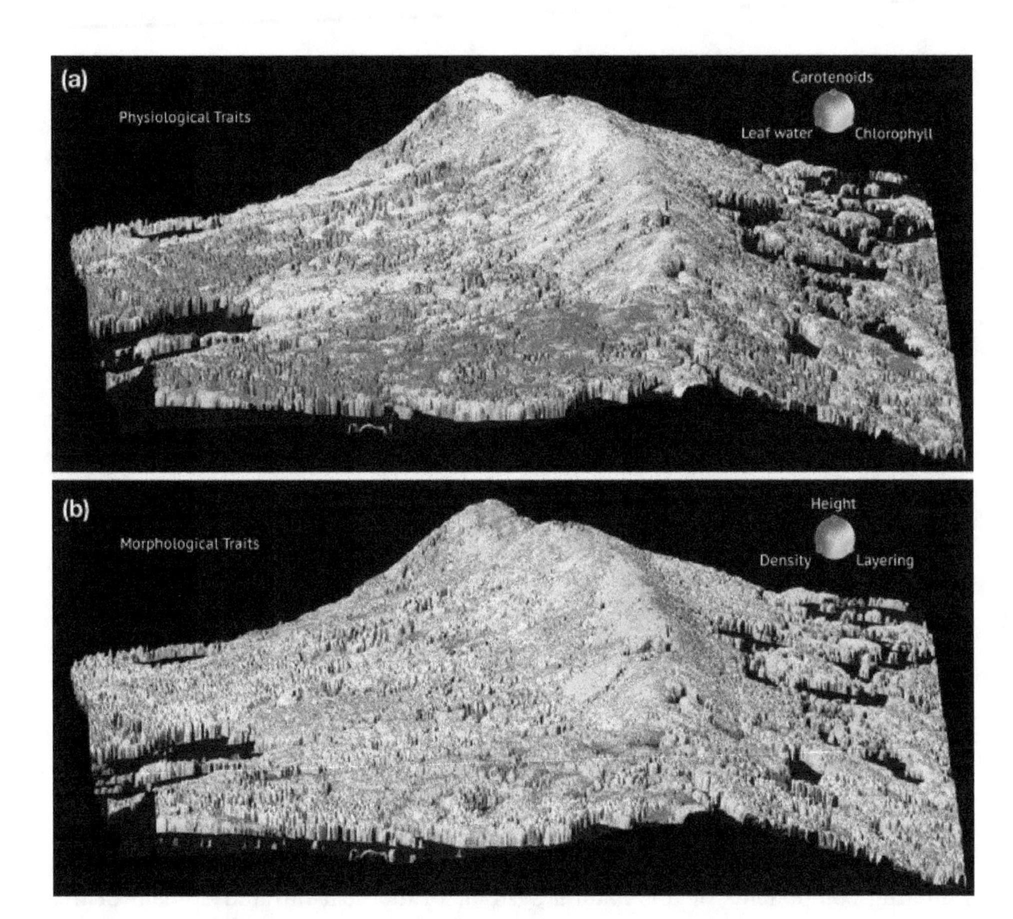

Fig. 4.7 Physiological (**a**) and morphological (**b**) traits derived from IS and ALS. For the ALS-based morphological traits, density is plant area index (PAI) and layering foliage height diversity (FHD)

4.4.4 Computation of Functional Richness

To showcasc how the RTM-validated EO traits can be used for spatially explicit diversity assessments, we compute the functional richness within the 3-D trait space using a spatial subset of pixels (e.g., in a 60 m × 60 m box containing 100 pixels). In the case of the morphological traits, the 3-D trait space is spanned by the axes CH, PAI, and FHD, whereas for the physiological traits the trait space is spanned by the axes CHL, CAR, and EWT. The richness within the trait space is based on volume of a 3-D convex hull of all pixels' trait values (i.e., the larger the variation of the respective traits, the larger the volume of the convex hull). As an example, if all trait values were the same, the richness would be zero because no volume would be spanned in the 3-D trait space. Computing the richness using a pixel-based approach has the advantage of resolving both inter- and intraspecific variation of the traits, with the latter being potentially as large as the former (e.g., as observed in our leaf spectra). For details on the definition and computation of richness and other diversity-related metrics in the scope of this work, please refer to Schneider et al. (2017).

4.5 Results and Discussion

4.5.1 Forward Simulation of Passive Optical Imagery and Comparison With EO Data

4.5.1.1 Spectral Validation

Figure 4.8 compares the spectral response of a 20 m × 20 m subplot within the Laegeren site simulated by the DART RTM with the average APEX spectrum of the same area. In contrast to Schneider et al. (2014), the improved version of the DART model used in this study shows good agreement (within the standard deviation for the 10 × 10 pixel areas) for all wavelengths, including the visible domain. The version of DART used in this study (5.6.0, v739) has a more sophisticated parameterization of the atmosphere than the older version, improving the spectral response in the visible domain (Grau and Gastellu-Etchegorry 2013; Yin et al. 2013; Gastellu-Etchegorry et al. 2015). The very good agreement of simulated and measured spectra across all bands shows that our approach of combining a 3-D reconstruction of the forest and LOPs of leaves and needles was successful in capturing the dominant scattering components of this natural system. In the near-infrared domain of the spectra, this is likely due to ALS and TLS providing accurate physical representations of 3-D canopy structure, whereas in the visible domain the quality of the LOPs

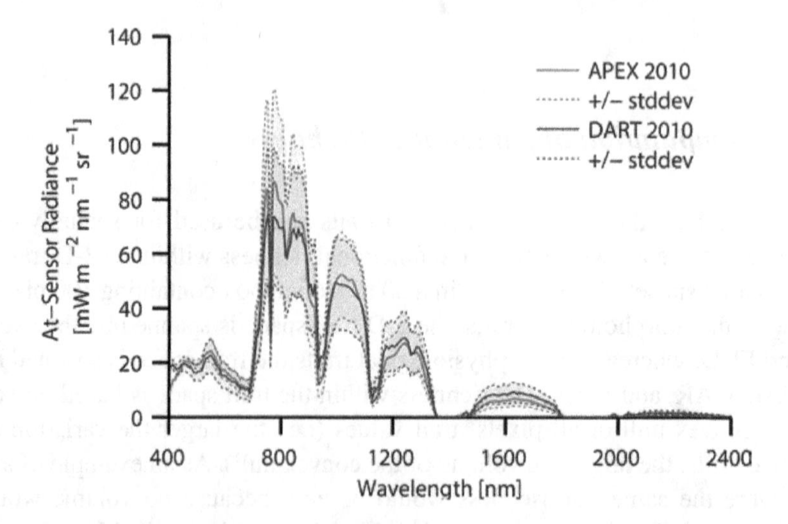

Fig. 4.8 Simulated spectral response by DART for subplot S1 in comparison with APEX data acquired over the same area. The standard deviation is computed from the single pixels in the 20 m × 20 m plot

and the representation of the atmosphere are contributing more to this excellent result. Thus, we have shown that the measured physiological trait variation at leaf level can be upscaled to canopy level (as observed by IS instruments). We used this to forward validate IS-derived physiological traits that are the basis of the functional richness computation in Sect. 4.5.2. This RTM-based link is a key component of our validation framework because typically field-measured spectra and the traits based on spectral indices cannot be assumed to be representative for the respective signals measured at the imaging sensor above the canopy.

4.5.1.2 Spatial Validation

Figure 4.9 shows a comparison of the spatial patterns in both simulated and actual RapidEye imagery obtained over the Laegeren site in leaf-on conditions. When subsampled to the 5 m resolution of RapidEye, our approach produces very similar spatial patterns, properly resolving shadows and highlights due to forest structure and underlying topography effectively contained in the ALS data and transferred to

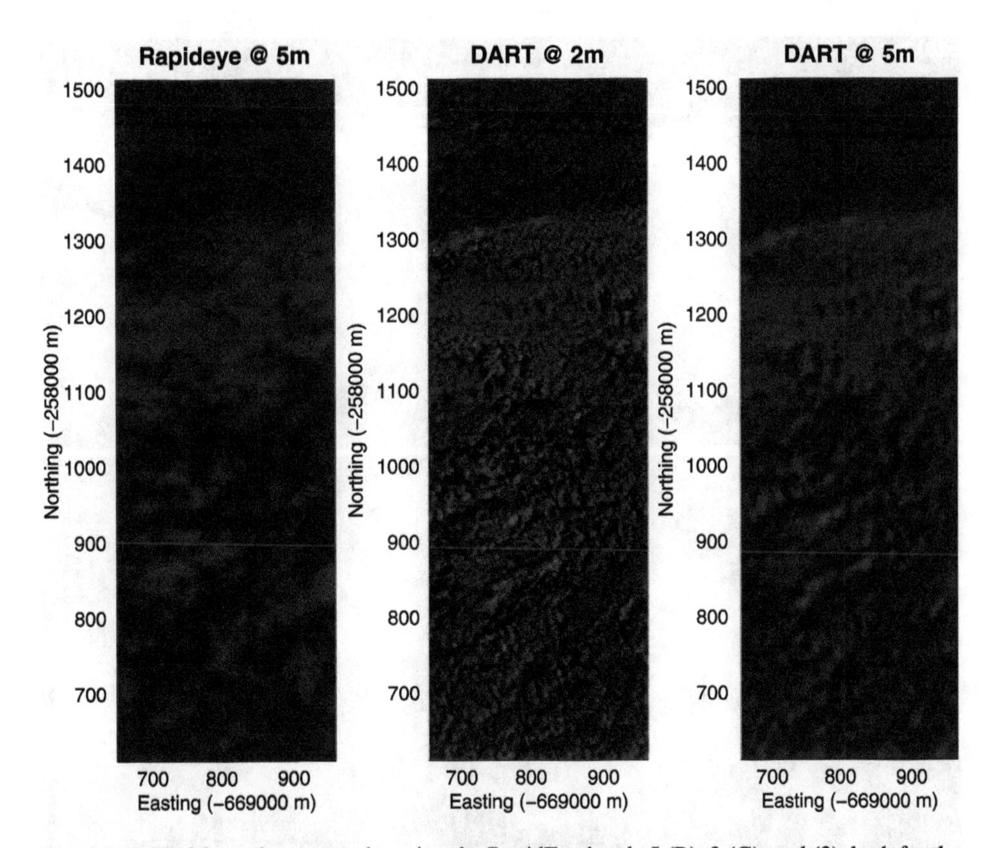

Fig. 4.9 RGB false-color composite using the RapidEye bands 5 (R), 3 (G), and (2), both for the simulated image using DART (center panel) and actual RapidEye data acquired over the site (left panel). The right panel shows a smoothed version of the DART simulation to accommodate for the lower resolution (5 m) of the RapidEye imaging sensor

the RT model by the PAI voxel grid. If LOPs can be assumed to be uniform across a site or region, the presented approach can be used for larger areas, only relying on ALS data to parameterize the RT model and using the LOPs measured at a subset of the site or taken from spectral libraries. ALS data are generally available at regional and national scales, effectively bridging the gap between point-based field inventories and global scale satellite imagery. Using such larger-scale simulated EO data, retrieval methods for RS-EBV primary observables such as CHL and vegetation height can be tested and validated. The 3-D simulation environment we established explicitly or implicitly contains all these variables in an easily retrievable format.

4.5.2 Functional Diversity of Laegeren Site

Figure 4.10 shows the spatial distribution of the richness as computed from the three morphological and physiological traits. The most prominent pattern is the strong topographic effect of the Laegeren mountain ridge, which is equally present in both

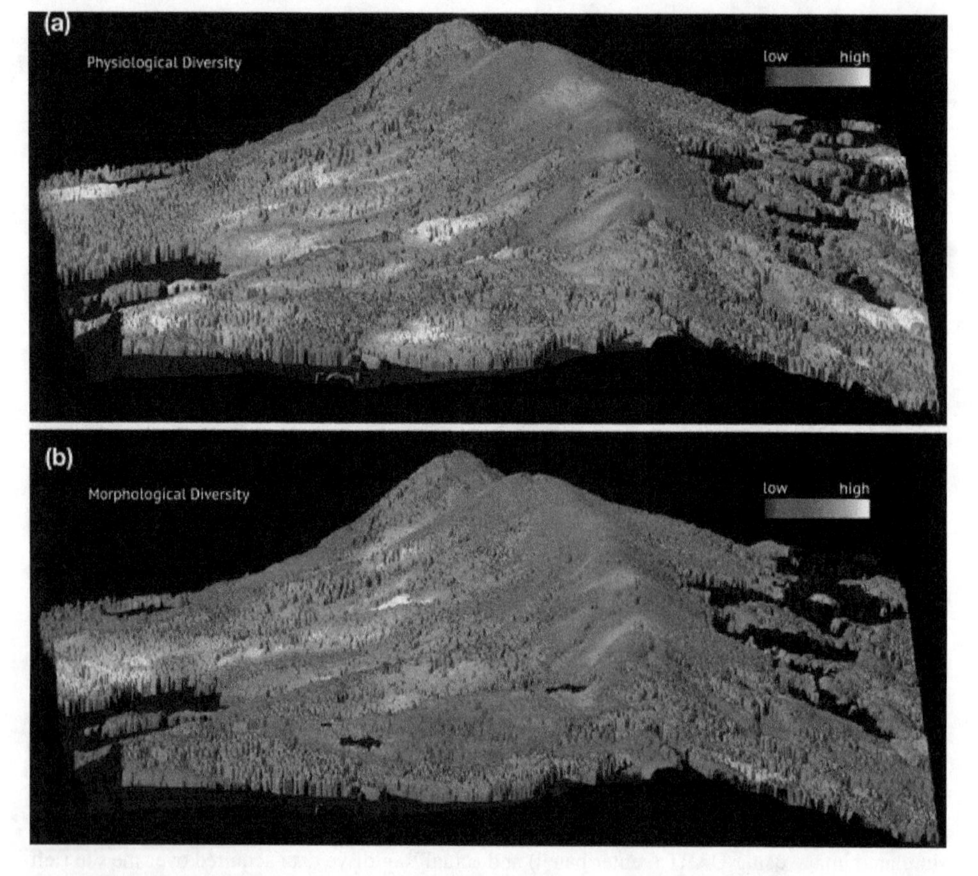

Fig. 4.10 Physiological (**a**) and morphological (**b**) richness determined within the trait space spanned by three traits

richness maps. The ridge and the associated steep slopes affect many environmental variables, which could act as filter for niche space and thus diversity. Higher altitude is linked with decreased temperature, whereas the slope is contributing to lesser soil depths and water availability and increased incoming radiation, at least for the part south of the ridge. Thus, the environmental conditions are harsher close to the ridge, which might explain the decrease of functional richness we observe in this context. In the lower regions, morphological and physiological richness exhibit differing spatial patterns. We assume that changes of the morphological richness in lower parts parallel to ridge are caused by the different stand management regimes and associated stand ages and structures. For the physiological richness, differences in species composition seem to be a dominant effect, with the conifer-dominated stands having a lower functional richness and the old-growth mixed stands being functionally richer.

4.6 Conclusion and Outlook

Modern RS technologies increasingly face a validation paradox—i.e., it is very difficult to provide ground-based validation data that match the spatial (resolution and extent), temporal, and thematic characteristics of modern EO data sets. As an example, ALS-derived tree height is assumed to be more accurate than field measurements, but it cannot be proved using field data alone. By using laser scanning-derived 3-D structure together with LOPs in an RT model approach, we have shown a way to overcome such mismatches and provide a framework that could be established across a range of sites around the globe to prototype and validate EO-based data and products in the future. Such a forward validation will as well pave the way for products that are not measurable in the field, but still might be relevant in the context of ecosystem function and diversity. The RTM approach provides a physical and mechanistic way to learn about the information content of EO data, and a combination of this approach with recent developments in the machine learning domain could provide interesting perspectives.

With the trait-based functional richness assessment, we demonstrated how a spatially extended monitoring using the complementary technologies imaging spectroscopy and lidar would work and what kind of insights into ecosystem functioning it could generate. In addition, the trait maps and the derived functional richness could be used for spatiotemporal gap filling of in-situ observational networks such as the global forest biodiversity initiative, complementing the diversity information that these provide.

In the future, these data streams in conjunction with the EBV concept (Fernández et al., Chap. 18) will give policy-makers around the world useful tools to assess and report on the biodiversity. To speed up this process, the European Space Agency funded the GlobDiversity project starting in 2017 in the tradition of similar projects for some of the essential climate variables. The project's goal is to demonstrate the capability and utility of producing a set of selected RS-EBV data sets in different regions and biomes around the globe and with high spatial resolution (10–30 m) using the newest-generation satellite data, such as Sentinel-2 and Landsat 8. In addition, the project shall suggest in a reference document how to describe RS-EBVs and how they could be engineered and validated. We believe that the 3-D reconstruction and RT modeling approach highlighted in this chapter could be applied across a global range of sites to fulfill this task.

Acknowledgments The contributions of F.M., F.D.S., and M.E.S. are supported by the University of Zurich Research Priority Program on Global Change and Biodiversity, and the contribution of D.K. was with support from the European Union's 7th Framework Program (FP7/ 2014–2018) under EUFAR2 contract no. 312609.

References

Asner GP, Warner AS (2003) Canopy shadow in ikonos satellite observations of tropical forests and savannas of. Environment 87(4):521–533

Bar-On YM, Phillips R, Milo R (2018) The biomass distribution on earth. Proc Natl Acad Sci 115(25):6506

Chapin FS III, Zavaleta ES, Eviner VT, Naylor RL, Vitousek PM, Reynolds HL, Hooper DU, Lavorel S, Sala OE, Hobbie SE, Mack MC, Dıaz S (2000) Consequences of changing biodiversity. Nature 405(6783):234–242

Chave J, Réjou-Méchain M, Burquez A, Chidumayo E, Colgan MS, Delitti WB, Duque A, Eid T, Fearnside PM, Goodman RC, Henry M, Martínez-Yrízar A, Mugasha WA, Muller-Landau HC, Mencuccini M, Nelson BW, Ngomanda A, Nogueira EM, Ortiz-Malavassi E, Pelissier R, Ploton P, Ryan CM, Saldarriaga JG, Vieilledent G (2014) Improved allometric models to estimate the aboveground biomass of tropical trees. Glob Chang Biol 20(10):3177–3190

Cote JF, Widlowski JL, Fournier RA, Verstraete MM (2009) The structural and radiative consistency of three-dimensional tree reconstructions from terrestrial lidar. Remote Sens Environ 113(5):1067–1081

Disney M, Lewis P, Saich P (2006) 3d modeling of forest canopy structure for remote sensing simulations in the optical and microwave domains. Remote Sens Environ 100(1):114–132

Eugster W, Zeyer K, Zeeman M, Michna P, Zingg A, Buchmann N, Emmeneg- ger L (2007) Nitrous oxide net exchange in a beech dominated mixed forest in Switzerland measured with a quantum cascade laser spectrometer. Biogeosci Discuss 4:1167–1200

Feret JB, Francois C, Asner GP, Gitelson AA, Martin RE, Bidel LP, Ustin SL, le Maire G, Jacquemoud S (2008) Prospect-4 and 5: advances in the leaf optical properties model separating photosynthetic pigments. Remote Sens Environ 112(6):3030–3043

Gardner T (2010) Monitoring forest biodiversity: improving conservation through ecologically responsible management, vol 9781849775106. Earthscan, London

Gastellu-Etchegorry J, Zagolski F, Romier J (1996) A simple anisotropic reflectance model for homogeneous multilayer canopies. Remote Sens Environ 57(1):22–38

Gastellu-Etchegorry JP, Yin T, Lauret N, Cajgfinger T, Gregoire T, Grau E, Feret JB, Lopes M, Guilleux J, Dedieu G, Malenovský Z, Cook BD, Morton D, Rubio J, Durrieu S, Cazanave G, Martin E, Ristorcelli T (2015) Discrete anisotropic radiative transfer (DART 5) for modeling airborne and satellite spectroradiometer and lidar acquisitions of natural and urban landscapes. Remote Sens 7(2):1667–1701

Grau E, Gastellu-Etchegorry JP (2013) Radiative transfer modeling in the earth-atmosphere system with {DART} model. Remote Sens Environ 139:149–170

Hansen MC, Potapov PV, Moore R, Hancher M, Turubanova SA, Tyukavina A, Thau D, Stehman SV, Goetz SJ, Loveland TR, Kommareddy A, Egorov A, Chini L, Justice CO, Townshend JRG (2013) High-resolution global maps of 21st-century forest cover change. Science 342(6160):850–853

Hilker T, Coops NC, Hall FG, Black TA, Wulder MA, Nesic Z, Krishnan P (2008) Separating physiologically and directionally induced changes in pri using brdf models. Remote Sens Environ 112(6):2777–2788

Homolov́a L, Malenovský Z, Clevers JGPW, Garćıa-Santos G, Schaepman ME (2013) Review of optical-based remote sensing for plant trait mapping. Ecol Complex 15:1–16

Huemmrich KF (2001) The geosail model: a simple addition to the sail model to describe discontinuous canopy reflectance. Remote Sens Environ 75(3):423–431

Hueni A, Biesemans J, Meuleman K, Dell'Endice F, Schlapfer D, Odermatt D, Kneubuehler M, Adriaensen S, Kempenaers S, Nieke J, Itten K (2009) Structure, components, and interfaces of the airborne prism experiment (apex) processing and archiving facility. IEEE Trans Geosci Remote Sens 47(1):29–43

Hyyppa J, Kelle O, Lehikoinen M, Inkinen M (2001) A segmentation-based method to retrieve stem volume estimates from 3-d tree height models produced by laser scanners. IEEE Trans Geosci Remote Sens 39:969–975

Isbell F, Gonzalez A, Loreau M, Cowles J, Dıaz S, Hector A, Mace GM, Wardle DA, O'Connor MI, Duffy JE, Turnbull LA, Thompson PL, Larigauderie A (2017) Linking the influence and dependence of people on biodiversity across scales. Nature 546:65–72

Jacquemoud S (1993) Inversion of the prospect + sail canopy reflectance model from aviris equivalent spectra: theoretical study. Remote Sens Environ 44(2–3):281–292

Jacquemoud S, Baret F (1990) Prospect: a model of leaf optical properties spectra. Remote Sens Environ 34(2):75–91

Jetz W, Cavender-Bares J, Pavlick R, Schimel D, Davis F, Asner G, Guralnick R, Kattge J, Latimer A, Moorcroft P, Schaepman M, Schildhauer M, Schneider F, Schrodt F, Stahl U, Ustin S (2016) Monitoring plant functional diversity from space. Nat Plants 2(3):16024

Kaartinen H, Hyyppa J, Yu X, Vastaranta M, Hyyppa H, Kukko A, Holopainen M, Heipke C, Hirschmugl M, Morsdorf F, Næsset E, Pitkanen J, Popescu S, Solberg S, Wolf BM, Wu JC (2012) An international comparison of individual tree detection and extraction using airborne laser scanning. Remote Sens 4:950–974

Knyazikhin Y, Lewis P, Disney MI, Mottus M, Rautiainen M, Stenberg P, Kaufmann RK, Marshak A, Schull MA, Latorre Carmona P, Vanderbilt V, Davis AB, Baret F, Jacquemoud S, Lyapustin A, Yang Y, Myneni RB (2013) Reply to Ollinger et al.: remote sensing of leaf nitrogen and emergent ecosystem properties. Proc Natl Acad Sci 110(27):E2438

Kotz B, Schaepman M, Morsdorf F, Bowyer P, Itten K, Allg̈ower B (2004) Radiative transfer modeling within a heterogeneous canopy for estimation of forest fire fuel properties. Remote Sens Environ 92(3):332–344

Lefsky MA, Cohen WB, Acker SA, Parker GG, Spies TA, Harding D (1999) Lidar remote sensing of the canopy structure and biophysical properties of Douglas-fir western hemlock forests. Remote Sens Environ 70:339–361

Leiterer R, Mücke W, Morsdorf F, Hollaus M, Pfeifer N, Schaepman M (2013) Operational forest structure monitoring using airborne laser scanning. Photogrammetrie Fernerkundung Geoinformation 3:173–184

Lewis P, Disney M (2007) Spectral invariants and scattering across multiple scales from within-leaf to canopy. Remote Sens Environ 109(2):196–206

Meroni M, Colombo R, Panigada C (2004) Inversion of a radiative transfer model with hyperspectral observations for LAI mapping in poplar plantations. Remote Sens Environ 92(2):195–206

Morsdorf F, Meier E, Kotz B, Itten KI, Dobbertin M, Allgower B (2004) Lidar-based geometric reconstruction of boreal type forest stands at single tree level for forest and wildland fire management. Remote Sens Environ 92(3):353–362, forest Fire Prevention and Assessment

Morsdorf F, Kotz B, Meier E, Itten K, Allgower B (2006) Estimation of LAI and fractional cover from small footprint airborne laser scanning data based on gap fraction. Remote Sens Environ 104(1):50–61

Morsdorf F, Marell A, Koetz B, Cassagne N, Pimont F, Rigolot E, Allgower B (2010) Discrimination of vegetation strata in a multi-layered mediterranean forest ecosystem using height and intensity information derived from airborne laser scanning. Remote Sens Environ 114(7):1403–1415

Morsdorf F, Eck C, Zgraggen C, Imbach B, Schneider F, Kükenbrink D (2017) UAV-based LiDAR acquisition for the derivation of high-resolution forest and ground information. Lead Edge 36(7):566–570

Morsdorf F, Kükenbrink D, Schneider FD, Abegg M, Schaepman ME (2018) Close-range laser scanning in forests: towards physically based semantics across scales. Interface Focus 8(2):20170046

Myneni R, Nemani R, Running S (1997) Estimation of global leaf area index and absorbed PAR using radiative transfer models. IEEE Trans Geosci Remote Sens 35:1380–1393

Myneni RB, Maggion S, Iaquinta J, Privette JL, Gobron N, Pinty B, Kimes DS, Verstraete MM, Williams DL (1995) Optical remote sensing of vegetation: Modeling, caveats, and algorithms. Remote Sens Environ 51(1):169–188

Næsset E (2002) Predicting forest stand characteristics with airborne scanning laser using a practical two-stage procedure and field data. Remote Sens Environ 80(1):88–99

Nelson R (1997) Modeling forest canopy heights: the effects of canopy shape. Remote Sens Environ 60:327–334

Nelson R (2013) How did we get here? An early history of forestry lidar. Can J Remote Sens 39(s1):S6–S17

Ni W, Li X, Woodcock C, Caetano M, Strahler A (1999) An analytical hybrid gort model for bidirectional reflectance over discontinuous plant canopies. IEEE Trans Geosci Remote Sens 37:987–999

Niinemets Ü, Kull O, Tenhunen J (1998) An analysis of light effects on foliar morphology, physiology, and light interception in temperate deciduous woody species of contrasting shade tolerance. Tree Physiol 18(10):681–696

North P (1996) Three-dimensional forest light interaction model using a monte carlo method. IEEE Trans Geosci Remote Sens 34(4):946–956

O'Connor B, Secades C, Penner J, Sonnenschein R, Skidmore A, Burgess ND, Hutton JM (2015) Earth observation as a tool for tracking progress towards the aichi biodiversity targets. Remote Sens Ecol Conserv 1(1):19–28

Parmesan C, Yohe G (2003) A globally coherent fingerprint of climate change impacts across natural systems. Nature 421(6918):37–42

Pereira H, Ferrier S, Walters M, Geller G, Jongman R, Scholes R, Bruford M, Brummitt N, Butchart S, Cardoso A, Coops N, Dulloo E, Faith D, Freyhof J, Gregory R, Heip C, Hoft R, Hurtt G, Jetz W, Karp D, McGeoch M, Obura D, Onoda Y, Pettorelli N, Reyers B, Sayre R, Scharlemann J, Stuart S, Turak E, Walpole M, Wegmann M (2013) Essential biodiversity variables. Science 339(6117):277–278

Pettorelli N, Wegmann M, Skidmore A, Mücher S, Dawson T, Fernandez M, Lucas R, Schaepman M, Wang T, O'Connor B, Jongman R, Kempeneers P, Sonnenschein R, Leidner A, Bohm M, He K, Nagendra H, Dubois G, Fatoyinbo T, Hansen M, Paganini M, de Klerk H, Asner G, Kerr J, Estes A, Schmeller D, Heiden U, Rocchini D, Pereira H, Turak E, Fernandez N, Lausch A, Cho M, Alcaraz-Segura D, McGeoch M, Turner W, Mueller A, St-Louis V, Penner J, Vihervaara P, Belward A, Reyers B, Geller G (2016) Framing the concept of satellite remote sensing essential biodiversity variables: challenges and future directions. Remote Sensing in Ecology and Conservation 2(3):122–131

Popescu SC, Wynne RH, Nelson RF (2002) Estimating plot-level tree heights with lidar: local filtering with a canopy-height based variable window size. Comput Electron Agric 37(1–3):71–95

Raumonen P, Kaasalainen M, Akerblom M, Kaasalainen S, Kaartinen H, Vastaranta M, Holopainen M, Disney M, Lewis P (2013) Fast automatic precision tree models from terrestrial laser scanner data. Remote Sens 5(2):491

Running S, Nemani R, Heinsch F, Zhao M, Reeves M, Hashimoto H (2004) A continuous satellite-derived measure of global terrestrial primary production. Bioscience 54(6):547–560

Schaepman ME, Ustin SL, Plaza AJ, Painter TH, Verrelst J, Liang S (2009) Earth system science related imaging spectroscopy - an assessment. Remote Sens Environ 113(Suppl. 1):S123–S137

Schaepman ME, Jehle M, Hueni A, D'Odorico P, Damm A, Weyermann J, Schnei- d FD, Laurent V, Popp C, Seidel FC, Lenhard K, Gege P, Küchler C, Brazile J, Kohler P, Vos LD, Meuleman K, Meynart R, Schlapfer D, Kneubühler M, Itten KI (2015) Advanced radiometry measurements and earth science applications with the airborne prism experiment (apex). Remote Sens Environ 158:207–219

Schaepman-Strub G, Schaepman M, Painter T, Dangel S, Martonchik J (2006) Reflectance quantities in optical remote sensing–definitions and case studies. Remote Sens Environ 103(1):27–42

Schneider FD, Leiterer R, Morsdorf F, Gastellu-Etchegorry JP, Lauret N, Pfeifer N, Schaepman ME (2014) Simulating imaging spectrometer data: 3d forest modeling based on lidar and in situ data. Remote Sens Environ 152:235–250

Schneider FD, Morsdorf F, Schmid B, Petchey OL, Hueni A, Schimel DS, Schaepman ME (2017) Mapping functional diversity from remotely sensed morphological and physiological forest traits. Nat Commun 8(1):1441

Skidmore A, Pettorelli N, Coops N, Geller G, Hansen M, Lucas R, Mücher C, O'Connor B, Paganini M, Pereira H, Schaepman M, Turner W, Wang T, Weg- mann M (2015) Environmental science: agree on biodiversity metrics to track from space. Nature 523(7561):403–405

Turner M (1989) Landscape ecology: the effect of pattern on process. Annu Rev Ecol Syst 20:171–197

Verhoef W, Bach H (2007) Coupled soil-leaf-canopy and atmosphere radiative transfer modeling to simulate hyperspectral multi-angular surface reflectance and {TOA} radiance data. Remote Sens Environ 109(2):166–182

Wang Y, Hyyppa J, Liang X, Kaartinen H, Yu X, Lindberg E, Holmgren J, Qin Y, Mallet C, Ferraz A, Torabzadeh H, Morsdorf F, Zhu L, Liu J, Alho P (2016) International benchmarking of the individual tree detection methods for modeling 3-d canopy structure for silviculture and forest ecology using airborne laser scanning. IEEE Trans Geosci Remote Sens 54(9):5011–5027

Widlowski JL, Robustelli M, Disney M, Gastellu-Etchegorry JP, Lavergne T, Lewis P, North P, Pinty B, Thompson R, Verstraete M (2008) The rami online model checker (romc): a web-based benchmarking facility for canopy reflectance models. Remote Sens Environ 112(3):1144–1150

Widlowski JL, Mio C, Disney M, Adams J, Andredakis I, Atzberger C, Brennan J, Busetto L, Chelle M, Ceccherini G, Colombo R, Cote JF, Eenmae A, Essery R, Gastellu-Etchegorry JP, Gobron N, Grau E, Haverd V, Homolova L, Huang H, Hunt L, Kobayashi H, Koetz B, Kuusk A, Kuusk J, Lang M, Lewis PE, Lovell JL, Malenovsky Z, Meroni M, Morsdorf F, Mottus M, Ni-Meister W, Pinty B, Rautiainen M, Schlerf M, Somers B, Stuckens J, Verstraete MM, Yang W, Zhao F, Zenone T (2015) The fourth phase of the radiative transfer model intercomparison (rami) exercise: actual canopy scenarios and conformity testing. Remote Sens Environ 169:418–437

Wilson J, Peet R, Dengler J, Partel M (2012) Plant species richness: the world records. J Veg Sci 23(4):796–802

Wulder M, Coops N, Hudak A, Morsdorf F, Nelson R, Newnham G, Vastaranta M (2013) Status and prospects for lidar remote sensing of forested ecosystems. Can J Remote Sens 39(s1):S1–S5

Wulder MA, White JC, Nelson RF, Næsset E, Orka HO, Coops NC, Hilker T, Bater CW, Gobakken T (2012) Lidar sampling for large-area forest characterization: a review. Remote Sens Environ 121:196–209

Yin T, Gastellu-Etchegorry JP, Lauret N, Grau E, Rubio J (2013) A new approach of direction discretization and oversampling for 3d anisotropic radiative transfer modeling. Remote Sens Environ 135:213–223

5

Spectranomics: An Emerging Approach

Roberta E. Martin

5.1 Introduction

One of the major challenges for biodiversity science is how to measure biodiversity at spatial scales relevant for conservation and management (Turner 2014). Supported by technological, computational, and modeling advances, along with increased data availability, remote sensing (RS) has become an essential tool for ecologists and land managers because it provides data on the optical properties of the Earth's surface at landscape to global scales (Jetz et al. 2016). At the same time, increasing awareness of how little we know about the species inhabiting our planet has led to a surge in ground-based activities to catalog what's out there and establish baselines such as Conservation International's Rapid Assessment Program and/or community aggregated information needed for biodiversity assessment (Myers et al. 2000). In addition, advances in genetic analysis, physiological experiments, and trait-based studies have advanced our understanding of functional biodiversity (Cavender-Bares et al. 2006; Kress et al. 2009; Baraloto et al. 2012). Despite these knowledge gains, linking the information from these disparate sources in a useful manner presented a new hurdle. In 2007 the Spectranomics approach was launched to address this challenge using canopy functional traits and their resultant spectral properties.

R. E. Martin (✉)
School of Geographical Sciences and Urban Planning, Arizona State University, Tempe, AZ, USA

Center for Global Discovery and Conservation Science, Arizona State University, Tempe, AZ, USA
e-mail: Roberta.Martin@asu.edu

Plants play a foundational role in establishing and maintaining ecosystem function, biogeochemical cycling, hydrological cycling, and biodiversity (Mooney et al. 1996; Schimel et al. 2013). More specifically, canopy plants (those that occupy the sun-facing portion of a landscape) serve as dominant primary producers through the capture and utilization of light. Their structures also provide habitat for vast numbers of species living in the shadows. To maintain this premier position in a forest ecosystem, plants have evolved a vast array of strategies for growth, defense, and longevity, largely manifested as chemical and/or structural adjustments in their leaves (Reich et al. 2003; Wright et al. 2004; Diaz et al. 2016). The molecular arrangement of these foliar properties generates an optical reflectance spectrum that can be measured at a variety of scales with spectroscopy (Curran 1989; Jacquemoud and Ustin 2001; Ustin et al. 2009; Ustin and Jacquemoud, Chap. 14). The ultimate result is a massive number of tree species coalescing into forest communities of varying complexity, with unique taxonomic compositions and functional roles that can potentially be mapped across a forested landscape (Reichstein et al. 2014 and others).

Despite understanding the important role different canopy species and communities of species play in creating and maintaining biodiversity, the measurement, mapping, and monitoring of forest canopy composition and functional diversity has remained a challenge. Current Earth-observing satellite technology is limited to detecting changes in vegetation cover as well as major differences in vegetation type and photosynthesis (Running et al. 1994; Tucker and Townshend 2000) and does not easily reveal compositional differences or changes over time (Turner et al. 2003). Tropical forest canopy diversity is especially underexplored because spatial and temporal variation often exceeds our ability to adequately utilize field-based approaches (Marvin et al. 2014). Airborne imaging spectroscopy can provide an intermediate solution; however, a fundamental prerequisite for determining whether species diversity or a particular species might be successfully mapped is an assessment of chemical uniqueness and diversity among plant taxa. This is important because the spectroscopy of canopies is driven primarily by the chemical composition of the foliage (Curran 1989; Asner et al. 2015).

5.2 Spectranomics Approach

The Spectranomics approach was developed to link plant canopy functional traits to their spectral properties with the objective of providing time-varying, scalable methods for remote sensing (RS) of forest biodiversity (Asner and Martin 2009). In the pool of potentially important plant functional traits, foliar chemicals stand out as core physiologically based predictors of plant adaptation to environmental conditions (Díaz et al. 1998; Wright et al. 2010). We selected a suite of 23 canopy chemical traits based on their strong ecological and evolutionary relevance, spatial variation in species and communities, and measurable spectral properties. These traits consist of those that (i) mediate or are indicative of photosynthesis and carbon uptake (chlorophyll *a* and *b*, carotenoids, nitrogen, $\delta^{13}C$, and $\delta^{15}N$; non-soluble carbohydrates); (ii) are related to structure (leaf mass per area and water content, lignin, cellulose, and hemicellulose) and chemical defense (phenols and tannins); and (iii) are defining general metabolic processes (macro- and micronutrients; here calcium, magnesium, phosphorus, potassium and boron, iron, manganese, zinc) (Table 5.1). The distribution and variation of these traits in plant canopy leaves evolve as a function of stoichiometric relationships among constituents in response to biotic and abiotic pressures and are often formulated differently at the species level (Díaz et al. 1998). This evolved chemical makeup of plant canopies and its similarity and uniqueness among species, which we call *chemical phylogeny,* is an essential component of Spectranomics (Fig. 5.1a).

Table 5.1 Summary statistics for 22 foliar chemical traits and leaf mass per area (LMA) from top-of-canopy leaves collected from 12,012 individual trees at sites across the wet tropics as part of the Spectranomics Program

	Mean	Standard deviation	Minimum	Maximum	Median	Skew	Kurtosis
Light capture and growth							
Chlorophyll *a* (mg g^{-1})	4.67	1.96	0.01	26.71	4.40	2.47	1.76
Chlorophyll *b* (mg g^{-1})	1.73	0.78	0.01	11.32	1.62	0.86	0.58
Carotenoids (mg g^{-1})	1.38	0.53	0.01	7.87	1.31	0.80	0.59
Nitrogen (%)	2.01	0.70	0.35	6.15	1.91	1.21	0.91
NSC (%)	46.20	11.56	12.73	86.33	45.72	31.40	25.61
$\delta^{13}C$ (‰)	−30.59	1.89	−36.40	−19.90	−30.70	−32.90	−34.00
$\delta^{15}N$ (‰)	1.38	2.70	−10.30	10.50	1.40	−2.00	−4.00
Structure and defense							
LMA (g m^{-2})	113.63	44.44	15.65	622.36	105.33	66.92	51.68
Water (%)	58.40	8.28	9.17	90.79	57.57	48.92	44.87
Carbon (%)	49.30	3.31	31.60	65.00	49.70	45.00	41.80
Lignin (%)	24.15	9.94	0.25	81.08	23.47	11.67	7.65
Cellulose (%)	17.82	5.75	1.08	56.60	17.30	10.89	8.41
Hemicellulose (%)	11.63	5.03	0.00	49.31	11.25	5.72	2.84
Phenols (mg g^{-1})	101.11	53.63	0.00	358.19	101.71	27.02	10.54
Tannins (mg g^{-1})	46.45	26.82	−0.64	238.79	43.14	15.27	5.66
Macronutrients							
Calcium (%)	0.96	0.81	0.00	8.36	0.74	0.18	0.08
Magnesium (%)	0.26	0.15	0.02	2.71	0.23	0.11	0.08
Phosphorus (%)	0.12	0.07	0.02	0.86	0.10	0.05	0.04
Potassium (%)	0.76	0.45	0.13	5.64	0.65	0.35	0.25
Micronutrients							
Boron (μg g^{-1})	27.19	23.48	1.16	321.89	20.03	8.35	5.31
Iron (μg g^{-1})	80.58	206.63	7.13	9470.68	47.78	26.55	19.47
Manganese (μg g^{-1})	304.32	512.14	3.03	7331.67	103.80	19.75	11.55
Zinc (μg g^{-1})	17.08	44.47	1.65	2535.98	11.77	6.49	4.62

NSC nonstructural carbohydrates

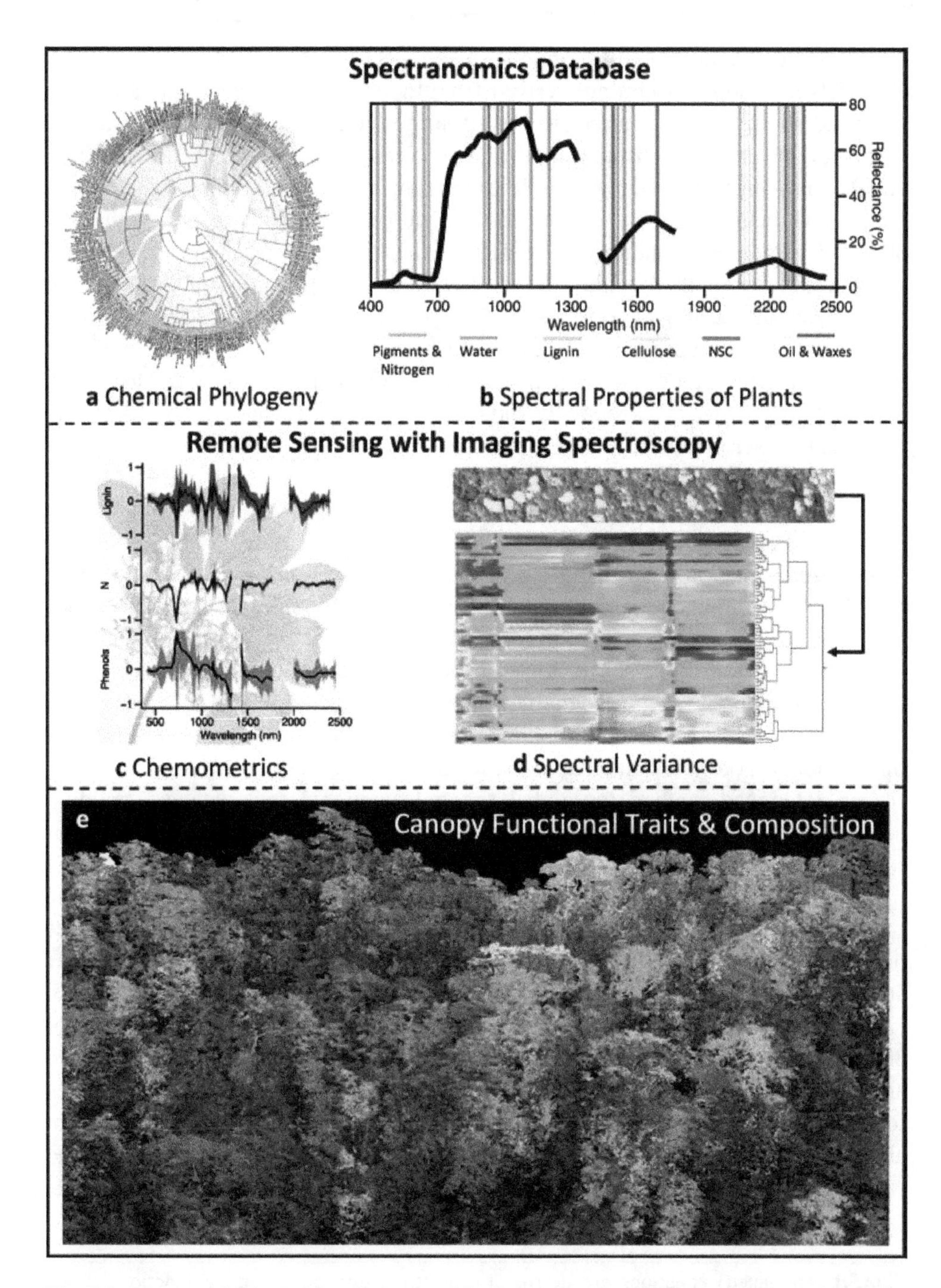

Fig. 5.1 The essential interactive elements of the Spectranomics Database include phylogenetic, chemical, and spectral information on canopy species. (**a**) Assays of 23 foliar chemical traits combined are collected, organized, and analyzed phylogenetically, producing a new tree of life based on the relatedness of functional trait signatures. This generic phylogeny shows the chemical relatedness of thousands of species in the Spectranomics Database. (**b**) An example of a remotely sensed canopy reflectance spectrum of one species is shown along with indicators of key chemical contributions to the spectrum (Curran 1989; Ustin et al. 2009; Kokaly et al. 2009). (**c**) Chemometric

Another component of Spectranomics is the spectral properties of plant canopies (Fig. 5.1b). Canopy spectra are derived from the way plant foliage interacts with solar radiation, and this interaction is strongly determined by foliar chemicals. Across the full solar spectrum, from the ultraviolet to the visible to the near-infrared and the shortwave-infrared regions of the electromagnetic spectrum (350–3500 nm), plants have many common and yet also unique patterns of interaction with solar energy. Chemometric studies determine how these chemicals relate to reflectance spectra, and the methods today range from traditional spectroscopic assays and newer machine learning approaches (Wold et al. 2001; Serbin et al. 2014; Feilhauer et al. 2015). Spectral properties also provide a tantalizing pathway forward to scale up from leaves to landscapes (Ustin et al. 2004) to the planetary level (Jetz et al. 2016), but only if we can accurately and repeatedly measure and interpret the spectra of plants over increasingly larger portions of Earth (Fig. 5.1c–e).

The realization of Spectranomics rests in a number of choices made early on to attempt to reduce unwanted sources of variation combined with extensive sampling. We focused on humid tropical forests for their high diversity and relative freedom from extreme phenological changes brought about by seasonal cycles such as those experienced in temperate regions but may not completely eliminate smaller phenological variation that might arise in reaction to drought or solar variations. We targeted only mature, fully sunlit, top-of-canopy leaves (trees and lianas) to limit variation attributable to intra-canopy shade and ontogeny and to best relate leaf properties to airborne and satellite-based spectral measurements. Prior to Spectranomics, our work and that of many others did not follow a strategically consistent, integrated method for global spectral-functional trait database building needed to reveal canopy plant functional spectral-chemical patterns at the biospheric scale.

We have collected, cataloged, and stored more than 13,000 canopy tree and liana specimens, in over 3 million tissue samples, representing about 10,000 species biased to humid tropical ecosystems (Fig. 5.2a). For perspective, this number approaches the total number of tree species in the Amazon basin (roughly 11,000; Hubbell et al. 2008), a value that would put the global tropical tree inventory at 30,000 species if we liberally extrapolate to the entire Neotropics plus the African and Asian-Oceanic tropics. The Spectranomics database focuses only on species found in the canopy, meaning they are in full sunlight and are observable from above. Since roughly 30–60% of tree species in a tropical forest plot makes it to the canopy (e.g., Bohlman 2015), the current Spectranomics database contains at least

Fig. 5.1 (continued) equations are derived to quantitatively relate canopy functional traits (chemicals) to spectral data. Example relationships are shown for foliar lignin, nitrogen (N), and polyphenols. The x-axis indicates spectral wavelengths of 400–2500 nm; the y-axes indicate relative importance of the spectrum to each example chemical constituent shown. (**d**) An example of spectra from individual crowns clustered based on their spectral variation. (**e**) A 3-D view of a portion of lowland Amazonian forest canopy. Different colors indicate different species detected based on 15 chemical traits using airborne imaging spectroscopy

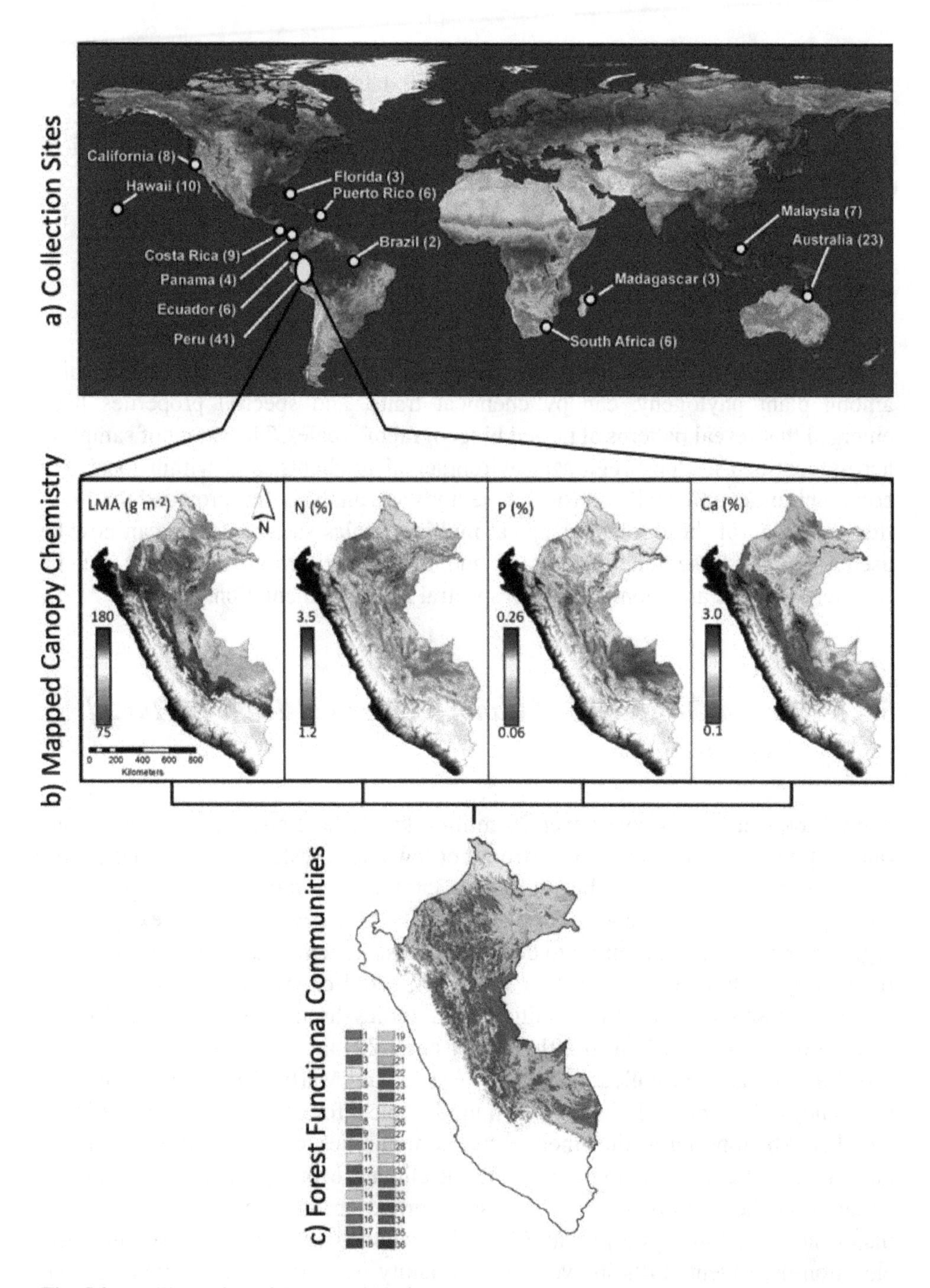

Fig. 5.2 An illustration of functional biodiversity mapping from foliar traits. (**a**) The 2018 global distribution of 128 forest landscapes contributing to the Spectranomics database. (**b**) Example maps of four foliar traits generated for the Andes-to-Amazon region of Peru using airborne imaging spectroscopy and modeling (Asner et al. 2017). (**c**) Map of 36 forest functional communities derived from a classification based on seven forest canopy traits derived from airborne imaging spectroscopy

half of the known tropical forest canopy species worldwide with measured foliar traits (Table 5.1). From investigations of these data and the fundamental patterns they uncover, Spectranomics has evolved into a new pathway to biological and ecological discovery, as well as a new tool for conservation-relevant mapping, particularly in high-diversity tropical forests.

5.3 Lessons Learned from Spectranomics

As the Spectranomics database has grown through the years, new relationships among plant phylogeny, canopy chemical traits, and spectral properties have emerged that reveal patterns at nested biogeographic scales. The extent of sampling across continents, along regional environmental gradients, and within local tree communities, coupled with consistent methods and analysis, has provided for quantitative testing of these relationships at multiple scales such that they can now be used to forecast the functional traits and biodiversity components that can be remotely mapped and monitored with spectral RS instrumentation.

5.3.1 Nested Geography of Canopy Chemical Traits in Humid Tropical Forest

Humid tropical forests cover over 20 million km of land area, span an enormous range of environmental conditions from hot lowland forests to cool montane rainforests along equatorial tree line at almost 3500 m on a variety of geological substrates, and support thousands of tree species. The high degree of complexity of this region provided an ideal setting to develop and use Spectranomics to test how environment and phylogeny interact to sort the spectral-chemical diversity of forest canopies. Based on results from multiple field studies throughout this region (Martin et al. 2007; Asner and Martin 2011; Asner et al. 2014b; McManus Chauvin et al. 2018) as well as their collective analysis (Asner and Martin 2016), we discovered that canopy chemical trait diversity of humid tropical forests occurs in a nested pattern driven by long-term adjustment of tree communities to large-scale environmental factors, particularly geologic substrate and climate. More specifically, geographic variation at the soil order level, expressing broad changes in fertility, underpins major shifts in foliar phosphorus (P) and calcium (Ca) (Fig. 5.2). Additionally, elevation-dependent shifts in average community leaf dry mass per area (LMA), chlorophyll, and carbon allocation (including nonstructural carbohydrates) are most strongly correlated with changes in foliar Ca. We also found that chemical diversity within communities is driven by differences between species rather than by plasticity within species. Finally, elevation- and soil-dependent changes in nitrogen (N), LMA, and leaf carbon allocation are mediated by canopy compositional turnover, whereas foliar P and Ca are driven more by changes in site conditions than by phylogeny. In short, Spectranomics led us to understand that canopy functional traits can be nested regionally by environmental setting but expressed locally within any given environment by their evolutionary origin.

5.3.2 Spectral Properties of Humid Tropical Forest Canopies

In concert with chemical trait collections, we measured the spectral properties of canopy foliage from thousands of humid tropical tree canopies and determined that all 23 chemical traits can be remotely sensed to varying degrees (Asner et al. 2011; Chadwick and Asner 2016; Martin et al. 2018). Utilizing leaf-level spectral-chemical relationships, we discovered that the spectral properties of canopy foliage closely tracked canopy functional trait responses to macro-environmental changes such as broad differences in soil fertility (Asner and Martin 2011; Asner et al. 2012b). Similar to the functional trait findings, we discovered that the spectral properties of foliage within communities along elevation gradients were largely determined by phylogenetic identity (Asner et al. 2014a). Consequently, canopy functional traits and spectral properties tracked one another at nested ecological scales, a result that suggests what we might find if we collected map-based spectral data over a much larger geographic area using RS instrumentation.

When coupled with DNA analyses, Spectranomics data indicate that forest canopies show strong phylogenetic organization of their foliar spectral properties, particularly in the shortwave-infrared (1500–2500 nm) wavelength region (McManus et al. 2016). This finding suggests that mapping of forest canopies with airborne imaging spectroscopy may provide spatial insight to the genetic distribution and genealogy of forest canopy taxa. Growth-form-specific studies using the Spectranomics approach revealed that lianas (woody vines) maintain functional traits and spectral properties unique from their host tree canopies (Asner and Martin 2012). Lianas are important drivers and limiters of biodiversity and carbon cycling in tropical forests (Schnitzer and Bongers 2011), and these measured differences predicted and underpinned the subsequent mapping of lianas in tropical forests using airborne imaging spectroscopy (Marvin et al. 2016).

Spectranomics data have been collected and archived under stringent field and analytical standards, which has facilitated the development new quantitative linkages between canopy foliar spectroscopy and canopy functional traits (Feilhauer et al. 2010, 2015; Féret et al. 2011, 2017). Spectral modeling studies showed that full-spectrum (350–3500 nm) data provided retrieval capability for three times the number of chemicals as 350–1300 nm data from less expensive, more common visible to near-infrared spectrometers. These studies also pointed to the need for sampling fully sunlit foliage in higher-density portions of tree crowns to minimize the effect of canopy structure on chemical trait retrievals. These findings were key guiding components in the development of laser-guided imaging spectroscopy that links Spectranomics field surveys to remotely sensed spectra to generate consistent canopy chemical trait retrieval at multiple geographic scales.

5.3.3 Spectranomics for Biodiversity Mapping

The Spectranomics fieldwork pointed toward two particular forecasts. First, Spectranomics suggested that spectral mapping from current aircraft and future satellites will reveal where whole forest communities are functionally similar and where they are unique. Second, Spectranomics suggested that spectral RS will reveal the presence and patterning of specific canopy species, within communities and across environmental gradients, based on their functional trait "signatures."

Both forecasts were subsequently proven correct during mapping studies. Numerous landscape-scale studies now show that location of particular forest canopy species and their evolved canopy functional traits mirror soil nutrient resources mediated by topography, parent material, and climate (Higgins et al. 2014; Chadwick and Asner 2016; Balzotti et al. 2016). These findings demonstrate that Spectranomics directly connects plants to ecosystem processes such as biogeochemical cycles, which form an essential link to the rest of the Earth system. At a larger scale, a 2016 report on Andean and Amazonian forests mapped with airborne imaging spectroscopy confirmed the forecasted ecological shifts in forest canopy functional composition, sorted geographically by large-scale environmental factors including elevation, geology, soils, and climate (Fig. 5.2b, c; Asner et al. 2017). While the Spectranomics database provided a field-based preview of how communities of species would differ from one another, the mapping step provided a first synoptic view of the geographic distribution. Importantly, the mapping phase also revealed numerous new combinations of functional traits that had not been detected in the field program. The new canopy functional trait maps are a key stepping-stone to biogeographic assembly, not only of the functional diversity of the Andes-to-Amazon but also of the biological diversity of the region. The approach from Peru is currently being applied in Ecuador as well as Malaysian Borneo.

The second forecast from Spectranomics—which coexisting species within communities can maintain relatively unique canopy functional traits and spectral properties—has been explored and confirmed in a series of studies using airborne and space-based imaging spectroscopy. From Hawaii to Panama, and from Africa to the Amazon, hundreds of target species have been mapped based on their spectral signatures, underpinned by a knowledge of their functional traits (Fig. 5.3; Carlson et al. 2007; Papeş et al. 2010; Colgan et al. 2012; Baldeck and Asner 2014; Baldeck et al. 2015; Graves et al. 2016). Further, the new concept of "spectral species" was developed to map species richness (alpha diversity) and compositional turnover (beta diversity) in forest landscapes without the need to detect individual species (Féret and Asner 2014).The separability of the spectral species is determined by their canopy functional traits.

More broadly, Spectranomics has enabled a different kind of interaction between field or laboratory studies of plants and RS of functional and biological diversity of ecosystems. The forecasting capability made possible with the Spectranomics database has been central to planning whether and how to undertake spectral mapping activities in different regions and under what environmental conditions the RS technology will yield new insight. In turn, this has transformed the interaction

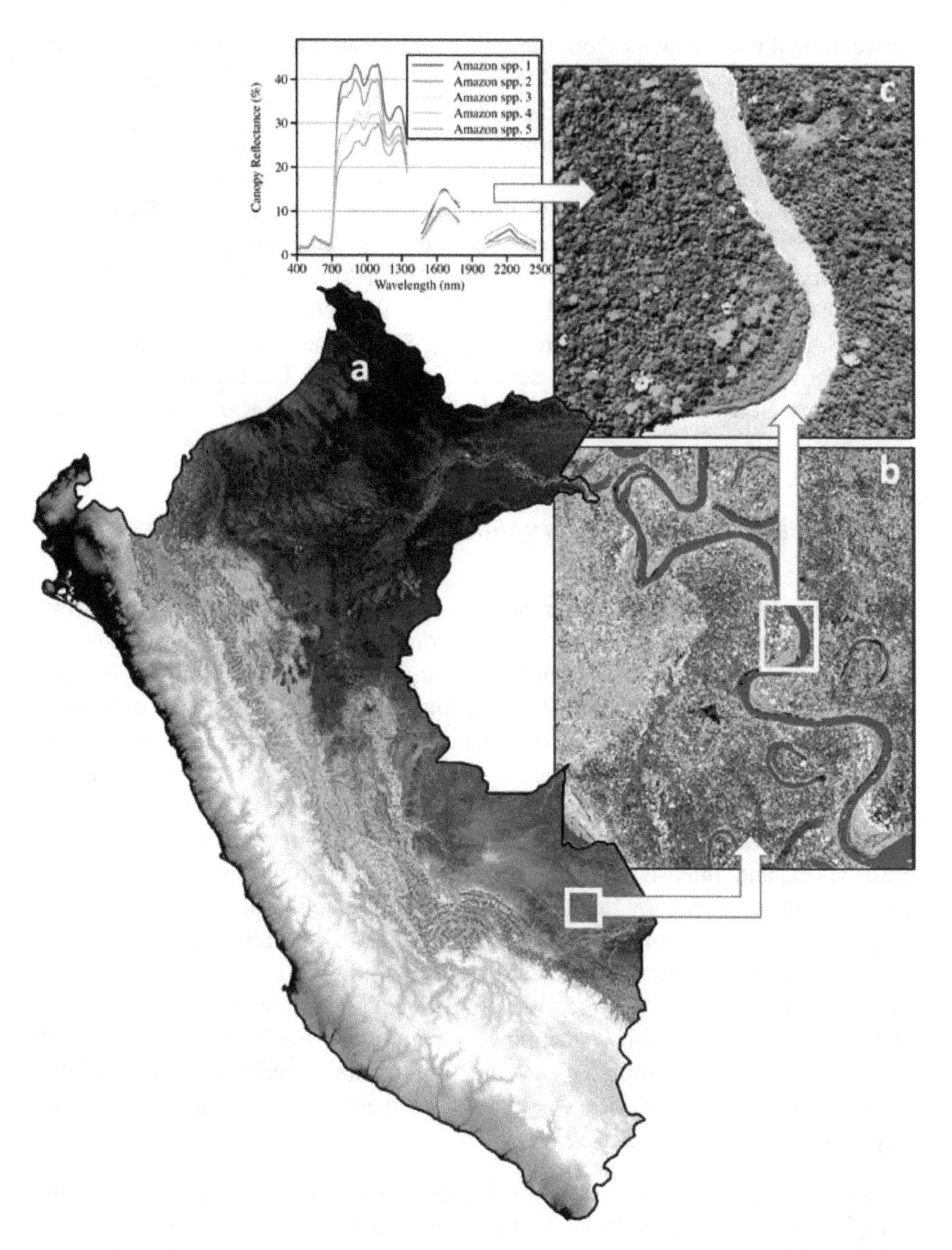

Fig. 5.3 Three scale-dependent views of the Peruvian Andes-Amazon region derived from airborne imaging spectroscopy using data and information from Spectranomics. (**a**) Peru-wide map shows the distribution of functionally distinct forests. Different colors indicate varying combinations of remotely sensed canopy foliar nitrogen (N), phosphorus (P), and leaf mass per area (LMA) (Asner et al. 2016). (**b**) Zoom image from the Peru-wide map indicates major changes in canopy N, P, and LMA with a lowland Amazonian forest (Asner et al. 2015a). Red indicates higher N + P and lower LMA relative to yellow and blue. (**c**) Individual species detections within the zoom box of panel b, derived using species-specific canopy spectra (Féret and Asner 2013; Baldeck et al. 2015)

between field and RS work from the traditional approach of mapping and ground truthing to one based on botanical, ecological, and biophysical knowledge in the interpretation of remotely sensed data.

This interaction between Spectranomics and RS also provided the scientific guidance, and initial funding, for a new class of mapping instruments, starting with a next-generation, high-fidelity visible-to-shortwave infrared (VSWIR) imaging spectrometer, built by the California Institute of Technology's Jet Propulsion Laboratory (JPL) for the Global Airborne Observatory, formerly the Carnegie Airborne Observatory (Asner et al. 2012a). JPL then built an identical instrument for NASA's Airborne Visible/Infrared Imaging Spectrometer (AVIRIS; http://aviris.jpl.nasa.gov) program, as well as several copies for the US National Ecological Observatory Network (NEON, https://www.neonscience.org; Kampe et al. 2011).

5.3.4 Scientific and Conservation Opportunities

An important outgrowth of Spectranomics is an emerging opportunity to partner discovery-based science with applied environmental conservation at large geographic scales. Conservation and management actions are usually limited in scope and effectiveness by numerous interacting financial, logistical, cultural, and political factors. An increasing ability to map canopy diversity may provide an avenue to identify the location and essential components of high-value conservation targets. Moreover, near-real-time scientific discovery from spectral RS can lead to more tactical conservation decision-making. Our specific experience is that, as land use pressures expand, intensify, and change over time, a mapping capability built upon the details of forest canopy function and composition, rather than just forest cover, supports improved conservation discussions and planning. This type of approach is needed to identify current and potential threats to, as well as current protections and opportunities for new protection of, species, communities, and ecosystems. The evolving biodiversity mapping capabilities made possible through Spectranomics are providing a tool set to support the current portfolio of Global Airborne Observatory activities (e.g., http://www.theborneopost.com/2016/04/06/3d-mapping-to-decide-on-land-use/).

The Spectranomics approach is starting to catch on in the scientific community, as highlighted in chapters throughout this book as well as new programs such as NEON and Canada's recently announced Spectranomics program for boreal forests (the Canadian Airborne Biodiversity Observatory; http://www.caboscience.org/), but there is much more to do to bring our approach to the global level. First, more scientists could get involved through building plant canopy trait laboratories and databases, paired with a specific style of leaf-level spectral measurements in the field. Currently, many functional trait and spectral measurement protocols are incompatible with the Spectranomics approach. For example, many foliar trait studies have involved the collection of samples in understory or shaded settings, in part because this foliage is easier to reach, yet spectral RS is most sensitive to canopy-level foliar chemical and structural traits (Jacquemoud et al. 2009). Additionally, most field-based trait studies do not include the use of a high-fidelity field spectrometer, which must be applied on fresh foliage to ensure connectivity to biotic and environmental conditions. Moreover, high-fidelity imaging spectrometers needed for mapping, such as the Global Airborne Observatory or AVIRIS, demand stringent and consistent field and laboratory trait measurement practices. Most of these issues can be remedied by incorporating one or more of the protocols provided on the Spectranomics website (https://gdcs.asu.edu/labs/martinlab/spectranomics). More could be done to boost capacity throughout the science community to generate data suitable for Spectranomics-type applications. Community-wide efforts to develop a global biodiversity monitoring system (Geller, Chap. 20) will greatly enhance humanity's ability to monitor and manage biodiversity for sustainability in the Anthropocene.

Acknowledgments I thank the Spectranomics scientific co-founder and programmatic funder, Dr. Greg Asner, for the inspiration to write this piece. I also thank the numerous scientists, engineers, technicians, students, and supporters of the Spectranomics Project, which has been made possible by the John D. and Catherine T. MacArthur Foundation.

References

Asner GP, Martin RE (2009) Airborne spectranomics: mapping canopy chemical and taxonomic diversity in tropical forests. Front Ecol Environ 7:269–276. https://doi.org/10.1890/070152

Asner GP, Martin RE (2011) Canopy phylogenetic, chemical and spectral assembly in a lowland Amazonian forest. New Phytol 189:999–1012. https://doi.org/10.1111/j.1469-8137.2010.03549.x

Asner GP, Martin RE (2012) Contrasting leaf chemical traits in tropical lianas and trees: implications for future forest composition. Ecol Lett 15:1001–1007

Asner GP, Martin RE (2016) Convergent elevation trends in canopy chemical traits of tropical forests. Glob Chang Biol 22:2216–2227. https://doi.org/10.1111/gcb.13164

Asner GP, Martin RE, Knapp DE, Tupayachi R, Anderson C, Carranza L, Martinez P, Houcheime M, Sinca F, Weiss P (2011) Spectroscopy of canopy chemicals in humid tropical forests. Remote Sens Environ 115:3587–3598. https://doi.org/10.1016/j.rse.2011.08.020

Asner GP, Knapp DE, Boardman J, Green RO, Kennedy-Bowdoin T, Eastwood M, Martin RE, Anderson C, Field CB (2012a) Carnegie Airborne Observatory-2: increasing science data dimensionality via high-fidelity multi-sensor fusion. Remote Sens Environ 124:454–465. https://doi.org/10.1016/j.rse.2012.06.012

Asner GP, Martin RE, Bin SA (2012b) Sources of canopy chemical and spectral diversity in lowland bornean forest. Ecosystems 15:504–517. https://doi.org/10.1007/s10021-012-9526-2

Asner GP, Martin RE, Carranza-Jiménez L, Sinca F, Tupayachi R, Anderson CB, Martinez P (2014a) Functional and biological diversity of foliar spectra in tree canopies throughout the Andes to Amazon region. New Phytol 204:127–139. https://doi.org/10.1111/nph.12895

Asner GP, Martin RE, Tupayachi R, Anderson CB, Sinca F, Carranza-Jimenez L, Martinez P (2014b) Amazonian functional diversity from forest canopy chemical assembly. Proc Natl Acad Sci 111:5604–5609. https://doi.org/10.1073/pnas.1401181111

Asner GP, Anderson CB, Martin RE, Tupayachi R, Knapp DE, Sinca F (2015a) Landscape biogeochemistry reflected in shifting distributions of chemical traits in the Amazon forest canopy. Nat Geosci 8:567–575. https://doi.org/10.1038/ngeo2443

Asner GP, Martin RE, Anderson CB, Knapp DE (2015b) Quantifying forest canopy traits: imaging spectroscopy versus field survey. Remote Sens Environ 158:15–27. https://doi.org/10.1016/j.rse.2014.11.011

Asner GP, Ustin SL, Townsend PA, Martin RE and Chadwick KD (2015) Forest biophysical and biochemical properties from hyperspectral and LiDAR remote sensing. In: Remote sensing handbook. Land resources monitoring, modeling, and mapping with remote sensing, pp 429–448. https://doi.org/10.1201/b19322-22

Asner GP, Knapp DE, Anderson CB, Martin RE, Vaughn N (2016) Large-scale climatic and geophysical controls on the leaf economics spectrum. Proc Natl Acad Sci 113:E4043–E4051. https://doi.org/10.1073/pnas.1604863113

Asner GP, Martin RE, Knapp DE, Tupayachi R, Anderson CB, Sinca F, Vaughn NR, Llactayo W (2017) Airborne laser-guided imaging spectroscopy to map forest trait diversity and guide conservation. Science 355:385–389. https://doi.org/10.1126/science.aaj1987

Baldeck C, Asner G (2014) Single-species detection with airborne imaging spectroscopy data: a comparison of support vector techniques. IEEE J Sel Top Appl Earth Obs Remote Sens 8:2501–2512

Baldeck CA, Asner GP, Martin RE, Anderson CB, Knapp DE, Kellner JR, Wright SJ (2015) Operational tree species mapping in a diverse tropical forest with airborne imaging spectroscopy. PLoS One 10:e0118403. https://doi.org/10.1371/journal.pone.0118403

Balzotti CS, Asner GP, Taylor PG, Cleveland CC, Cole R, Martin RE, Nasto M, Osborne BB, Porder S, Townsend AR (2016) Environmental controls on canopy foliar nitrogen distributions in a Neotropical lowland forest. Ecol Appl 26:2449–2462. https://doi.org/10.1002/eap.1408

Baraloto C, Hardy OJ, Paine CET, Dexter KG, Cruaud C, Dunning LT, Gonzalez M-A, Molino J-F, Sabatier D, Savolainen V, Chave J (2012) Using functional traits and phylogenetic trees to examine the assembly of tropical tree communities. J Ecol 100:690–701. https://doi.org/10.1111/j.1365-2745.2012.01966.x

Bohlman SA (2015) Species diversity of canopy versus understory trees in a Neotropical Forest: implications for Forest structure, function and monitoring. Ecosystems 18:658–670. https://doi.org/10.1007/s10021-015-9854-0

Carlson KM, Asner GP, Hughes RF, Ostertag R, Martin RE, Hughes FR, Ostertag R, Martin RE (2007) Hyperspectral remote sensing of canopy biodiversity in Hawaiian lowland rainforests. Ecosystems 10:536–549. https://doi.org/10.1007/s10021-007-9041-z

Cavender-Bares J, Keen A, Miles B (2006) Phylogenetic structure of Floridian plant communities depends on taxonomic and spatial scale. Ecology 87:S109–S122

Chadwick KD, Asner GP (2016) Organismic-scale remote sensing of canopy foliar traits in lowland tropical forests. Remote Sens 8:87. https://doi.org/10.3390/rs8020087

Colgan MS, Baldeck CA, baptiste FJ, Asner GP (2012) Mapping savanna tree species at ecosystem scales using support vector machine classification and BRDF correction on airborne hyperspectral and LiDAR data. Remote Sens 4:3462–3480. https://doi.org/10.3390/rs4113462

Curran PJ (1989) Remote sensing of foliar chemistry. Remote Sens Environ 30:271–278. https://doi.org/10.1016/0034-4257(89)90069-2

Díaz S, Cabido M, Casanoves F, Diaz S, Cabido M, Casanoves F (1998) Plant functional traits and environmental filters at a regional scale. J Veg Sci 9:113–122. https://doi.org/10.2307/3237229

Diaz S, Kattge J, Cornelissen JH, Wright IJ, Lavorel S, Dray S, Reu B, Kleyer M, Wirth C, Prentice IC, Garnier E, Bonisch G, Westoby M, Poorter H, Reich PB, Moles AT, Dickie J, Gillison AN, Zanne AE, Chave J, Wright SJ, Sheremet'ev SN, Jactel H, Baraloto C, Cerabolini B, Pierce S, Shipley B, Kirkup D, Casanoves F, Joswig JS, Gunther A, Falczuk V, Ruger N, Mahecha MD, Gorne LD (2016) The global spectrum of plant form and function. Nature 529:167–171. https://doi.org/10.1038/nature16489

Feilhauer H, Asner GP, Martin RE, Schmidtlein S (2010) Brightness-normalized partial least squares regression for hyperspectral data. J Quant Spectrosc Radiat Transf 111:1947–1957. https://doi.org/10.1016/j.jqsrt.2010.03.007

Feilhauer H, Asner GP, Martin RE (2015) Multi-method ensemble selection of spectral bands related to leaf biochemistry. Remote Sens Environ 164:57–65. https://doi.org/10.1016/j.rse.2015.03.033

Féret JB, Asner GP (2013) Tree species discrimination in tropical forests using airborne imaging spectroscopy. IEEE Trans Geosci Remote Sens 51:73–84

Feret J-B, Asner GP (2014) Microtopographic controls on lowland Amazonian canopy diversity from imaging spectroscopy. Ecol Appl 24:1297–1310

Féret J-B, Asner GP (2014) Mapping tropical forest canopy diversity using high-fidelity imaging spectroscopy. Ecol Appl 24:1289–1296. https://doi.org/10.1890/13-1824.1

Féret J-B, François C, Gitelson A, Asner GP, Barry KM, Panigada C, Richardson AD, Jacquemoud S (2011) Optimizing spectral indices and chemometric analysis of leaf chemical properties using radiative transfer modeling. Remote Sens Environ 115:2742–2750. https://doi.org/10.1016/j.rse.2011.06.016

Féret J-B, Gitelson AA, Noble SD, Jacquemoud S (2017) PROSPECT-D: towards modeling leaf optical properties through a complete lifecycle. Remote Sens Environ 193:204–215. https://doi.org/10.1016/J.RSE.2017.03.004

Graves S, Asner G, Martin R, Anderson C, Colgan M, Kalantari L, Bohlman S (2016) Tree species abundance predictions in a tropical agricultural landscape with a supervised classification model and imbalanced data. Remote Sens 8:161. https://doi.org/10.3390/rs8020161

Higgins MA, Asner GP, Martin RE, Knapp DE, Anderson C, Kennedy-Bowdoin T, Saenz R, Aguilar A, Joseph Wright S (2014) Linking imaging spectroscopy and LiDAR with floristic composition and forest structure in Panama. Remote Sens Environ 154:358–367. https://doi.org/10.1016/j.rse.2013.09.032

Hubbell SP, He F, Condit R, Borda-de-Água L, Kellner J, ter Steege H (2008) How many tree species are there in the Amazon and how many of them will go extinct? Proc Natl Acad Sci 105:11498–11504

Jacquemoud S, Ustin SL (2001) Leaf Optical properties: a state of the art. In: 8th International Symposium on Physical Measurements & Signatures inRemote Sensing, Aussois, France, pp 223–232

Jacquemoud S, Verhoef W, Baret F, Bacour C, Zarco-Tejada PJ, Asner GP, Francois C, Ustin SL (2009) PROSPECT plus SAIL models: a review of use for vegetation characterization. Remote Sens Environ 113:S56–S66. https://doi.org/10.1016/j.rse.2008.01.026

Jetz W, Cavender-Bares J, Pavlick R, Schimel D, Davis FW, Asner GP, Guralnick R, Kattge J, Latimer AM, Moorcroft P, Schaepman ME, Schildhauer MP, Schneider FD, Schrodt F, Stahl U, Ustin SL (2016) Monitoring plant functional diversity from space. Nat Plants 2:16024. https://doi.org/10.1038/nplants.2016.24

Kampe TU, Johnson BR, Kuester M, McCorkel J (2011) Airborne remote sensing instrumentation for NEON: status and development. 2011 Aerospace conference

Kokaly RF, Asner GP, Ollinger SV, Martin ME, Wessman CA (2009) Characterizing canopy biochemistry from imaging spectroscopy and its application to ecosystem studies. Remote Sens Environ 113:S78–S91. https://doi.org/10.1016/j.rse.2008.10.018

Kress WJ, Erickson DL, Jones A, Swenson NG, Perez R, Sanjur O, Bermingham E (2009) Plant DNA barcodes and a community phylogeny of a tropical forest dynamics plot in Panama. Proc Natl Acad Sci 106:18621. https://doi.org/10.1073/pnas.0909820106

Martin RE, Asner GP, Sack L (2007) Genetic variation in leaf pigment, optical and photosynthetic function among diverse phenotypes of Metrosideros polymorpha grown in a common garden. Oecologia 151:387–400. https://doi.org/10.1007/s00442-006-0604-z

Martin RE, Chadwick K, Brodrick PG, Carranza-Jimenez L, Vaughn NR, Asner GP, Dana Chadwick K, Brodrick PG, Carranza-Jimenez L, Vaughn NR, Asner GP (2018) An approach for foliar trait retrieval from airborne imaging spectroscopy of tropical forests. Remote Sens 10:199. https://doi.org/10.3390/rs10020199

Marvin DC, Asner GP, Knapp DE, Anderson CB, Martin RE, Sinca F, Tupayachi R (2014) Amazonian landscapes and the bias in field studies of forest structure and biomass. Proc Natl Acad Sci 111:E5224–E5232. https://doi.org/10.1073/pnas.1412999111

Marvin DC, Koh LP, Lynam AJ, Wich S, Davies AB, Krishnamurthy R, Stokes E, Starkey R, Asner GP (2016) Integrating technologies for scalable ecology and conservation. Glob Ecol Conserv 7:262–275. https://doi.org/10.1016/J.GECCO.2016.07.002

McManus Chauvin K, Asner GP, Martin RE, Kress WJ, Wright SJ, Field CB (2018) Decoupled dimensions of leaf economic and anti-herbivore defense strategies in a tropical canopy tree community. Oecologia 186:765–782. https://doi.org/10.1007/s00442-017-4043-9

McManus KM, Asner GP, Martin RE, Dexter KG, Kress WJ, Field CB (2016) Phylogenetic structure of foliar spectral traits in tropical forest canopies. Remote Sens 8:196. https://doi.org/10.3390/rs8030196

Mooney HA, Cushman JH, Medina E, Sala OE, Schulze E-D (1996) Functional roles of biodiversity: a global perspective. p 493

Myers N, Mittermeier RA, Mittermeier CG, da Fonseca GAB, Kent J (2000) Biodiversity hotspots for conservation priorities. Nature 403:853–858. https://doi.org/10.1038/35002501

Papeş M, Tupayachi R, Martínez P, Peterson a TT, Powell GVNVN (2010) Using hyperspectral satellite imagery for regional inventories: a test with tropical emergent trees in the Amazon Basin. J Veg Sci 21:342–354. https://doi.org/10.1111/j.1654-1103.2009.01147.x

Reich PB, Wright IJ, Cavender-Bares J, Craine JM, Oleksyn J, Westoby M, Walters MB (2003) The evolution of plant functional variation: traits, spectra, and strategies. Int J Plant Sci 164:S143–S164. https://doi.org/10.1086/374368

Reichstein M, Bahn M, Mahecha MD, Kattge J, Baldocchi DD (2014) Linking plant and ecosystem functional biogeography. Proc Natl Acad Sci 111:13697–13702. https://doi.org/10.1073/pnas.1216065111

Running SW, Justice CO, Salomonson V, Hall D, Barker J, Kaufmann YJ, Strahler AH, Huete AR, Muller J-P, Vanderbilt V, Wan ZM, Teillet P, Carneggie D (1994) Terrestrial remote sensing science and algorithms planned for EOS/MODIS. Int J Remote Sens 15:3587–3620

Schimel DS, Asner GP, Moorcroft PR (2013) Observing changing ecological diversity in the Anthropocene. Front Ecol Environ 11:129

Schnitzer SA, Bongers F (2011) Increasing liana abundance and biomass in tropical forests: emerging patterns and putative mechanisms. Ecol Lett 14:397–406

Serbin SP, Singh A, McNeil BE, Kingdon CC, Townsend PA (2014) Spectroscopic determination of leaf morphological and biochemical traits for northern temperate and boreal tree species. Ecol Appl 24:1651–1669. https://doi.org/10.1890/13-2110.1

Tucker CJ, Townshend JRG (2000) Strategies for monitoring tropical deforestation using satellite data. Int J Remote Sens 21:1461–1471

Turner W (2014) Sensing biodiversity. Science 346:301–302. https://doi.org/10.1126/science.1256014

Turner W, Spector S, Gardiner N, Fladeland M, Sterling E, Steininger M (2003) Remote sensing for biodiversity science and conservation. Trends Ecol Evol 18:306–314. https://doi.org/10.1016/S0169-5347(03)00070-3

Ustin SL, Roberts DARA, Gamon JA, Gregory P, Green RO (2004) Using imaging spectroscopy to study ecosystem processes and properties. Bioscience 54:523–534

Ustin SL, Gitelson AA, Jacquemoud S, Schaepman M, Asner GP, Gamon JA, Zarco-Tejada P (2009) Retrieval of foliar information about plant pigment systems from high resolution spectroscopy. Remote Sens Environ 113(Suppl):S67–S77. https://doi.org/10.1016/j.rse.2008.10.019

Wold S, Sjöström M, Eriksson L (2001) PLS-regression: a basic tool of chemometrics. Chemom Intell Lab Syst 58:109–130. https://doi.org/10.1016/S0169-7439(01)00155-1

Wright IJ, Reich PB, Westoby M, Ackerly DD, Baruch Z, Bongers F, Cavender-Bares J, Chapin T, Cornelissen JHC, Diemer M, Flexas J, Garnier E, Groom PK, Gulias J, Hikosaka K, Lamont BB, Lee T, Lee W, Lusk C, Midgley JJ, Navas M-LL, Niinemets U, Oleksyn J, Osada N, Poorter H, Poot P, Prior L, Pyankov VI, Roumet C, Thomas SC, Tjoelker MG, Veneklaas EJ,

Villar R (2004) The worldwide leaf economics spectrum. Nature 428:821–827. https://doi.org/10.1038/nature02403

Wright SJ, Kitajima K, Kraft NJB, Reich PB, Wright IJ, Bunker DE, Condit R, Dalling JW, Davies SJ, DíAz S, Engelbrecht BMJ, Harms KE, Hubbell SP, Marks CO, Ruiz-Jaen MC, Salvador CM, Zanne AE (2010) Functional traits and the growth-mortality trade-off in tropical trees. Ecology 91:3664–3674. https://doi.org/10.1890/09-2335.1

6

Forest Ecosystem and Biodiversity: Detection and Protection

Jennifer Pontius, Paul Schaberg, and Ryan Hanavan

6.1 Introduction

In many ways, biodiversity is a foundational component of healthy, productive forests and maintenance of the many ecosystem services that they provide (e.g., carbon sequestration, nutrient cycling, water filtration and provisioning, wildlife habitat). Forested landscapes are often characterized by a mosaic of species, age classes, and structural characteristics that results from natural patterns of disturbance. This diversity within stands and across forested landscapes increases resilience of larger forested ecosystems, enabling them to recover and maintain ecological function following disturbance (Thompson et al. 2009). But many pests and pathogens, particularly exotic invasive insects, as well as various abiotic stresses (e.g., pollution impacts or increases in climate extremes), have the potential to alter native populations, reduce biodiversity, and impact ecosystem function and service provisioning. This is particularly true for ecosystems dominated by keystone or foundational species, which exert a relatively large impact on community stability and ecosystem function (Ellison et al. 2010).

There are many examples of the impacts of pests and pathogens on biodiversity and ecological function in forested ecosystems. Dutch elm disease was introduced in the United States in the 1930s and the United Kingdom in the 1970s, with

---

J. Pontius (✉)
Rubenstein School of Environment and Natural Resources, University of Vermont, Burlington, VT, USA

USDA Forest Service, Northern Research Station, Burlington, VT, USA
e-mail: Jennifer.pontius@uvm.edu

P. Schaberg
USDA Forest Service, Northern Research Station, Burlington, VT, USA

R. Hanavan
USDA Forest Service, Forest Health Protection, Northeastern Area, Durham, NH, USA

Fig. 6.1 Ancient whitebark pines killed by the recent mountain pine beetle outbreak stand on a windy ridge in Yellowstone National Park.

profound impacts on the biodiversity of rural landscapes (Harwood et al. 2011). The mountain pine beetle has impacted large swaths of coniferous and mixed forests in British Columbia, with severe impacts to avian biodiversity (Martin et al. 2006). In the western United States, pine blister rust has impacted biodiversity and ecological processes, particularly at high elevation sites where whitebark pine is a keystone species (Tomback and Achuff 2010, Fig. 6.1). Recent cases, such as the introduction of the Asian long-horned beetle and emerald ash borer to the United States, demonstrate the ongoing biosecurity challenges that currently face forested ecosystems.

Similarly, abiotic stresses can lead to declines that alter competition and biodiversity in the broader forest. For example, acid deposition that resulted from elevated inputs of sulfur and nitrogen pollution in the 1950s through 1980s led to declines in red spruce (*Picea rubens* Sarg.) (Schaberg et al. 2011) and sugar maple (*Acer saccharum* Marsh.) (Huggett et al. 2007) and increases in less sensitive species such as American beech (*Fagus grandifolia* Ehrh.) (Schaberg et al. 2001; Pontius et al. 2016). In another example, warming temperatures were associated with reductions in winter snowpacks, increased soil freezing, and root mortality that resulted in the broad-scale decline of yellow cedar (*Callitropsis nootkatensis*) but not sympatric species (Hennon et al. 2012). Warmer climates have also resulted in range expansion of native insects and disease with potential to further alter the landscape. For example, the southern pine beetle (*Dendroctonus frontalis*) continues to move north from the loblolly forests of the southern United States to pitch pine in the north.

Many resource managers cite the need for early detection of forest decline to minimize impacts of emergent stress agents (Genovesi et al. 2015; Sitzia et al. 2016). Research has shown that the earlier you can detect forest decline, the more successful management and control efforts will be (Epanchin-Niell and Hastings 2010). For invasive pests and pathogens, identifying the locations of incipient infestations is critical to minimizing spread, reducing ecosystem impacts, and targeting management and control (Mumford 2017).

But early detection also benefits the sustainable management of forested ecosystems responding to lower-level, chronic stress agents such as climate change and acid deposition. Such chronic stress agents often manifest in more subtle decline symptoms over many years. This slow and highly variable decline (some good years, some bad years) limits the ability to identify causal relationships, understand potential impacts to ecosystem function, and develop management strategies. As a result, we need to be able to *quantify* decline symptoms with greater detail and sensitivity to subtle changes, from the gradual loss of photosynthetic apparatus in response to initial stress, to reductions in canopy density, dieback, and ultimate mortality across the landscape.

Remote sensing (RS) has long been used to assess relative vegetation density, decline, and mortality. But landscape-scale assessment of small-scale or subtle decline symptoms has been more difficult. The spatial patial resolution of many sensors has limited our ability to detect small-scale decline in highly mixed pixels, while spectral resolution has limited our ability to detect early biogeochemical precursors to more severe decline symptoms. But as new sensors and modeling algorithms have come on board, there is a growing list of successful early decline detection efforts.

Here we present the science behind RS for the assessment of vegetation condition, with a focus on using these tools for more detailed and accurate monitoring of forest decline and disturbance. We also highlight the importance of this approach to inform the sustainable management of forested ecosystems and preservation of forest biodiversity.

6.2 The Basics of Forest Decline

In order to better understand how RS instruments can detect vegetation stress, and be used to quantify forest decline, it is important to understand the structural and physiological response of vegetation to stress. Any RS effort to detect or monitor decline is based on the sensor's ability to detect these biophysical changes that manifest following stress.

Trees adjust their physiology and form in response to environmental stimuli (e.g., light, temperature, moisture). Stress occurs when environmental conditions fall outside of the normal or optimal levels to which plants are adapted. As sessile organisms that cannot flee from the many stresses that they are routinely exposed to over their long life spans, trees have evolved enumerable mechanisms to avoid, mitigate, or rebound from stress. Some of these adaptations (e.g., protective pigments such as the yellow/orange carotenoids and red anthocyanins in leaves) can directly influence RS spectral measurements. Other stress adaptations (e.g., changes in carbohydrate storage and lipid and protein metabolism; Strimbeck et al. 2015) influence spectral characteristics indirectly through changes in leaf retention and life span. Here we walk through some of these physiological and structural changes relevant to RS efforts in more detail.

Leaf Size Small, emerging leaves can be difficult to detect via RS (e.g., White et al. 2014). Therefore, factors that delay or expedite bud break and leaf expansion, or lead to leaf wilting, curling, and folding can influence spectral signatures

Fig. 6.2 Leaf curl, wilt, and stunted expansion can result in decreased leaf area index that is commonly quantified in RS applications. (Credit: Eiku [CC BY-SA 4.0] from Wikimedia Commons)

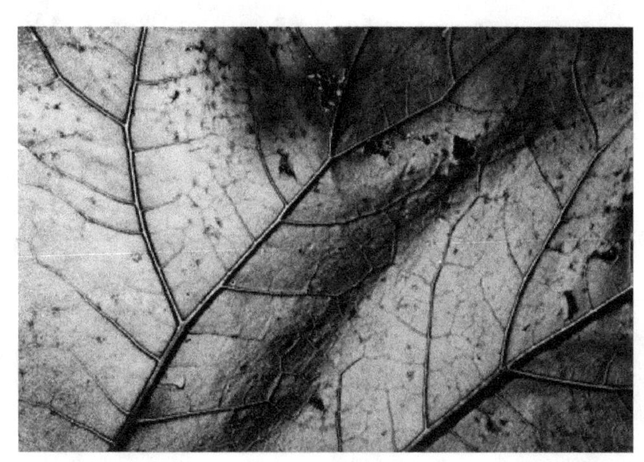

Fig. 6.3 Many sensors can detect changes in leaf pigment concentration and function before chlorosis is visible to the human eye. (Credit: [CC0] https://pxhere.com/en/photo/575928)

(Fig. 6.2). In addition, leaves that develop after episodic leaf mortality are often stunted, diminishing overall leaf area. Reduced leaf size can also result from carbohydrate losses associated with sucking insects, e.g., pear thrips (*Taeniothrips inconsequens*; Kolb and Teulon 1991), and insect herbivory can reduce the functional area of leaves through leaf consumption.

Leaf Chemistry and Physiology Plant pigments (chlorophylls essential in photosynthesis, xanthophylls that assist with light capture and protect leaves from photooxidation, and anthocyanins that have numerous protective capacities) are all spectrally responsive (Fig. 6.3). Therefore, environmental factors that influence their development and turnover (e.g., cold temperatures that can speed chlorophyll catabolism and trigger anthocyanin expression; Schaberg et al. 2017) can influence associated spectral signatures. Similarly, because leaf water content and chemistry have identifiable spectral features, environmental factors such as droughts, fertilization, and soil acidification can also influence spectral signatures.

Leaf Quantity and Longevity Despite remarkable and diverse capacities for stress response and protection, numerous biological and abiotic factors can reduce

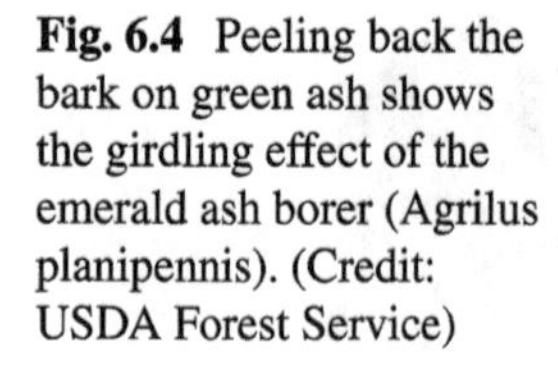

Fig. 6.4 Peeling back the bark on green ash shows the girdling effect of the emerald ash borer (Agrilus planipennis). (Credit: USDA Forest Service)

leaf longevity or lead to significant defoliation. The most prominent factors causing foliar reductions vary across ecoregions (e.g., drought is a dominant factor in the western United States, whereas insect defoliation is prominent in the eastern states) and over time (e.g., episodic drought, cyclic insect outbreaks). However, numerous anthropogenic factors (e.g., ozone pollution, acid deposition, introduction of exotic pests and pathogens) have expanded the list of stress agents that can lead to significant defoliation. Some stress agents directly result in defoliation, but many stress agents impact other organs that crowns rely on, for example, insects such as bark beetles and the emerald ash borer (Fig. 6.4) and pathogens such as chestnut blight girdle stems. Invasive pests such as hemlock woolly adelgid extract photosynthate directly from phloem. Root freezing injury (e.g., yellow cedar decline; Hennon et al. 2012) can limit resource uptake. All of these stress agents can manifest as reduced leaf area index and canopy density.

Branch Dieback, Tree Decline, and Mortality Repeated or severe direct damage to tree canopies or chronic imbalances in tree carbohydrate and/or stress response systems can lead to branch dieback. This dieback is typically first evident as mortality of the most distal portions of the crown (tip dieback) and can lead to significant carbon imbalances as the photosynthetic capacity of trees is outstripped by

Fig. 6.5 Dieback typically results in changes to spectral characteristics as pixels become dominated by understory or bark and soil surface features. (Credit: Joseph O'Brien, USDA Forest Service, Bugwood.org)

carbohydrate use associated with maintenance respiration as well as compensatory growth (e.g., epicormic branching), and seed production, which are often associated with decline. Significant crown loss exacerbates negative carbon balances, ultimately resulting in tree mortality. Temporary or partial crown dieback may be difficult to detect if it is not widespread, but protracted dieback, especially if it results in significant tree mortality, could dramatically alter spectral measurements in the near (during the decline event) and long terms if elevated mortality leads to significant changes in canopy density, gap fraction, species composition, or forest cover (Fig. 6.5).

6.3 RS Approaches to Forest Decline Detection

Aerial Sketch Mapping In the United States, federal and state forestry agencies have been conducting aerial detection surveys of forest decline for many decades (Fig. 6.6; Johnson and Wittwer 2008; Johnson and Ross 2008; McConnell 1999). This manual RS technique involves an observer mapping polygons by identifying host trees by crown shape and causal agent by damage signature from an aircraft. In the early decades (1950s–1980s), this was often deployed only in response to severe or widespread forest disturbance events, with limited flight lines and rough delineation of impacted stands onto paper maps. Now, organized by the national Forest Health Monitoring (FHM) program, many states are flown in their entirety each year to survey impacts from a suite of potential biotic and abiotic stressors and various disturbance types (e.g., defoliation, mortality, dieback), with mapping captured on digital, global positioning system (GPS)-enabled touchscreen tablets. Like other RS methods, ground validation adds confidence in the final map products. Aerial sketch mapping is currently the most widespread approach to forest condition mapping across the United States, and because of direct cooperation among federal and states agencies collecting and using the resulting maps, it also has the most direct link to land managers and decision-makers.

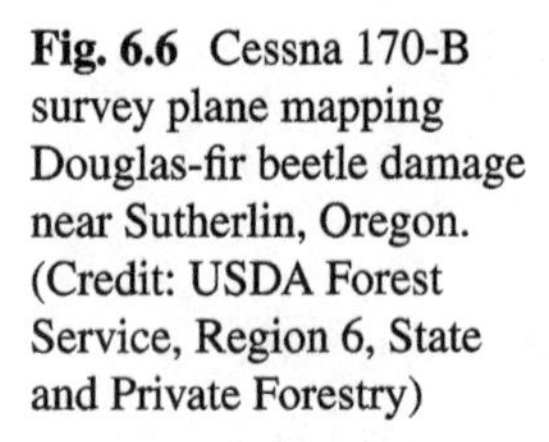

Fig. 6.6 Cessna 170-B survey plane mapping Douglas-fir beetle damage near Sutherlin, Oregon. (Credit: USDA Forest Service, Region 6, State and Private Forestry)

However, the mapping products generated vary based on differences in the base map scale used, observer bias, or agency emphasis (Kosiba et al. 2018). Products also vary year to year based on timing of flight and the visibility of different stress symptoms (e.g., early season vs. late season defoliators). Further, only decline symptoms that are severe enough, and in large enough patches to be visible to an observer in an aircraft traveling approximately 100 knots from an altitude of 1000–3000 feet above ground level, are mapped. As such, aerial sketch mapping can be highly subjective and should only be regarded as a coarse "snapshot" of landscape-level forest health.

Multispectral Sensors Terrestrial satellite RS began with the launch of the Landsat mission (then called the Earth Resources Technology Satellite (ERTS)) in 1972. Designed to supply regular images of Earth's surface, with multispectral bands designed to capture biospheric processes at medium-high spatial resolution, Landsat-1 enabled a revolution in terrestrial research (Williams et al. 2006). With continuous coverage since the 1972 launch, the family of Landsat sensors is particularly useful for studying forest change over time across regional to global scales (Fig. 6.7).

Initially, the broad, multispectral bands on the Landsat sensors were used to assess relative vegetative density, or "greenness." This was made possible by targeting the near-infrared (NIR) portions of the electromagnetic spectrum in addition to visible wavelengths. This "near-infrared plateau" is a region of strong reflectance in vegetation and is distinct from many other surface features such as soil, rock, and water, making it particularly useful for distinguishing vegetation from non-vegetative land cover types or assessing the relative amount of vegetation within mixed pixels. It is also highly responsive to common stress symptoms such as defoliation, chlorosis, and decreases in canopy density. Over the decades, scientists have developed a suite of vegetation indices to quantify vegetation condition and biophysical attributes (Table 6.1) that have been commonly used to assess changes in canopy cover (e.g., deforestation) and widespread defoliation or mortality.

Fig. 6.7 Landsat images from 1984 and 2010 show clear-cutting and forest regrowth in Washington State, highlighting the utility of multispectral sensors in detecting vegetation density and disturbance. (Credit: NASA image by Robert Simmon)

The use of multispectral sensors to identify more subtle or early decline symptoms is typically limited by the spectral resolution (few, broad-bands of spectral information to work with), spatial resolution (mixes of healthy and stressed vegetation in one pixel often mask the spectral stress signature of stressed individual trees), and temporal resolution (inability to acquire cloud-free images at intervals sufficient to detect change).

As the interest in RS products has grown, along with the range of applications, many commercial vendors have expanded access to multispectral products with both aerial and satellite platforms. We now have over 100 active satellite sensors with visible and NIR capabilities listed in the International Inst. for Aerospace Survey and Earth Sciences (Netherlands; formerly International Training Centre for

Table 6.1 A selection of broadband and narrowband vegetation indices useful for vegetation stress detection

Broadband indices	Abbreviation	Associated vegetation characteristics	Index formula (Landsat 7 ETM bands)	Reference
Difference vegetation index	DVI	Canopy cover, greenness	*B4 – B3*	Jordan (1969); Tucker (1979)
Enhanced vegetation index	EVI	Canopy cover, greenness	*2.5 * ((B4 – 33)/ (B4 + (6 * 33) – (7.5 * B1) + 1))*	Huete et al. (1994); Huete et al. (2002)
Greenness index		Canopy cover, greenness	*B4/B3*	Sivanpillai et al. (2006)
	B7	Vegetation & soil water content	*B7*	Lillesand and Kiefer (1994)
	EL	Green biomass	*B5/B7*	Elvidge and Lyon (1985)
Modified soil adjusted vegetation index	MSAVI	Canopy cover, greenness	*0.5 *(2*34 + 1 – (Sqr(((2 * B4 + 1) * (2 * B4 + 1))) – (8 * (B4 – B3))))*	Qi et al. (1994)
Moisture stress index	MSI	Canopy water content	*B5/B4*	Rock et al. (1986)
Normalized difference infrared index 5	$NDII_5$		*(B4 – B5)/(B4 + B5)*	Hardisky et al. (1983)
Normalized difference infrared index 7	$NDII_7$		*(B4–B7)/(34+ B7)*	Hunt and Rock (1989)
Normalized difference vegetation index	NDVI	Canopy greenness, condition, density	*(B4 – B3)/(B4 + B3)*	Deblonde and Cihlar (1993); Gamon et al. (1997); Myneni et al. (1995a, 1995b); Rouse et al. (1974)
Optomized soil adjusted vegetation index	OSAVI	Canopy density	*(B4 – b3)/ (B4 + B3 + 0.16)*	Rondeaux et al. (1996)
Renormalized difference vegetation index	RDM	Leaf area index; chlorophyll	*Sqr(((B4 – B3)/ (B4 + B3)) * (B4 – B3)))*	Roujean and Breon (1995)
Ratio vegetation index	RM	Chlorophyll	*B4/B3*	Pearson and Miller (1972)

(continued)

Table 6.1 (continued)

Broadband indices	Abbreviation	Associated vegetation characteristics	Index formula (Landsat 7 ETM bands)	Reference
Soil and atmospherically resistant vegetation index	SARM	Canopy density	*1.5 * ((B4 – (B3 – (B1 – B3)))/(B4 – (B3 + (B1 – B3) + 0.5)))*	Kaufman and Tanre (1992); Huete et al. (1994)
Soil adjusted vegetation index	SAM	Soil adjusted canopy cover	*1.5 * ((B4 – B3)/ (B4 + B3 + 0.5))*	Huete (1988)
	$SDRE_a$	General stress	*(B5 – B3) – (B4 – B2)*	Pontius et al. (2006)
	$SDRE_b$	General stress	*(B4 – B3) – (B3 – B2)*	Pontius et al. (2005a, 2005b)

Narrowband indices	Narrowband indices	Associated vegetation characteristics	Index formula narrowband formula broadband approximation	Landsat 7 ETM approximation	Reference
	Aoki	Chlorophyll	*R550/R800*	*B2/B4*	Aoki et al. (1981)
	BN_a	Chlorophyll	*R800 – R550*	*BB – B2*	Buschman and Nagel (1993)
	BN_b	Chlorophyll	*R800/R550*	*BB/B2*	Buschman and Nagel (1994)
Chlorophyll fluorescence ratio	CFR	Chlorophyll fluorescence	*F690/F735*	*X*	D'Ambrosio, et al. (1992)
	$Chap_A$	$Chlorophyll_a$	*R675/R700*	*X*	Chappelle et al. (1992)
	$Chap_B$	$Chlorophyll_a$	*R675/(R700 * R650)*	*X*	Chappelle et al. (1992)
Curvature index	CI	Chlorophyll and chlorophyll florescence	$(R683^2)/(R675 * R691)$	*X*	Zarco-Tejada et al. (2002)
	CS_a	Chlorophyll	*R694/R760*	*BB/B4*	Carter and Miller (1994)
	CS_b	General stress	*R694/R420*	*BB/B1*	Carter (1994)
	CS_c	$Chlorophyll_a$	*R605/R760*	*X*	Carter (1994); Carter and Miller (1994)

	CS_d	General stress	*R710/R760*	*X*	Carter (1994)
	CS_e	Chlorophyll	*R695/R760*	*BB/B1*	Carter (1994)
	$Datt_A$	Chlorophyll	*R672 * (R550 * R708)*	*BB * (B2 *BB)*	Datt (1998)
	$Datt_B$	Chlorophyll	*FD754/FD704*	*X*	Datt (1999)
	DattCS	Chlorophyll content	*(R672/R550) * (R695/R 760)*	*X*	Datt (1998)
Derivative chlorophyll index	DCI	$Chlorophyll_a$; $chlorophyll_b$. chlorophyll florescence	*FD705/FD723*	*X*	Zarco-Tejada et al. (2002)
	EZ	Chlorophyll	*Sum FD 625 to 795*	*X*	Elvidge and Chen (1995)
	FD_{703}	$Chlorophyll_a$	*FD703*	*X*	Boochs et al. (1990)
	FD_{720}	$Chlorophyll_b$	*FD720*	*X*	Boochs et al. (1990)
	Flo	Chlorophyll florescence; photo synthetic activity	*FD690/FD735*	*X*	Mohammed et al. (1995)
	FP	Leaf area index; chlorophyll	*Sum FD 680 to 780*	*X*	Filella and Penuelas (1994)
	Git_C	Chlorophyll	*1/R700*	*1/B3*	Gitelson et al. (1999); Gitelson et al. (2001)
	GM_a	Chlorophyll	*R750/R550*	*B4/B2*	Gitelson and Merzlyak (1994)
	GM_b	Chlorophyll	*R750/R700*	*B4/B3*	Gitelson and Merzlyak (1994)
	Mac	Chlorophyll	*(R780 – R710)/(R780 – R680)*	*X*	Maccioni et al. (2001)
Modified chlorophyll absorption in reflectance index	MCARI	Chlorophyll	*((R700 – R670) – 0.2* (R700 – R550)) * (R700/R670)*	*X*	Daughtry et al. (2000)
Modified chlorophyll absorption in reflectance index 1	$MCARI_1$	Chlorophyll	*1.2 * ((2.5 * (R800 – R670)) – (1.3 * (R800 – R550)))*	*1.2 * ((2.5 * (B3)) – (1.3 * (B4 – B2)))*	Haboudane et al. (2004)
	McM	Chlorophyll	*R700/R760*	*X*	McMurtrey III et al. (1994)
	mND_{705}	Chlorophyll	*(R750 – R705)/(R750 + RR705 + 2R445)*	*(B4 – B3)/(B4 + B3 + (2 * B1))*	Sims and Gamon (2002)

(continued)

Table 6.1 (continued)

Narrowband indices	Narrowband indices	Associated vegetation characteristics	Index formula narrowband formula broadband approximation	Landsat 7 ETM approximation	Reference
Modified simple ratio	MSR	Leaf area index; fraction of photo synthetically active radiation	*((R800/R678) –1)/ (Sqr((R800/ R670) + 1))*	*((B4 /B3) –1)/ (Sqr((B4 / B3) + 1))*	Chen (1996)
Modified simple ratio$_{705}$	MSR$_{705}$	Chlorophyll	*(R750 – R445)/(R705 – R445)*	*(B4 – B1)/ (B4 + B1)*	Sims and Gamon (2002)
Normalized difference index	NDI	Chlorophyll	*(R750 – R705)/ R750 + R705*	*(B4 – B3)/ (B4 – B3)*	Gitelson and Merzlyak (1994)
Normalized difference water index	NDWI	Canopy water content	*(R860 – R1240)/(R860 – R1240)*	*(B4 – B5)/ (B4 – B5)*	Gao (1996)
Modified triangular vegetation index	MTVI	Leaf area index	*1.2 * ((1.2 * (R800 – R550)) – (2.5 * (R670–R550)))*	*1.2 *((1.2 *(B4 – B2)) – (2.5 *(B3-B2)))*	Haboudane et al. (2004)
Normalized pigments chlorophyll ratio index	NPCI	Chlorophyll; general stress	*(R680 – R430)/ (R680 + R430)*	*(B3 – B1)/ (B3 + B1)*	Penuelas et al. (1994)
Normalized phae ophytinization index	NPQI	Chlorophyll; general stress	*(R415 – R435)/ (R415 + R435)*	*X*	Barnes (1992)
Photochemical reflectance index	PRI	Xanthophyll cycle activity	*(R531 – R570)/ (R531 + R570)*	*X*	Gamon et al. (1997); Rahman et al. (2001)
Pigment-specific normalized difference	PSND	Chlorophyll$_b$	*(R680 – R635)/ (R800 + R635)*	*(B4 – B3)/ (B4 + B3)*	Blackburn (1998, 1999)
Plant senescence reflectance index	PSRI	Xanthophyll cycle activity	*(R680 – R500))/R750*		Merzlyak et al. (1999)
Pigment specific simple ratio	PSSR	Chlorophyll	*R800/R635*	*B4/B3*	Blackburn (1998, 1999)

	R_{550}	$Chlorophyll_a$	*R550*	*X*	Carter (1993)
	R_{678}	General stress	*R678*	*X*	Pietrzykowski et al. (2006)
	R_{680}	$Chlorophyll_a$; chlorophyll florescence	*R678*	*X*	Mohammed et al. (1995)
	R_{687}	Chlorophyll florescence	*R687*	*X*	Meroni et al. (2008)
	R_{702}	$Chlorophyll_a$	*R702*	*X*	Carter (1993); Carter and Knapp (2001)
	R_{760}	Chlorophyll florescence	*R760*	*X*	Carter and Miller (1994)
	R_{950}	Canopy water content	*R950*	*X*	Williams and Norris (2001)
Red edge inflection point	REIP	$Chlorophyll_a$; vegetation greenness; canopy density	*FD max near 700*	*X*	Horler et al. (1983); Gitelson and Merzlyak (1996); Rock et al. (1988); Vogelmann et al. (1993)
Structure insensitive pigment index	SIPI	Carotenoid:chlorophyll	*(R803 – R445)/(R800 – R680)*	*(B4 – B1)/(B4 – B3)*	Penuelas et al. (1995)
Simple ratio pigment index	SRPI	Chlorophyll; general stress	*R430/R680*	*B1/B3*	Penuelas et al. (1993)
Triangular vegetation index	TVI	Chlorophyll	*0.5 * (120 * (Ravg760to800 – Rcrvg530to570) – (200 * (Ravg650to680 – Ravg530to570)*	*0.5 * (120 * (B4 – B2) – 200 * (B3 – B2))*	Broge and Leblanc (2001)
	Vog_A	Chlorophyll	*R740/R720*	*X*	Vogelmann et al. (1993)
	Vog_B	Chlorophyll	*FD715/FD705*	*X*	Vogelmann et al. (1993)
Water band index	WBI	Canopy water content	*R900/R970*		Penuelas et al. (1993)

Note that this is not all-inclusive but intended to demonstrate the wide range of algorithms available and how these have been linked to specific leaf pigment, canopy structure, or condition metrics

Where no name is specified for a vegetation index by the author, we assign an abbreviation based on index function or author initials. *R* indicates reflectance values from narrow spectral bands (~10 nm), *FD* indicates a spectral first derivative centered on the specified wavelength, *X* indicates that a broadband equivalent does not exist for sensors such as Landsat 7 ETM.

Aerial Survey) ITC Satellite and Sensor Database: https://webapps.itc.utwente.nl/sensor/default.aspx?view=allsensors

One particularly promising sensor for improved forest health detection includes Sentinel 2 (A and B), recently launched by the European Space Agency. This is the first civil Earth observation sensor to include three bands in the red edge, providing additional information to quantify vegetation condition. Its 5-day repeat time and 10 m pixels also improve its ability to detect more subtle decline symptoms. This temporal resolution has proven useful in identifying forest decline based on detecting changes in the spectra of declining trees relative to healthy ones over time (Zarco-Tejada et al. 2018). Geostationary sensors like the GOES-R series also provide a unique opportunity to monitor forest condition at rapid time intervals across large landscapes. With two visible and four infrared bands useful to inform vegetation condition, the Advanced Baseline Imager on GOES-16 can provide images every 5 minutes with a spatial resolution of 0.5–2 km.

Improvements in computing technologies and modeling techniques have also increased the utility of multispectral sensors in early vegetation decline detection (Lausch et al. 2017). For example, Pontius (2014) demonstrated that using a multitemporal approach mimicking hyperspectral algorithms could successfully quantify a detailed decline scale using Landsat TM data. Over time, ongoing improvements in sensor resolution, computing capabilities, and modeling options will enable measurements of more subtle changes in reflectance associated with early decline detection.

Hyperspectral Sensors While multispectral sensors record electromagnetic radiation averaged over a broad "band" of wavelengths, a hyperspectral instrument records many adjacent narrow bands to image most of the spectrum within a set range. What makes these instruments so useful for vegetation assessment extends beyond the simple availability of *more* bands to work with. Typically, these bands record reflectance from much narrower regions of the electromagnetic spectrum. This narrowband design provides two key modeling capabilities that are not possible with broadband sensors: (1) narrow bands are able to target specific absorption features linked to specific physiological structures or processes that we can directly relate to plant stress response and (2) narrow, contiguous bands allow us to consider the overall shape of spectral signatures, including mathematical techniques (e.g., derivatives, area under the curve, slope of the line between key regions) that are not possible with broadband data.

Building off of the science of spectroscopy (the study of constituents and materials using specific wavelengths), RS analysts have used hyperspectral imagery to quantify specific vegetation constituents and processes. The best hyperspectral narrow bands to study vegetation are in the 400–2500 nm spectral range (Thenkabail et al. 2013; Fig. 6.8), enabling direct links to species composition, foliar chemistry, foliar function, and ecosystem characteristics (Smith et al. 2002; Williams and Hunt 2002; Kokaly et al. 2003; Asner and Heidebrecht 2003; Townsend et al. 2003; Carter et al. 2005; Cheng et al. 2006; Singh et al. 2015).

While it is generally believed that spectral changes in stressed vegetation are common across stress agents, the ability of hyperspectral sensors to target specific chemical, physiological, and morphological traits allows RS analysts to target and assess specific, early symptoms of decline and target detection efforts based on known physiological responses to a particular pest or pathogen. Lausch et al. (2013) targeted changes in chlorophyll absorption as an indicator of bark beetle-induced decline; Pontius et al. (2008) targeted chlorophyll fluoresce to map the invasive emerald ash borer (Pontius et al. 2008) and canopy density for detailed monitoring impacts of hemlock woolly adelgid (Pontius et al. 2005b).

Hyperspectral imagery has historically been limited in availability. NASA's Airborne Visible/Infrared Spectrometer (AVIRIS; Porter and Enmark 1987) hyperspectral sensor was the pioneer of airborne applications. But the launch of the NASA Hyperion Instrument (Pearlman et al. 2003) on the EO-1 satellite in 2000, and the addition of commercial vendors with aerial hyperspectral platforms (e.g., ITRES http://www.itres.com/; SPECIM http://www.specim.fi/hyperspectral-RS/), has increased the availability of hyperspectral imagery. The promise of new hyperspectral satellites such as the Environmental Mapping and Analysis Program (EnMAP http://www.enmap.org/mission.html) suggests there is potential for expanding applications in forest health monitoring and assessment. Recent examples include assessments of hemlock woolly adelgid-induced decline in the Catskills region of New York (Hanavan et al. 2015) and detection of drought-induced decline in the chaparral ecosystems of California (Coates et al. 2015). Fused hyperspectral and LiDAR imagery have also enabled the assessment of early decline at the canopy level in urban environments (e.g., Degerickx et al. 2018; Pontius et al. 2017).

6.4 Spectroscopy of Early Decline Detection

While different species have unique spectral signatures, there are similar changes in general spectral characteristics in response to stress (Buschmann and Nagel 1993). Many of these spectral features can be directly linked to the stress symptoms and physiological characteristics described above (Fig. 6.8). For example, changes in leaf chemistry and physiology are captured in the 480–520 nm (blue) and 600–680 nm (red) regions, where chlorophyll absorption is strong. But changes in this region are relatively small compared with the dramatic changes that can be seen with stress between 750 and 1300 nm. The sharp rise in reflectance between the red and NIR regions (red edge inflection point) can be used to quantify changes in both the slope of the spectral signature and the location of the inflection point of the slope in response to changes in leaf chemistry and canopy density. Spectral information at longer wavelengths (1650–2200 shortwave infrared) has also been useful in quantifying changes in leaf water content, often a key signal of early vegetation stress.

Often the most useful information about general canopy condition, density, and function is derived from combining bands from various regions in mathematical expressions referred to as vegetation indices (Elvidge and Chen 1995; Pinty et al. 1993). Sometimes these indices incorporate information from multiple wavelengths with known absorption features. But other times a nonresponsive "control" band

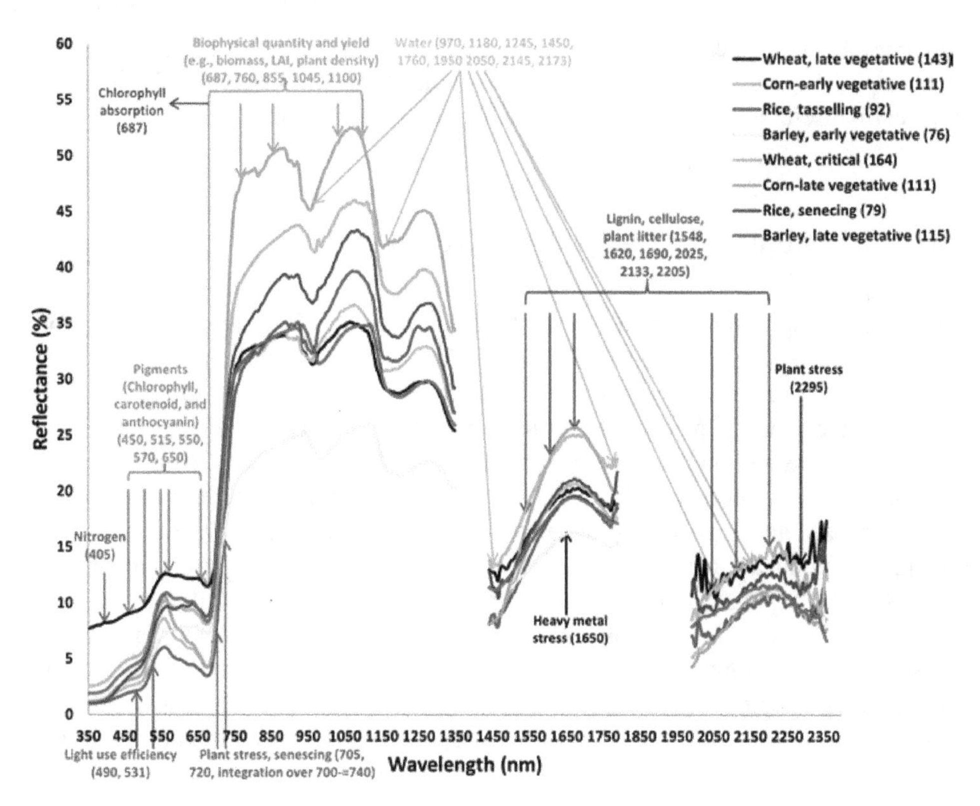

Fig. 6.8 Hyperspectral RS of vegetation condition is possible because of a suite of absorption and reflectance features across the visible and NIR spectra. (Credit: USGS by P. Thenkabail)

may be used to help account for differences in reflectance due to illumination or topography. Many vegetation indices have been designed for use with specific broadband sensors to assess general canopy characteristics such as relative "greenness," canopy density, or canopy condition (Table 6.1). But because of contributions in the field of spectroscopy, there is a wealth of literature that highlights specific regions of the electromagnetic spectrum (EMS) that are specifically associated with foliar chemistry, chlorophyll or carotenoid content, various metrics of photosynthetic activity, and other common stress markers (see Serbin et al. 2014, 2015; Singh et al. 2015).

Some of the vegetation indices listed in Table 6.1 are easily captured with widely available sensors. Others require reflectance information from narrow spectral regions that may only be accurately measured with hyperspectral sensors. Others may be located in regions that are outside of the EMS range of the imagery that is available. Thus, the number of available indices will depend on the imagery you have. Which index will prove most useful in detecting early canopy stress depends on the specific stress symptoms and the conditions of your study area. For example, in ecosystems with relatively sparse vegetation, a soil-adjusted vegetation index may work best to minimize the impact of background reflectance. Similarly, in ecosystems with very dense vegetation, you may need to select an index that does not saturate at high biomass levels. In most cases, you won't know which index, or set of indices and wavelengths, is best to use until you examine them as a part of your analyses. The best way to identify useful vegetation indices is detailed in the next section.

6.5 Techniques for Early Stress Detection

While mapping severe or widespread forest decline can be relatively straightforward using simple vegetation indices, it can be much more challenging to identify early or small-scale decline, particularly in mixed forests. For example, an insect outbreak may cause severe decline symptoms in the host tree species, but this signal may be washed out in a heterogeneous forest where reflectance from the larger canopy of other species dominates. Similarly, tree mortality is often accompanied by the release and ingrowth of understory vegetation. This can make detection of decline difficult as increased vegetation density from the understory masks the reduction in vegetation density in the upper canopy. Further, different species inherently have different chemical and structural characteristics, resulting in sometimes starkly different spectral signatures, even among healthy canopies. A healthy oak may be spectrally similar to a declining sugar maple. This underscores the importance of knowing the distribution of species across a landscape of interest and the characteristics of a "healthy" vs. "declining" spectral signature for a target forest type.

Because the identification of subtle stress characteristics relies on subtle changes in spectral characteristics, RS of early decline is very sensitive to anything that might alter spectral signatures. For example, an algorithm designed for early stress detection with one instrument may not be appropriate to apply to imagery from a different sensor. Even with a similar spectral, radiometric, and spatial configuration, differences in calibration may introduce differences that have nothing to do with the health of the canopy. Even when using the same instrument, atmospheric or illumination conditions may vary over time. For these reasons, it is important to calibrate each image to the specific conditions (atmospheric, illumination, canopy condition) at the time of acquisition.

There are several methodological approaches that can help to isolate and quantify decline symptoms, regardless of the sensor system (Pontius and Hallett 2014). Here we summarize the key components to identifying and quantifying early vegetation stress:

1. *Know the spectral characteristics of your baseline ecosystem.* While all vegetation has a common spectral curve, there are distinct differences in the spectral signature across different species and at different spatial resolutions. Because of inherent differences in foliar chemistry and canopy structure, a sugar maple has a spectral signature that is distinct from an eastern hemlock, even when both are in optimal health. Because of the spectral contribution from surrounding surface features, a healthy sugar maple in a heterogeneous forest will look different from

a healthy sugar maple grown in someone's front yard. Thus, it is important to know what the spectral signature for a pixel of your target ecosystem would look like in optimal condition.

There are many spectral libraries where "typical" spectra for a range of surface features can be downloaded and used for image calibration (e.g., ECOSTRESS Spectral Library https://speclib.jpl.nasa.gov/documents/jhu_desc or US Geological Survey (USGS) Spectral Library https://crustal.usgs.gov/speclab/QueryAll07a.php?quick_filter=vegetation). However, because of inherent differences between sensors, as well as atmospheric and illumination conditions at the time of image acquisition, it is best to also collect field spectra or identify homogeneous calibration pixels from across the imagery. Linking field data directly to the pixels will provide a spectral signature that is specific to the imagery you are using and ecosystem you are working in. This will serve as an important baseline and provide essential calibration data to model the species and stress condition of interest.

2. *Identify, quantify, and gather calibration data for the specific stress symptoms you expect to see.* While there are many common stress responses across vegetation types and stress agents, many symptoms can be species- or stress-specific. Of these, only some may be visible to the human eye. This is why it is important to identify the common stress symptoms you expect to see, from the earliest symptoms to the most obvious and severe decline, and design field data collection efforts that quantify each of those stress symptoms. Field calibration data should include measurements from locations across the imagery and cover the full range for each of these metrics that you would expect to manifest in the system you are studying and that you hope to quantify in your final product. These field data will provide valuable information as you analyze your imagery and model decline conditions across your study area.

   For example, hemlock woolly adelgid feed on photosynthate stored within hemlock twigs, limiting the ability of trees to put on new growth. This may not be visible in a broad assessment of canopy vigor, but can be quantified in the field by collecting multiple branches from across the canopy and assessing the proportion of terminal branchlets that have put on new growth. This serves as a relatively quick and low-tech way to quantify foliar productivity and the reductions in new growth that are often the first sign of infestation. Similarly, the most obvious visible sign of emerald ash borer infestation in ash trees is often scarified bark that results from increased woodpecker activity. Woodpeckers strip bark as they feed on larvae, leaving obvious white markings. These telltale signs of early infestation can serve as a proxy for subtle biophysical changes in the canopy that are not yet visible to field crews.

   Most often, decline manifests as many different concurrent stress symptoms (e.g., chlorosis and defoliation and dieback in various parts of the canopy) or a progression of decline symptoms that vary with the degree of impact (e.g., early decline manifests as chlorosis, later stages dominated by reductions in the live crown ratio, and ultimately mortality). In such cases, you may choose to develop an aggregate "field health" index that mathematically normalizes a suite of stress metrics into one summary metric (Pontius and Hallett 2014). This may be easier and more efficient than creating models to assess each of the various decline symptoms you expect to see in your target system or having to pick one decline metric to use.

3. *Calibrate imagery with field data.* In an ideal world, we would be able to develop one model that could be automated and applied to imagery over time and space regardless of sensor, acquisition condition, or location. Several automated RS tools currently available (see Sect. 6) have proven incredibly useful for monitoring large areas over time. But automated applications are limited in their ability to detect subtle, early decline, which requires careful calibration between the imagery acquired and ground conditions at the time that imagery was collected to make it possible to identify the targeted stress response while controlling for other sources of spectral variability. Ideally, field calibration data can be collected within several weeks of imagery acquisition (or at least before conditions on the ground change). GPS locations of field calibration sites link field data to the spectra of the associated pixel or pixels to calibrate the larger image.

   Various proprietary software modules exist for spectral calibration, modeling, and analyses. These modules can range from simple classification techniques that match pixels to various stages of decline based on your calibration spectra, to more complex spectral unmixing algorithms that approximate the proportion of "healthy to declining" spectra contained within each pixel. Even without specialized RS software, simple statistics can be used to quantify relationships between spectral reflectance and derived vegetation indices using field calibration data. A common approach is to use correlations between individual vegetation indices and decline metrics to qualitatively assess canopy condition across the landscape. Another approach uses multivariate statistical models to identify the best combination of bands or vegetation indices to quantify the decline metric of interest. Regardless of the mathematical approach, accuracy and detail are ultimately determined by the quality and range field calibration data available for model development. This type of targeted calibration to match the timing, location, and sensor characteristics for each decline assessment maximizes accuracy and detail of the final products.

4. *Validate and assess accuracy to inform interpretation.* One of the dangers inherent in linking RS products with management applications is overconfidence in the RS products. There is error inherent in each component of the RS process, from incorrect sensor calibration, to the variability introduced by atmospheric, topographic, and georegistration errors. However, when presented with a RS product, many end users develop their plans without consideration of how accurate the product may be or how inaccuracies can be avoided.

   Any RS product should include some measure of accuracy as well as any caveats that should be considered in its use. In some cases (e.g., the use of a vegetation index to qualitatively describe relative states of decline), it is sufficient to remind users that the scale presented is intended to be relative and does not necessarily identify stands in specific states of decline or resulting from specific stress agents. In other cases (e.g., the classification of pixels into levels of decline), we can use field data to present an accuracy assessment. Any accuracy assessment of classified image products should include overall accuracy as well as users' accuracy (percent of target pixels correctly classified; inverse = errors of omission) and producers' accuracy (percent of nontarget pixels that are *not* classified as the target class; inverse = errors of commission; Congalton 2001; Fassnacht et al. 2006). Splitting accuracy into users' and producers' values allows the end user to understand how false positives (saying a stand is dead when it is not) and false negatives (saying a stand is healthy when it is dead) can influence how the end product is used to inform management activities. For example, if overall accuracy in classifying forest mortality is 70% but almost all of the error results from false positives (many stands classified as dead when they are actually alive), end users may decide to limit management to locations with large clusters of predicted mortality or to clusters in higher-decline categories in order to avoid these common errors.

RS decline-detection products that result in ordinal classes of decline (e.g., healthy, degrees of decline, dead) can also be assessed for "fuzzy accuracy," which considers not only correct class assignments but also those within one ordinal class of the correct class. Products that provide a continuous decline metric can be used to produce more detailed accuracy metrics. Standard statistical regression techniques produce a coefficient of determination (r^2) to describe how well a statistical model fits the relationship between the input spectral variables and the output decline metric. Root mean square error, standard errors, and prediction errors can be used to place confidence bounds on predicted values. We can also examine how accuracy changes across the range of decline values predicted. For example, some models may be very good at quantifying severe decline but may not be able to detect early decline symptoms. Some models may overpredict early decline but underpredict severe decline. Standard statistical methods can be useful to examine how well your model works, which is critical to ensure that end users know how to best integrate your resulting RS products into their decision-making process.

A Nested Approach No one sensor, field methodology, or scale is appropriate for all applications. Different goals may require that you work at different scales (Fig. 6.9). The most detailed and accurate information about specific stress agents and response symptoms will always be obtained from on-the-ground field surveys (Tier 1). Such location-specific studies allow researchers to directly measure foliar chemistry, canopy structure, and spectral characteristics in situ. But these studies are limited in their utility to inform management across the broader landscape. Aerial sensors are often used to collect RS imagery at the local scale (Tier 2). Typically, this scale allows for the use of high spatial and spectral resolution imagery, ideally suited to detect forest stress conditions. However, such efforts may still be limited in geographic extent due to the high cost and computing needs. Most common is the use of broadband sensors at the regional-continental scale (Tier 3). Landsat sensors have been widely used for such applications, with sufficient spatial (30 m) and spectral resolution to prove useful in assessment of

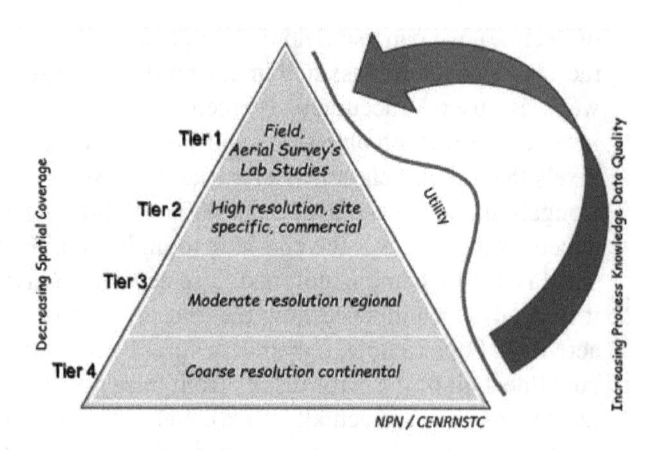

Fig. 6.9 RS work occurs at a variety of scales, with benefits and limitations at each level. Sometimes the best approach includes nesting your analyses across multiple scales to gain a comprehensive understanding of the forest health dynamics on the ground

relative levels of forest decline. The recent addition of improved satellite sensors (e.g., Sentinel 2) is rapidly increasing the capability to cover broad landscapes at higher spatial resolutions. Global assessments (Tier 4) of forest condition typically require a reduction in spatial resolution in order to process information over vast geographic extents. The much larger, mixed pixels often mask subtle changes in vegetation condition but can be useful in time series analyses when focused on relative changes in vegetation indices on continental scales.

The best approach to mapping and modeling forest decline depends on the scale of the investigation, level of detail needed, resources available, and time frame. For example, a regional assessment may have to forgo spatial and spectral resolution (and predictive detail and accuracy) in order to achieve the spatial coverage desired. In contrast, a municipality concerned about the spread of a recently detected invasive insect pest may forgo widespread spatial coverage to maximize the spatial and spectral resolution necessary to identify individual, newly infested trees. Sometimes you are limited by what is available in terms of imagery, time, or financial resources. For example, it is impossible to go back in time to collect high-resolution imagery, but you may be able to make use of historical broadband satellite imagery for a general assessment of past conditions. In most cases RS products, even when not exactly matched to the user's needs, can still provide insight that is not available through traditional monitoring.

Perhaps the most comprehensive approach to detecting novel forest health issues is to combine approaches. For example, a broad landscape assessment can be useful to identify localized areas for more detailed image acquisition. Even better, examining the relationship between spectral characteristics from higher-resolution imagery could be used to train coarser resolution imagery for a larger-scale assessment. The key is to recognize that there is no *one* right approach and that perhaps there are several RS approaches that can be used to achieve your objectives.

Stakeholder Engagement For each of the steps suggested above, stakeholder engagement is critical to success. RS specialists typically are not experts in entomology, invasive species, tree physiology, or forest ecology, and may not be aware of the specific stress symptoms to target for a given application. Because we work in various locations, we are rarely experts on the ecological specifics of a new study area. Where can we find target species or stands in various stages of decline? What key landscape features or characteristics should be covered in our calibration to best inform management? We also may not be sure how the products we develop could be most useful to land managers and practitioners. Would a classification product be most useful, with simple "healthy vs. dead" groupings, or would a range of decline condition be better? Do we need to develop a species map first to better target the declining stands end users hope to find? Are they looking for potential healthy "refugia" areas for conservation, newly declining stands for intervention, or high-mortality stands for salvage? Knowing what they need will allow us to design our modeling outputs to best suit their needs.

To maximize the impact of the products you develop, we suggest engaging a range of stakeholders throughout the entire process, for example:

- Go beyond simply obtaining letters of support to include end users and other key stakeholders in proposal development and experimental design from the outset of a new project.
- Find practitioners in your study area to identify and visit potential field sites.
- Present at local and regional meetings with the specific intent to introduce the project and solicit feedback on product format and delivery (prior to obtaining results).
- Include stakeholders in fieldwork, training them in field methodologies and learning from their expertise. Creating a sense of ownership or investment in a project improves the chances that your final products will actually be used.
- Meet with potential users as products are developed to gauge if the format (metric scale/range, spatial resolution, file format, etc.) are useful and, if not, how you might modify products to meet their needs.
- In addition to presenting your results at scientific meetings, target professional meetings and workshops to reach end users.
- Make your data products easily discoverable and available. This may include posting final products in online databases or web portals. Be sure the format is not limiting. Google Earth provides a useful platform for users without ARC or other proprietary geocomputing resources.

Including stakeholders in this way not only helps maximize the utility and impact of your efforts but also builds bridges between scientific and management communities. Historically, there have been limited collaborations among land managers, practitioners, decision-makers, and the RS scientific community. In some cases, there has even been mistrust as products are promised but delivered on a scientific timeline rather than a management timeline. But there has been a recent push to include stakeholders in RS and modeling efforts, exemplified by the recent "Voices from the Land" project led by researchers at Harvard Forest (McBride et al. 2017). This stakeholder-driven approach used interviews with New Englanders to identify key outcomes and likely scenarios for modeling. Such steps can build relationships that can serve all communities interested in sustaining forested ecosystems.

6.6 Using RS to Inform Forest Management

The application of RS for vegetation stress detection has advanced rapidly, evolving from classical aerial survey and photointerpretation techniques to digital image processing, where manual interpretation has been replaced with machine learning to identify subtle signatures humans are incapable of seeing with the naked eye. This technological evolution has effectively transferred these tools to the sustainable management of forest resources, but limitations remain in their widespread use. Monitoring, detecting, and reporting on forest health threats has always been a priority of federal and state forestry agencies. Conversion of forest land and changes in land use; climate change, intensified storms, higher frequency and intensity of forest fires and concerns of host range recession; and the threat of introduction and establishment from invasive insects and diseases have created an even more urgent demand for improved *near-real-time* tools and products. The capabilities of most sensors and the applications on which they have been tested are impressive, and more promising techniques and approaches continue to build on field application.

Recently, several programs have been developed with the goal of advancing and improving RS applications for forest management, including online tools developed to bring RS products to the forest health management community in near real time. Here we present some examples of online resources developed to transfer RS products to end users on time scales useful to inform management and planning.

World Vegetation Health Index https://www.star.nesdis.noaa.gov/smcd/emb/vci/VH/vh_browse.php The National Oceanic and Atmospheric Administration (NOAA)-National Environmental Satellite, Data, and Information Service (NESDIS) has developed several RS products designed specifically to assess vegetation health across the globe. Their Center for Satellite Applications and Research (STAR) Vegetation Health Index (Fig. 6.10) uses Advanced Very High-Resolution Radiometer (AVHRR) imagery produced from the NOAA/NESDIS Global Area Coverage (GAC) data set from 1981 to the present, with 4 km spatial and 7-day composite temporal resolution. Common vegetation indices are used to estimate vegetation health, moisture, and temperature and serve as a proxy to monitor vegetation cover, density, productivity, and drought conditions, as well as phenological stages such as the start/end of the growing season. Outputs are scaled to a range (0 to 100), providing a relative assessment of vegetation condition rather than a prediction of actual decline symptoms or identification of stress agents. However, these products are useful for examining short-term changes in vegetation that can be used to identify widespread decline events such as drought, land degradation, or fire.

ForWarn Online Mapper http://forwarn.forestthreats.org/; https://forwarn.forestthreats.org/fcav2/ ForWarn Satellite-Based Change Recognition and Tracking (Fig. 6.11) is a near-real-time product from the US Forest Service that uses 250 m MODIS data to compare current NDVI to seasonally similar historic NDVI values to identify disturbance such as wildfires, windstorms, insects, disease outbreaks,

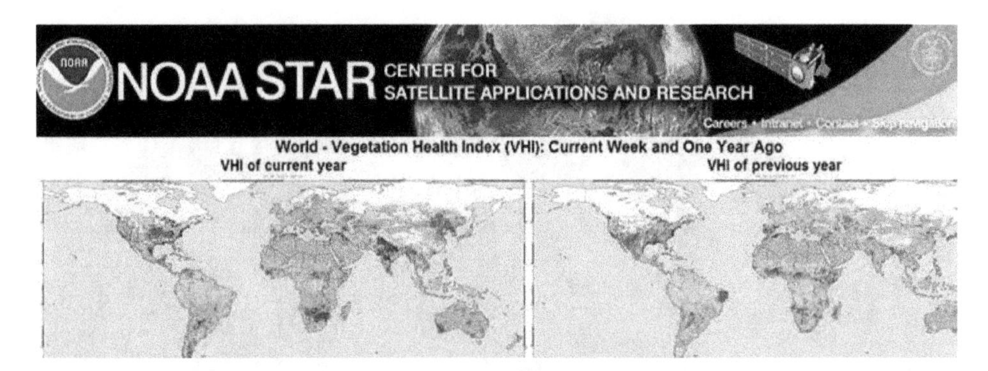

Fig. 6.10 The NOAA STAR World Vegetation Health Index visualization and data download portal

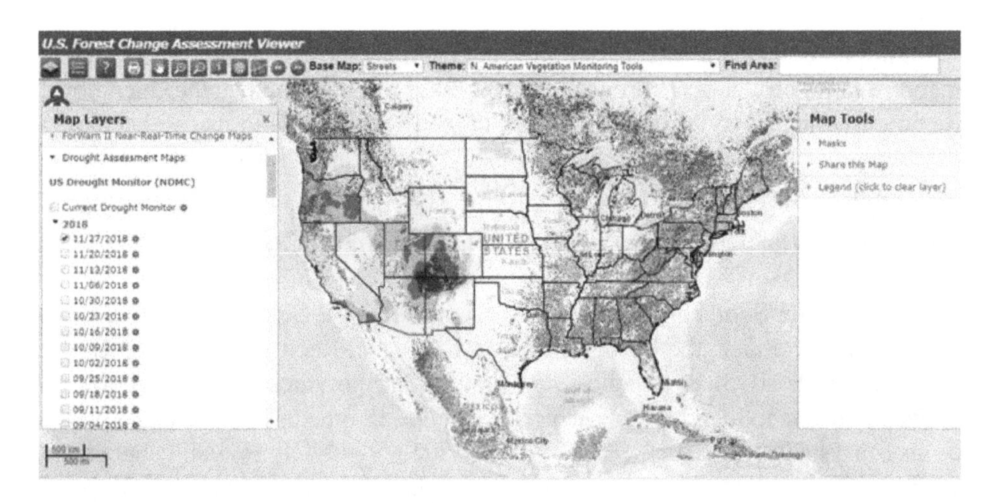

Fig. 6.11 The USFS ForWarn II online mapping portal provides weekly vegetation change and identification products dating back to 2003

logging, and land use change (Norman et al. 2013). Recent improvements in historical NDVI baseline data now provide the end user more tools to diagnose the severity and cause of changes in the mapping products.

Forest Disturbance Monitor (FDM) and Operational Remote Sensing (ORS) https://foresthealth.fs.usda.gov/FDM; http://foresthealth.fs.usda.gov/portal The US Forest Service Forest Disturbance Monitor (FDM; Fig. 6.12) is a forest disturbance web portal based on 16-day and 24-day MODIS composites that are updated every 8 days. FDM produces two forest disturbance products, 3-year Real-Time Forest Disturbance (RTFD) data and 5-year Trend Disturbance Data (TDD), providing near-real-time forest disturbance maps for land managers to target forest insect and disease events and complement aerial sketch mapping annual insect and disease surveys (IDSs; Chastain et al. 2015).

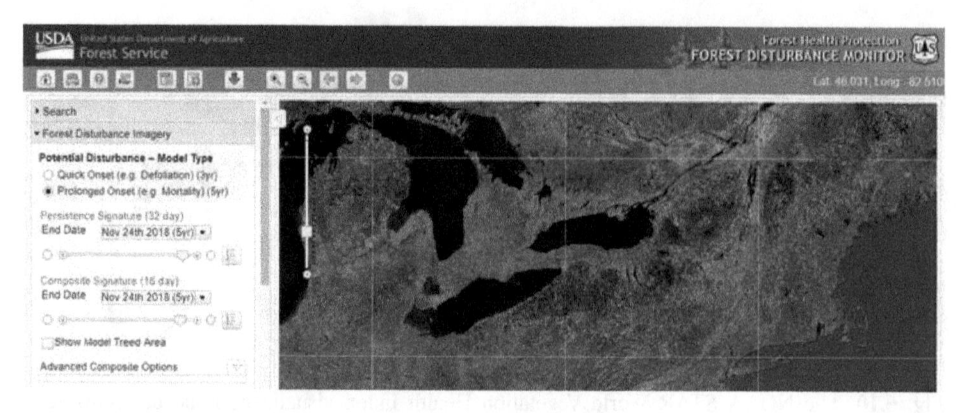

Fig. 6.12 The USFS Forest Disturbance Monitor online portal

To improve insect and disease surveys and facilitate the use of forest health information that RS products can provide, the USFS has recently initiated the Operational Remote Sensing (ORS) program. Similar to the FDM, ORS will use a phenology-based approach to intensifying surveys using 30 m Landsat and other moderate-resolution data.

Ecosystem Disturbance and Recovery Tracker (eDaRT) http://www.cstarsd3s.ucdavis.edu/systems/edart/ A collaboration among the University of California, Davis, Center for Southeastern Tropical Advanced Remote Sensing (CSTARS), and the US Forest Service, the Ecosystem Disturbance and Recovery Tracker (eDaRT; Koltunov et al. 2015) is an automated system that provides a suite of Landsat-derived products to identify and categorize changes in forest, shrubland, and herbaceous ecosystems. Currently, eDaRT products are not publicly available, but recent efforts are focused on expanding operations by the US Forest Service in California and elsewhere in the western United States in support of daily ecosystem management tasks.

Looking Ahead Because of the vast potential for RS to inform the sustainable management of terrestrial landscapes, there are several new Earth observation missions on the horizon. The European Space Agency (ESA) will launch Earth Explorer 7 in 2021 (https://www.esa.int/Our_Activities/Observing_the_Earth). This ecology mission, known as Biomass, is designed to characterize forests. The Biomass mission will be followed by the Earth Explorer 8 Fluorescence Explorer (FLEX) mission in 2022, with capabilities to quantify chlorophyll fluorescence in terrestrial vegetation. Landsat 9, part of the Earth observation data continuity mission from NASA (fast-tracked for December 2020 launch date), will maintain nearly 50 years of continuous Earth observation. This instrument is designed to simultaneously image 11 spectral bands, including a 15 m panchromatic band, with 12 bit radiometric resolution to increase sensitivity to small differences in reflectance. Such advances are critical to the early stress detection and detailed decline assessment that land managers need.

6.7 Management Applications: Limitations and Opportunities

Thanks to continuing advances in computing and software technologies, we are poised to bring near-real-time RS products to more stakeholders. Applications like Google Earth Engine (https://earthengine.google.com/) now have the ability to automate image acquisition, preprocessing, and more complex modeling algorithms to provide critical forest health information across large landscapes at regular time intervals. Similarly, the ESA's Grid Processing on Demand (G-POD) provides an online environment where scientists can build and automate RS applications (https://gpod.eo.esa.int/). While several organizations (see ForWarn, FDM, and eDaRT above) are making final products from this type of rapid analysis and assessment operational for coarse forest health assessments and disturbance mapping efforts, higher-level products (higher spatial resolution, low-level stress detection) are not yet publicly available for use by broad stakeholder groups.

Currently, most RS efforts to detect incipient stress factors or detailed vegetation condition are conducted by the research community with scientific journals as their primary outputs. The more widespread use of more advanced RS techniques in forest management is primarily limited by:

- *The cost of image acquisition and expertise required to accurately calibrate sensors and validate products.* This is particularly true for hyperspectral efforts, which generate large amounts of data and require specialized expertise for preprocessing corrections, calibration, and data management. Computing advances and the growing commercial sector promise improved access, but for many land managers, cost is still a strong deterrent. Some organizations are hoping to make cutting-edge imagery more accessible. For example, NASA's Goddard's LiDAR, Hyperspectral, and Thermal Imager (G-LiHT) (https://gliht.gsfc.nasa.gov/) is a portable, airborne imaging system that simultaneously maps composition, structure, and function of terrestrial ecosystems using multispectral LiDARs (3-D information about the vertical and horizontal distribution of foliage and other canopy elements), hyperspectral imaging spectrometer to discern species composition and variations in biophysical variables (photosynthetic pigments and nutrient and water content), and a thermal camera to measure surface temperatures to detect heat and moisture stress (Cook et al. 2013). Owned and operated by NASA Goddard, this instrument has proven to be more affordable and accessible than comparable commercial vendors and may greatly expand access to cutting-edge sensor technologies for a variety of applications (Fig. 6.13).
- *The turnaround time required to deliver final mapping products.* Typically, the more irruptive forest health issues require immediate attention in the current growing season (e.g., pest outbreaks, extreme climate events, wildfires), while turnaround from RS projects doesn't always occur in the same year. This disparity between product delivery and product need is especially evident in studies

Fig. 6.13 NASA's G-LiHT online data portal

where method development is necessary and limits the adoption of more advanced RS efforts by the forest management community. However, the increased use of automated image processing scripts that make satellite image products available in near real time is expanding the use of traditional (vegetation index-based) relative assessments available for a variety of applications. The resulting online tools described above are being adopted by a range of state and federal agencies to inform management decisions.

- *Integration of mapping products into decision-making processes.* Even when RS products are available, there is no clear path on how to *use* the information they provide to inform decision-making. Land managers may reference mapping products to target specific locations, but more complete integration of spatial products into management plans can be challenging for those not used to working with spatial data. Foresters are typically trained in making decisions based on generalized inventories of forest stands or management units, not pixelated rasters across a landscape with a high degree of variability. End users may not be aware that mapping products should come with an accuracy assessment that informs how the information can best be used and how it impacts the overall confidence in the product. Many of these limitations can be resolved by scientists working more closely with end users as outlined in the Stakeholder Engagement section above. By working together, both scientists and land managers can learn from each other and so better use RS technologies to manage critical environmental resources.

6.8 Conclusions

While historically RS has been successfully used to assess and monitor vegetation condition on a coarse, relative scale, recent advances and new analysis techniques now enable us to also use RS to identify and track early decline, disturbance, and stress conditions in vegetative systems. Considering the environmental challenges currently facing terrestrial systems, this information is critical to inform management, policy, and planning in order to maintain the structure and function of these systems.

The challenge is for scientists to look beyond traditional approaches to vegetation assessment and target earlier or more subtle decline response resulting from incipient or chronic environmental stress agents (e.g., climate change, pollution). Key challenges include linking hyperspectral data to specific stress agents, extending the availability of higher-resolution imagery, and operationalizing near-real-time monitoring of the forest resource (Senf et al. 2017). Scientists must work closely with land managers to bring these new technologies to application in order to harness RS's full potential to inform the management of critical ecological resources.

References

Aoki M, Yabuki K, Totsuka T (1981) An evaluation of chlorophyll content of leaves based on the spectral reflectivity in several plants. Res Rep Nat Inst Environ Stud Jpn 66:125–130

Asner GP, Heidebrecht KB (2003) Imaging spectroscopy for desertification studies: comparing AVIRIS and EO-1 Hyperion in Argentina drylands. IEEE Trans Geosci Remote Sens 41(6):1283–1296

Barnes JD (1992) A reappraisal of the use of DMSO for the extraction and determination of chlorophylls a and b in lichens and higher plants. Environ Experim Bot 32(2):85–100

Blackburn GA (1998) Quantifying chlorophylls and carotenoids at leaf and canopy scales: an evaluation of some hyperspectral approaches. Remote Sens Environ 66(3):273–285

Blackburn GA (1999) Relationships between spectral reflectance and pigment concentrations in stacks of deciduous broadleaves. Remote Sens Environ 70:224–237

Boochs F, Kupfer G, Dockter K, Kuhbauh W (1990) Shape of the red edge as vitality indicator for plants. *International Journal of Remote Sensing* 11(10):1741–1753

Broge NH, Leblanc E (2001) Comparing prediction power and stability of broadband and hyperspectral vegetation indices for estimation of green leaf area index and canopy chlorophyll density. Remote Sens Environ 76(2):156–172

Buschmann C, Nagel E (1993) In vivo spectroscopy and internal optics of leaves as a basis for remote sensing of vegetation. Int J Remote Sens 14:711–722

Carter GA (1993) Responses of leaf spectral reflectance to plant stress. Am J Bot 80(3):239–243

Carter GA (1994) Ratios of leaf reflectances in narrow wavebands as indicators of plant stress. Int J Remote Sens 15(3):697–703

Carter GA, Knapp AK (2001) Leaf optical properties in higher plants: linking spectral characteristics to stress and chlorophyll concentration. Am J Bot 88(4):677–684

Carter GA, Miller RL (1994) Early detection of plant stress by digital imaging within narrow stress-sensitive wavebands. Remote Sens Environ 50(3):295–302

Carter GA, Knapp AK, Anderson JE, Hoch GA, Smith MD (2005) Indicators of plant species richness in AVIRIS spectra of a mesic grassland. Remote Sens Environ 98:304–316

Chappelle EW, Kim MS, McMurtrey JE III (1992) Ratio analysis of reflectance spectra (RARS): an algorithm for the remote estimation of the concentrations of chlorophyll a, chlorophyll b, and carotenoids in soybean leaves. Remote Sens Environ 39(3):239–247

Chastain RA, Fisk H, Ellenwood JR, Sapio FJ, Ruefenacht B, Finco MV, Thomas V (2015) Near-real time delivery of MODIS-based information on forest disturbances. In: Time-sensitive remote sensing. Springer, New York, pp 147–164

Chen J (1996) Evaluation of vegetation indices and a modified simple ratio for boreal applications. Can J Remote Sens 22(3):229–242

Cheng Y-B, Zarco-Tejada PJ, Riano D, Rueda CA, Ustin SL (2006) Estimating vegetation water content with hyperspectral data for different canopy scenarios: relationships between AVIRIS and MODIS indexes. Remote Sens Environ 105(2006):354–366

Coates AR, Dennison PE, Roberts DA, Roth KL (2015) Monitoring the impacts of severe drought on southern California chaparral species using hyperspectral and thermal infrared imagery. Remote Sens 7(11):14276–14291

Congalton RG (2001) Accuracy assessment and validation of remotely sensed and other spatial information. Int J Wildland Fire 10(3–4):321–328

Cook BD, Corp LW, Nelson RF, Middleton EM, Morton DC, McCorkel JT, Masek JG, Ranson KJ, Ly V, Montesano PM (2013) NASA Goddard's Lidar, Hyperspectral and Thermal (G-LiHT) airborne imager. Remote Sens Environ 5:4045–4066

D'ambrosio N, Szabo K, Lichtenthaler H (1992) Increase of the chlorophyll fluorescence ratio F690/F735 during the autumnal chlorophyll breakdown. Radiat Environ Biophys 31(1):51–62

Datt B (1998) Remote sensing of chlorophyll a, chlorophyll b, chlorophyll a + b, and total carotenoid content in eucalyptus leaves. Remote Sens Environ 66:111–121

Datt B (1999) Visible/near infrared reflectance and chlorophyll content in Eucalyptus leaves. Int J Remote Sens 20(14):2741–2759

Daughtry CST, Walthall CL, Kim MS, de Colstoun EB, McMurtrey JE (2000) Estimating corn leaf chlorophyll concentration from leaf and canopy reflectance. Remote Sens Environ 74(2):229–239. https://doi.org/10.1016/s0034-4257(00)00113-9

Deblonde G, Cihlar J (1993) A multiyear analysis of the relationship between surface environmental variables and NDVI over the Canadian landmass. Remote Sens Rev 7:151–177

Degerickx J, Roberts DA, McFadden JP, Hermy M, Somers B (2018) Urban tree health assessment using airborne hyperspectral and LiDAR imagery. Int J Appl Earth Obs Geoinf 73:26–38

Ellison AM, Barker-Plotkin AA, Foster DR, Orwig DA (2010) Experimentally testing the role of foundation species in forests: the Harvard Forest Hemlock Removal Experiment. Methods Ecol Evol 1(2):168–179. https://doi.org/10.1111/j.2041-210X.2010.00025.x

Elvidge CD, Chen Z (1995) Comparison of broad-band and narrow-band red and near-infrared vegetation indices. Remote Sens Environ 54(1):38–48

Elvidge CD, Lyon RJ (1985) Estimation of the vegetation contribution to the 1 65/2 22 μm ratio in airborne thematic-mapper imagery of the Virginia Range, Nevada. Int J Remote Sens 6(1):75–88

Epanchin-Niell RS, Hastings A (2010) Controlling established invaders: integrating economics and spread dynamics to determine optimal management. Ecol Lett 13(4):528–541

Fassnacht KS, Cohen WB, Spies TA (2006) Key issues in making and using satellite-based maps in ecology: a primer. For Ecol Manag 222:167–181

Filella I, Penuelas J (1994) The red edge position and shape as indicators of plant chlorophyll content, biomass, and hydric status. Int J Remote Sens 15(7):1459–1470

Gamon JA, Serrano L, Surfus JS (1997) The photochemical reflectance index: an optical indicator of photosynthetic radiation use efficiency across species, functional types, and nutrient levels. Oecologia 112(4):492–501

Gao BC (1996) Ndwi - a normalized difference water index for remote sensing of vegetation liquid water from space. Remote Sens Environ 58(3):257–266

Genovesi P, Carboneras C, Vila M, Walton P (2015) EU adopts innovative legislation on invasive species: a step towards a global response to biological invasions? Biol Invasions 17(5):1307–1311

Gitelson A, Merzlyak MN (1994) Quantitative estimation of chlorophyll-a using reflectance spectra: experiments with autumn chestnut and maple leaves. J Photochem Photobiol B Biol 22(3):247–252

Gitelson AA, Merzlyak MN (1996) Signature analysis of leaf reflectance spectra: algorithm development for remote sensing of chlorophyll. J Plant Physiol 148(3–4):494–500

Gitelson AA, Buschmann C, Lichtenthaler HK (1999) The chlorophyll fluorescence ratio F-735/F-700 as an accurate measure of the chlorophyll content in plants. Remote Sens Environ 69(3):296–302

Gitelson AA, Merzlyak MN, Chivkunova OB (2001) Optical properties and non-destructive estimation of anthocyanin content in plant leaves. Photochem Photobiol 74(1):38–45

Haboudane D, Miller JR, Pattey E, Zarco-Tejada PJ, Strachan IB (2004) Hyperspectral vegetation indices and novel algorithms for predicting green LAI of crop canopies: modeling and validation in the context of precision agriculture. Remote Sens Environ 90(3):337–352. https://doi.org/10.1016/j.rse.2003.12.013

Hanavan RP, Pontius J, Hallett R (2015) A 10-year assessment of hemlock decline in the Catskill Mountain region of New York State using hyperspectral remote sensing techniques. J Econ Entomol 108(1):339–349

Hardisky MA, Klemas V, Smart RM (1983) The influence of soil salinity, growth form, and leaf moisture on the spectral radiance of *Spartina alterniflora* canopies. Photogramm Eng Remote Sens 49(1):77–83

Harwood T, Tomlinson I, Potter C, Knight J (2011) Dutch elm disease revisited: past, present and future management in Great Britain. Plant Pathol 60(3):545–555

Hennon PE, D'Amore DV, Schaberg PG, Wittwer DT, Shanley CS (2012) Shifting climate, altered niche, and a dynamic conservation strategy for yellow-cedar in the North Pacific coastal rainforest. Bioscience 62(2):147–158

Horler D, DOCKRAY M, Barber J (1983) The red edge of plant leaf reflectance. Int J Remote Sens 4(2):273–288

Huete AR (1988) A soil adjusted vegetation index (SAVI). Remote Sens Environ 25(3):295–309. https://doi.org/10.1016/0034-4257(88)90106-x

Huete A, Justice C, Liu H (1994) Development of vegetation and soil indices for MODIS-EOS. Remote Sens Environ 49(3):224–234

Huete A, Didan K, Miura T, Rodriguez EP, Gao X, Ferreira LG (2002) Overview of the radiometric and biophysical performance of the MODIS vegetation indices. Remote Sensing of Environment 83(1–2):195–213

Huggett BA, Schaberg PG, Hawley GJ, Eagar C (2007) Long-term calcium addition increases growth release, wound closure, and health of sugar maple (Acer saccharum) trees at the Hubbard Brook Experimental Forest. Can J For Res 37(9):1692–1700

Hunt ER, Rock BN (1989) Detection of changes in leaf water content using near infrared and middle infrared reflectances. Remote Sens Environ 30(1):43–54. https://doi.org/10.1016/0034-4257(89)90046-1

Johnson EW, Ross J (2008) Quantifying error in aerial survey data. Aust For 71(3):216–222

Johnson E, Wittwer D (2008) Aerial detection surveys in the United States. Aust For 71(3):212–215

Jordan CF (1969) Derivation of leaf area index from quality of light on the forest floor. Ecology (Washington DC) 50(4):663–666. https://doi.org/10.2307/1936256

Kaufman YJ, Tanre D (1992) Atmospherically resistant vegetation index (ARVI) for EOS-MODIS. IEEE Trans Geosci Remote Sens 30(2):261–270

Kokaly RF, Despain DG, Clark RN, Livo KE (2003) Mapping vegetation in yellowstone national park using spectral feature analysis of AVIRIS data. Remote Sens Environ 84(3):437–456

Kolb T, Teulon D (1991) Relationship between sugar maple budburst phenology and pear thrips damage. Can J For Res 21(7):1043–1048

Koltunov A, Ramirez C, Ustin SL (2015) eDaRT: the ecosystem disturbance and recovery tracking system prototype supporting ecosystem management in California. NASA Carbon Cycle and Ecosystems Joint Science Workshop College Park, MD, April, 19–24

Kosiba AM, Meigs GW, Duncan J, Pontius J, Keeton WS, Tait E (2018) Spatiotemporal patterns of forest damage in the Northeastern United States: 2000–2016. Forest Ecology and Management 430:94–104.

Lausch A, Heurich M, Gordalla D, Dobner HJ, Gwillym-Margianto S, Salbach C (2013) Forecasting potential bark beetle outbreaks based on spruce forest vitality using hyperspectral remote-sensing techniques at different scales. For Ecol Manag 308:76–89

Lausch A, Erasmi S, King D, Magdon P, Heurich M (2017) Understanding forest health with remote sensing-part II—a review of approaches and data models. Remote Sens 9(2):129

Lillesand TM, Kiefer RW (1994) Remote sensing and photo interpretation, 3rd edn. John Wiley& Sons, New York

Maccioni A, Agati G, Mazzinghi P (2001) New vegetation indices for remote measurement of chlorophylls based on leaf directional reflectance spectra. J Photochem Photobiol 61(1,2):52–61

Martin K, Norris A, Drever M (2006) Effects of bark beetle outbreaks on avian biodiversity in the British Columbia interior: implications for critical habitat management. J Ecosyst Manag 7(3):10–24

McBride MF, Lambert KF, Huff ES, Theoharides KA, Field P, Thompson JR (2017) Increasing the effectiveness of participatory scenario development through codesign. Ecol Soc 22(3):16

McConnell TJ (1999) Aerial sketch mapping surveys, the past, present and future. In: Paper from the north American science symposium, toward a unified framework for inventorying and monitoring forest ecosystem resources, Guadalajara, Mexico

McMurtrey J III, Chappelle EW, Kim M, Meisinger J (1994) Distinguishing nitrogen fertilization levels in field corn (Zea mays L.) with actively induced fluorescence and passive reflectance measurements. Remote Sens Environ 47(1):36–44

Meroni M, Rossini M, Picchi V, Panigada C, Cogliati S, Nali C, Colombo R (2008) Assessing steady-state fluorescence and PRI from hyperspectral proximal sensing as early indicators of plant stress: the case of ozone exposure. Sensors 8(3):1740–1754. https://doi.org/10.3390/s8031740

Merzlyak MN, Gitelson AA, Chivkunova OB, Rakitin VY (1999) Non-destructive optical detection of pigment changes during leaf senescence and fruit ripening. Physiol Plant 106(1):135–141. https://doi.org/10.1034/j.1399-3054.1999.106119.x

Mohammed GH, Binder WD, Gillies SL (1995) Chlorophyll fluorescence - a review of its practical forestry applications and instrumentation. Scand J For Res 10(4):383–410

Mumford R (2017) New approaches for the early detection of tree health pests and pathogens. Impact 1(7):47–49

Myneni RB, Hall FG, Sellers PJ, Marshak AL (1995a) The interpretation of spectral vegetation indexes. IEEE Trans Geosci Remote Sens 33(2):481–486

Myneni RB, Maggion S, Iaquinta J, Privette JL, Gobron N, Pinty B, Kimes DS, Verstraete MM, Williams DL (1995b) Optical remote sensing of vegetation: modeling, caveats, and algorithms. Remote Sens Environ 51:169–188

Norman SP, Hargrove WW, Spruce JP, Christie WM, Schroeder SW (2013) Highlights of satellite-based forest change recognition and tracking using the ForWarn System. Gen Tech Rep SRS-GTR-180 Asheville, NC: USDA-Forest Service, Southern Research Station, 30 p, 180:1–30

Pearlman JS, Barry PS, Segal CC, Shepanski J, Beiso D, Carman SL (2003) Hyperion, a space-based imaging spectrometer. IEEE Trans Geosci Remote Sens 41:1160–1173

Pearson L, Miller LD (1972) Remote mapping of standing crop biomass for estimation of the productivity of the short-grass prairie, Pawnee National Grasslands, Colorado. In: Proceedings of the 8th international symposium on remote sensing of the environment, Ann Arbor, MI, 1972. ERIM, Ann Arbor, pp 1357–1381

Penuelas J, Filella I, Biel C, Serrano L, Save R (1993) The reflectance at the 950–970 nm region as an indicator of plant water status. Int J Remote Sens 14(10):1887–1905

Penuelas J, Gamon JA, Fredeen AL, Merino J, Field CB (1994) Reflectance indices associated with physiological changes in nitrogen-limited and water-limited sunflower leaves. Remote Sens Environ 48(2):135–146. https://doi.org/10.1016/0034-4257(94)90136-8

Penuelas J, Baret F, Filella I (1995) Semi-empirical indices to assess carotenoids/chlorophyll a ratio from leaf spectral reflectance. Photosynthetica 31:221–230

Pietrzykowski E, Stone C, Pinkard E, Mohammed C (2006) Effects of Mycosphaerella leaf disease on the spectral reflectance properties of juvenile Eucalyptus globulus foliage. For Pathog 36(5):334–348

Pinty B, Leprieur C, Verstraete MM (1993) Towards a quantitative interpretation of vegetation indices. Part I. Biophysical canopy properties and classical indices. Remote Sens Rev 7:127–150

Pontius J (2014) A new approach for forest decline assessments: maximizing detail and accuracy with multispectral imagery. Int J Remote Sens 35(9):3384–3402. https://doi.org/10.1080/01431161.2014.903439

Pontius J, Hallett R (2014) Comprehensive methods for earlier detection and monitoring of forest decline. For Sci 60(6):1156–1163. https://doi.org/10.5849/forsci.13-121

Pontius J, Hallett R, Martin M (2005a) Assessing hemlock decline using visible and near-infrared spectroscopy: indices comparison and algorithm development. Appl Spectrosc 59(6):836–843. https://doi.org/10.1366/0003702054280595

Pontius J, Hallett R, Martin M (2005b) Using AVIRIS to assess hemlock abundance and early decline in the Catskills, New York. Remote Sens Environ 97(2):163–173. https://doi.org/10.1016/j.rse.2005.04.011

Pontius JA, Hallett RA, Jenkins JC (2006) Foliar chemistry linked to infestation and susceptibility to hemlock woolly adelgid (Homoptera: Adelgidae). Environ Entomol 35(1):112–120

Pontius J, Martin M, Plourde L, Hallett R (2008) Ash decline assessment in emerald ash borer-infested regions: a test of tree-level, hyperspectral technologies. Remote Sens Environ 112(5):2665–2676. https://doi.org/10.1016/j.rse.2007.12.011

Pontius J, Halman JM, Schaberg PG (2016) Seventy years of forest growth and community dynamics in an undisturbed northern hardwood forest. Can J For Res 46(7):959–967. https://doi.org/10.1139/cjfr-2015-0304

Pontius J, Hanavan RP, Hallett RA, Cook BD, Corp LA (2017) High spatial resolution spectral unmixing for mapping ash species across a complex urban environment. Remote Sens Environ 199:360–369

Porter WM, Enmark HT (1987) A system overview of the airborne visible/infrared imaging spectrometer (AVIRIS). In: *Imaging Spectroscopy II*, vol 834. International Society for Optics and Photonics, Bellingham, WA. pp 22–32

Qi J, Chehbouni A, Huete AR, Kerr YH, Sorooshian S (1994) A modified soil adjusted vegetation index. Remote Sens Environ 48(2):119–126. https://doi.org/10.1016/0034-4257(94)90134-1

Rahman AF, Gamon JA, Fuentes DA, Roberts DA, Prentiss D (2001) Modeling spatially distributed ecosystem flux of boreal forest using hyperspectral indices from AVIRIS imagery. J Geophys Res-Atmos 106(D24):33579–33591

Rock BN, Vogelmann AF, Williams DL, Vogelmann DL, Hoshizaki T (1986) Remote detection of forest damage. Bioscience 36(7):439–445

Rock BN, Hoshizaki T, Miller JR (1988) Comparison of in situ and airborne spectral measurements of the blue shift associated with forest decline. Remote Sens Environ 24(1):109–127

Rondeaux G, Steven M, Baret F (1996) Optimization of soil-adjusted vegetation indices. Remote Sens Environ 55(2):95–107. https://doi.org/10.1016/0034-4257(95)00186-7

Roujean J-L, Breon F-M (1995) Estimating PAR absorbed by vegetation from bidirectional reflectance measurements. Remote Sens Environ 51(3):375–384

Rouse J, Hass R, Schell J, Deering D, Harlan J (1974) Monitoring the vernal advancement and retrogradation of natural vegetation. NASA Report, Greenbelt, MD

Schaberg PG, Dehayes DH, Hawley GJ (2001) Anthropogenic calcium depletion: a unique threat to forest ecosystem health? Ecosyst Health 7(4):214–228

Schaberg PG, Minocha R, Long S, Halman JM, Hawley GJ, Eagar C (2011) Calcium addition at the Hubbard Brook Experimental Forest increases the capacity for stress tolerance and carbon capture in red spruce (Picea rubens) trees during the cold season. Trees 25(6):1053–1061

Schaberg PG, Murakami PF, Butnor JR, Hawley GJ (2017) Experimental branch cooling increases foliar sugar and anthocyanin concentrations in sugar maple at the end of the growing season. Can J For Res 47(5):696–701

Senf C, Seidl R, Hostert P (2017) Remote sensing of forest insect disturbances: current state and future directions. Int J Appl Earth Obs Geoinf 60:49–60

Serbin SP, Singh A, McNeil BE, Kingdon CC, Townsend PA (2014) Spectroscopic determination of leaf morphological and biochemical traits for northern temperate and boreal tree species. Ecol Appl 24(7):1651–1669

Serbin SP, Singh A, Desai AR, Dubois SG, Jablonski AD, Kingdon CC et al (2015) Remotely estimating photosynthetic capacity, and its response to temperature, in vegetation canopies using imaging spectroscopy. Remote Sens Environ 167:78–87

Sims DA, Gamon JA (2002) Relationships between leaf pigment content and spectral reflectance across a wide range of species, leaf structures and developmental stages. Remote Sens Environ 81(2–3):337–354

Singh A, Serbin SP, McNeil BE, Kingdon CC, Townsend PA (2015) Imaging spectroscopy algorithms for mapping canopy foliar chemical and morphological traits and their uncertainties. Ecol Appl 25(8):2180–2197

Sitzia T, Campagnaro T, Kowarik I, Trentanovi G (2016) Using forest management to control invasive alien species: helping implement the new European regulation on invasive alien species. Biol Invasions 18(1):1–7

Sivanpillai R, Smith CT, Srinivasan R, Messina MG, Ben Wu X (2006) Estimation of managed loblolly pine stand age and density with Landsat ETM+ data. For Ecol Manag 223(1–3):247–254. https://doi.org/10.1016/j.foreco.2005.11.013

Smith ML, Ollinger SV, Martin ME, Aber JD, Hallett RA, Goodale CL (2002) Direct estimation of aboveground forest productivity through hyperspectral remote sensing of canopy nitrogen. Ecol Appl 12(5):1286–1302

Strimbeck GR, Schaberg PG, Fossdal CG, Schröder WP, Kjellsen TD (2015) Extreme low temperature tolerance in woody plants. Front Plant Sci 6:884

Thenkabail PS, Mariotto I, Gumma MK, Middleton EM, Landis DR, Huemmrich KF (2013) Selection of hyperspectral narrowbands (HNBs) and composition of hyperspectral twoband vegetation indices (HVIs) for biophysical characterization and discrimination of crop types using field reflectance and Hyperion/EO-1 data. IEEE J-STARS 6(2):427–439

Thompson I, Mackey B, McNulty S, Mosseler A (2009) Forest resilience, biodiversity, and climate change. In secretariat of the convention on biological diversity, Montreal. Technical Series No. 43. Vol 43, pp 1–67

Tomback DF, Achuff P (2010) Blister rust and western forest biodiversity: ecology, values and outlook for white pines. For Pathol 40(3–4):186–225

Townsend PA, Foster JR, Chastain RA, Currie WS (2003) Application of imaging spectroscopy to mapping canopy nitrogen in the forests of the central appalachian mountains using hyperion and aviris. IEEE Trans Geosci Remote Sens 41(6):1347–1354

Tucker CJ (1979) Red and photographic infrared linear combinations for monitoring vegetation. Remote Sens Environ 8:127–150

Vogelmann JE, Rock BN, Moss DM (1993) Red edge spectral measurements from sugar maple leaves. Int J Remote Sens 14(8):1563–1575

White K, Pontius J, Schaberg P (2014) Remote sensing of spring phenology in northeastern forests: a comparison of methods, field metrics and sources of uncertainty. Remote Sens Environ 148:97–107. https://doi.org/10.1016/j.rse.2014.03.017

Williams AP, Hunt ER (2002) Estimation of leafy spurge cover from hyperspectral imagery using mixture tuned matched filtering. Remote Sens Environ 82(2–3):446–456

Williams P, Norris K (2001) Near-infrared Technology in the Agricultural and Food Industries. American Association of Cereal Chemists, Inc., St. Paul

Williams DL, Goward S, Arvidson T (2006) Landsat. Photogramm Eng Remote Sens 72(10):1171–1178

Zarco-Tejada PJ, Miller JR, Mohammed GH, Noland TL, Sampson PH (2002) Vegetation stress detection through chlorophyll $a + b$ estimation and fluorescence effects on hyperspectral imagery. J Environ Qual 31(5):1433–1441

Zarco-Tejada PJ, Hornero A, Hernández-Clemente R, Beck PSA (2018) Understanding the temporal dimension of the red-edge spectral region for forest decline detection using high-resolution hyperspectral and sentinel-2a imagery. ISPRS J Photogramm Remote Sens 137:134

7

Leaf Spectra: An Integrated Study

José Eduardo Meireles, Brian O'Meara, and Jeannine Cavender-Bares

7.1 Introduction

Evolution is the engine behind the diversity in leaf structure and chemistry that is captured in their spectral profiles, and, therefore, leaf spectra are inexorably linked to the tree of life. Our ability to distinguish species using spectra is a consequence of trait differences that arise and accumulate over evolutionary time. By the same token, the amount of variation that exists in different spectral regions is ultimately determined by the pace of evolution, convergence, and other evolutionary dynamics affecting the underlying leaf traits. There is an increasing interest in understanding leaf spectra through the lens of evolution and in the context of phylogenetic history (Cavender-Bares et al. 2016; McManus et al. 2016). Advances on this front will require, however, a good understanding of how evolutionary biologists leverage the tree of life to make inferences about evolution.

J. E. Meireles (✉)
Department of Ecology, Evolution and Behavior, University of Minnesota, Saint Paul, MN, USA

School of Biology & Ecology, University of Maine, Orono, ME, USA

B. O'Meara
Department of Ecology & Evolutionary Biology, University of Tennessee, Knoxville, TN, USA

J. Cavender-Bares
Department of Ecology, Evolution and Behavior, University of Minnesota, Saint Paul, MN, USA

7.2 Evolutionary Trees

We refer to phylogenies in many different ways, and all of these terms appear in the literature. The terms *phylogeny*, *phylogenetic tree*, and *evolutionary tree* can be used interchangeably. We also use the term *tree of life* to refer to *the tree of life* (the evolutionary tree for all of life) or to the phylogeny of a really large group (or lineage) of organisms, such as the plant tree of life or the vertebrate tree of life.

7.2.1 *How to Read Phylogenies*

The idea that species descend from a common ancestor is at the very core of the theory of evolution. Evolutionary trees represent the branching structure of life and describe how species are related to each other similarly to how a genealogical tree recounts how people are related. A branch on a phylogenetic tree is a species; when it speciates, two (typically) descendant species arise. The two lineages coming from the same ancestor are known as sisters. These lineages can continue to branch, leading to more descendants. An ancestor and all its descendants are known as a clade: since these descendants all came from the same species, they share many inherited traits. Relatedness among organisms is encoded in the phylogeny's structure—its topology—which defines a series of lineages that are hierarchically nested. The branch lengths also usually convey information, such as the time since divergence, amount of molecular similarity, or number of generations (Fig. 7.1). Dated fossil information can be used to calibrate the age of some of the nodes in a phylogeny. This is one means by which branch lengths can be made to represent time fairly accurately. Typically, the spacing between tip nodes (the y-axis in Fig. 7.1a, b, and d) has no meaning, but it can sometimes be used to display information about the trait values of a species (Fig. 7.1c). Because no one has been taking notes of how lineages split over the last 4 billion years, phylogenetic trees must be estimated by analyzing current species data, generally DNA sequences, using models of evolution. This means that phylogenies are statistical inferences that have uncertainty about their topology and their branch lengths (Fig. 7.1d).

7.2.2 *Why Care About Phylogenetic Accuracy?*

An accurate phylogeny is key for understanding life. A phylogeny in which dandelions were more closely related to ferns than to roses would tell us a very different story about the evolution of flowering plants than would the true phylogeny, in which all flowering plants belong to a single lineage. In other words, the accuracy of the estimated tree topology—the structure of the relationships between species—matters to how we understand trait evolution. Accurately inferring divergence times

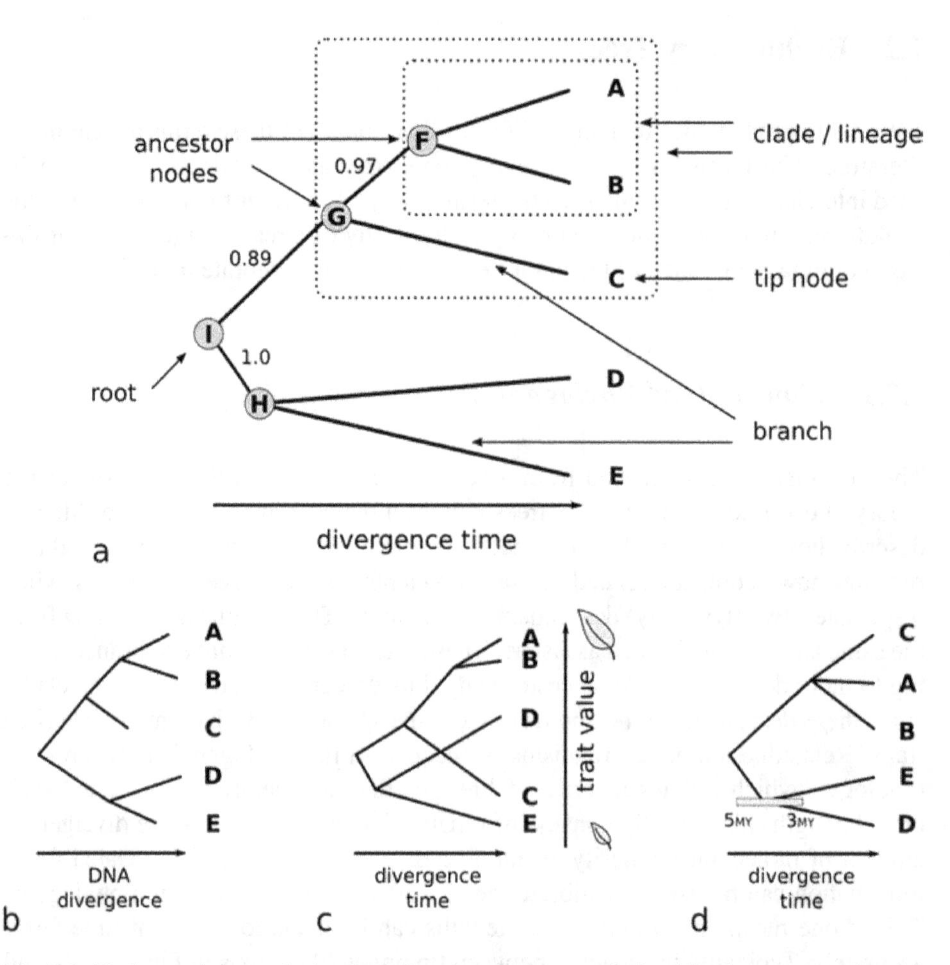

Fig. 7.1 Phylogenetic trees depict the inferred evolutionary relationships between species. (**a**) Clades (or lineages) are defined by a common ancestor and all of its descendants. Nodes are the branching points between descendants as well as tips, which are typically species. All nodes—tips and ancestors alike—share a common ancestor. The ancestral node from which all subsequent nodes of the tree descend is called the root. Confidence in the evolutionary relationships is shown above internal branches (maximum possible is 1). (**b**) Branch lengths (here shown along the *x*-axis) may represent divergence times, number of generations, or amount of molecular divergence. (**c**) In some cases the *y*-axis is used to display information about a quantitative trait—such as leaf size—in a tree known as a phenogram or a traitigram. (**d**) Unresolved relationships can be represented as three or more descendants stemming from the same ancestor, which is known as a polytomy. Uncertainty in divergence times are generally depicted with error bars at the internal nodes, if indicated at all

among species and lineages is also critical for making meaningful inferences about evolution. As we will discuss in the next section, estimates of the pace of trait evolution depends on the amount of change in a trait that occurs over a unit of time.

There are today several resources to help generate a good phylogenetic tree for a set of species. A common approach is to trim the whole plant tree of life—taken from the Open Tree of Life (Hinchliff et al. 2015) or Phylomatic (Webb and Donoghue 2005), for example—to the set of species of interest. A second option is to reconstruct the phylogeny from scratch using DNA sequences and then by time-calibrating the tree using fossil information and molecular clock models. Tree reconstruction is tricky and laborious, but there are many tools that can help (e.g., Antonelli et al. 2016; Pearse and Purvis 2013). Cobbling together a phylogenetic tree by manually assembling branches is not recommended for analysis of spectra or other traits.

Finally, as seen in the previous section, phylogenies are estimates, and systematists have means of assessing uncertainty in their topology and their branch lengths, which are together referred to as phylogenetic uncertainty. For example, the divergence between two lineages may have a mean of 20 million years and a confidence interval or 95% highest posterior density of 18–22 million years. That uncertainty can (and should) be carried over to downstream statistical analyses.

7.3 The Evolution of Quantitative Traits

The study of evolution is fundamentally concerned with describing how organisms change through time and with understanding the processes driving change. Evolutionary change, however, can be thought about at different phylogenetic and temporal scales. Because we are interested in understanding spectra in light of phylogenies, we will not discuss microevolutionary processes that occur at the population level such as genetic drift and natural selection. Instead, we will focus on describing macroevolution and how traits—such as leaf structure and chemical composition—change across entire lineages over long timescales (usually millions of years).

7.3.1 Macroevolutionary Models of Trait Evolution

Macroevolutionary models of trait evolution describe the long-term consequences of short timescale evolution. At any given time step, a trait value can increase or decrease due to mechanisms like selection, drift, and migration. For example, the reflectance in one spectral region may decrease due to selection for higher levels of a particular pigment, while reflectance in another spectral region may decrease due to a random change in leaf hair density. Many such changes occur over long evolutionary time in each lineage.

7.3.1.1 Brownian Motion

Most models for evolution of quantitative traits leverage the central limit theorem from statistics, which states that the sum of many random changes leads to a normal distribution. Because trait evolution at macroevolutionary scales integrates over many random changes in trait values (due to varied processes), it may be described by a normal distribution. This model of evolution is known as Brownian motion (Felsenstein 1985). The pace at which those changes accumulate is at the core of what we call the rate of evolution, and it is captured by the variance of the normal distribution (whose mean is the trait value at the root).

When lineages split, they start out with the same trait value and then diverge independently. It is easy then to see that the expected amount of trait variation between lineages depends on both the rate of evolution and on the divergence time. This leads to the expectation that trait values should be on average more similar among closely related taxa—which had little time to diverge—than among distantly related taxa. Such expectation is at the core of the concept of phylogenetic signal (see Sect. 7.3.2) and the idea that phylogenetic relatedness can be used as a proxy for functional similarity (Webb et al. 2002), particularly when integrating across a large number of traits (Cavender-Bares et al. 2009).

7.3.1.2 Ornstein–Uhlenbeck

With Brownian motion, an increase or decrease in a trait is equally likely, regardless of the current value of a trait (Fig. 7.2a). However, it could be more realistic to think of a trait as being pulled toward some optimum (or, similarly but not quite the same, away from extreme values). This force or "pull" could be due to many processes: it is often considered to be a pull toward some evolutionary optimum due to natural selection, but it could instead result from a bias in mutation toward a particular trait value, repulsion from extremes, or other factors that lead to a pattern that resembles a pull toward an optimum. The placement of the optimum, the strength of the pull, and the basic underlying rate of evolution are all parameters of this model, which is known as an Ornstein–Uhlenbeck process (Butler and King 2004). The degree of the pull toward the optimum is analogous to the strength of a rubber band linking

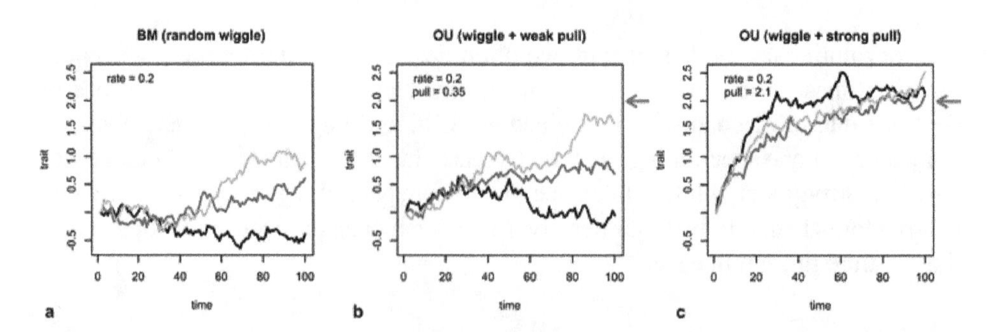

Fig. 7.2 Three independent realizations of the Brownian motion (BM) and Ornstein–Uhlenbeck (OU) processes. (**a**) In a BM model, trait values are equally likely to increase or decrease at each time step. (**b, c**) In contrast, traits in an OU model are more likely to move toward an optimum (represented by the red arrows). (**b**) When the evolutionary pull is weak, traits move slowly toward their optimum. (**c**) When the pull is strong, however, traits converge quickly toward their optimum

the evolving trait on one end and the optimum trait value on the other end. A weak rubber band will provide enough slack for the trait to wiggle around the optimum (Fig. 7.2b), whereas a strong rubber band will keep the evolving trait close to the optimum (Fig. 7.2c). The strength of the rubber band also affects how quickly the trait is pulled toward its optimum (Fig. 7.2b,c). The time a trait is expected to take to get halfway to the optimum is called the phylogenetic half-life, and this is an alternative way to think about the strength of the evolutionary pull.

Variation in traits within species, populations, and even individuals may result from responses to environmental conditions or have a genetic basis. Until recently, phylogenetic comparative methods largely ignored intraspecific variation and used species means instead. Ives et al. (2007) and Felsenstein (2008) devised methods to account for within species variation, which typically enters the model as the standard errors about the mean trait value of each species.

7.3.2 Phylogenetic Signal

Phylogenetic signal can be thought as the degree to which closely related species resemble each other. Two different metrics have been widely used to assess phylogenetic signal: Pagel's lambda (Pagel 1999) and Blomberg's *K* (Blomberg et al. 2003).

7.3.2.1 Pagel's Lambda

Pagel's lambda is a scalar for the correlation between the phylogenetic similarity matrix and the trait matrix. It has the effect of shrinking the internal branches (as opposed to the branches that lead to the tips) of a phylogeny, thereby reducing the expected species correlation due to shared evolutionary history (Fig. 7.3a–d). A lambda value of 0 indicates that trait correlations between species are independent from evolutionary history (Fig. 7.3d), whereas a lambda of 1 suggests that trait correlations are equal to the species correlation imposed by their shared evolutionary history (Fig. 7.3a), assuming a Brownian motion model of evolution.

7.3.2.2 Blomberg's *K*

Blomberg's *K* measures the degree to which trait variance lies within clades versus among clades. Brownian motion is used as an expectation. *K* values greater than 1 indicate that there is more variance among clades than expected by Brownian motion (Fig. 7.3e), while *K* values smaller than 1 imply that more variance is found within clades than expected under a Brownian motion model (Fig. 7.3f).

It is important to note that both Pagel's lambda and Blomberg's *K* are treewide metrics, meaning that they do not explicitly account for the heterogeneity in trait values among lineages. For example, an estimate of low phylogenetic signal in fruit

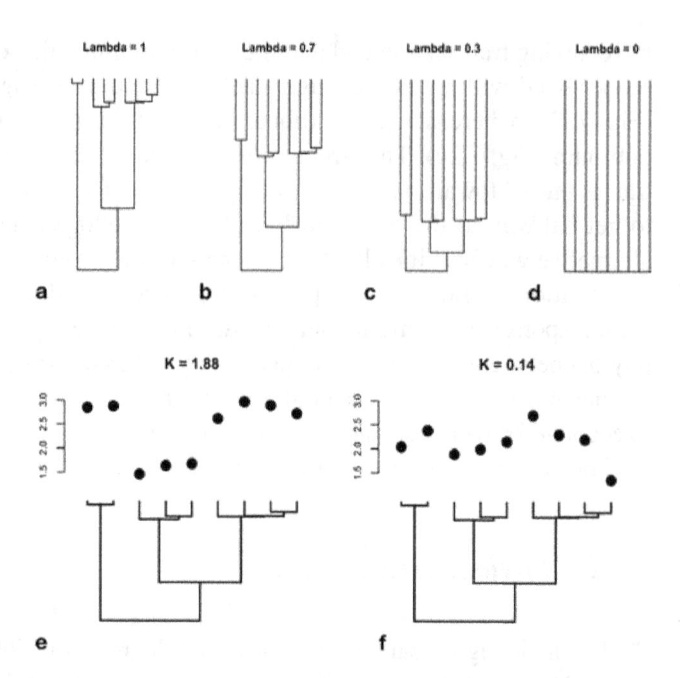

Fig. 7.3 This is how phylogenetic signal is inferred. (**a–d**) Pagel's lambda is equivalent to scaling the internal branches of the phylogeny, which reduces the expected covariance between species due to evolutionary history. (**e, f**) Blomberg's *K* measures phylogenetic signal by estimating the degree of variation between and within clades. (**e**) *K* value is high when most trait variation is found between clades instead of within them. (**f**) *K* values are low when trait variation is mostly within clades

shape across all flowering plants does not imply the lack of phylogenetic signal in fruit shape within the oaks. Therefore, assessments of phylogenetic signal should be seen as indicators that are contingent on the scale of analysis and the particular species sampled instead of as general, hard truths.

It is also important to recognize that every calculation of the phylogenetic signal of a trait involves fitting an evolutionary model that comes with a series of assumptions. For example, most procedures to estimate phylogenetic signal using Blomberg's *K* are based on a single-rate Brownian motion model; using and reporting the measure of phylogenetic signal implicitly requires accepting the Brownian assumptions.

7.4 Evolution and Spectra

Chemical and structural leaf attributes that underlie plant spectra evolve through time. Because leaf spectra integrate over these evolved leaf attributes, they can carry information about phylogenetic relationships and leaf evolution. Given this, how would one go about analyzing spectra in a phylogenetic context?

One approach is to subject the spectra directly to an evolutionary analysis, essentially taking reflectance values at different bands across the spectrum to be a set of "traits." For example, McManus et al. (2016) estimated Pagel's lambda on spectra from Amazonian plants, assuming each band to be an independent trait. Cavender-Bares et al. (2016) used principal component analysis (PCA) to reduce the dimensionality of the spectral data before estimating phylogenetic signal on the resulting principal component axes using Blomberg's *K*.

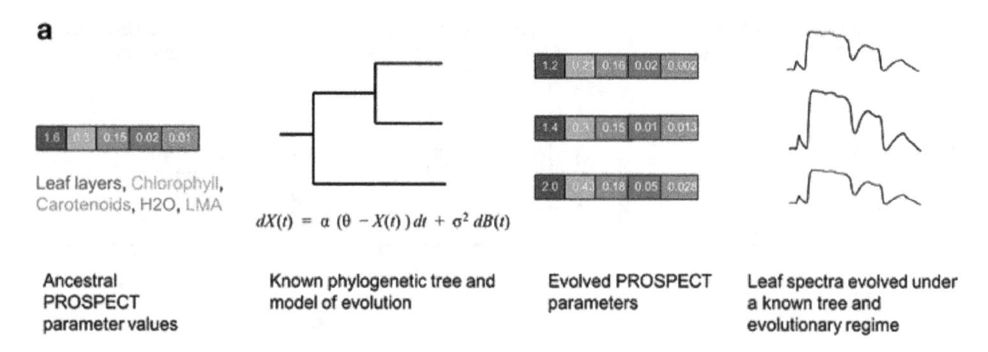

Fig. 7.4 Integration of trait evolution and leaf spectral models enables estimation of evolutionary parameters from spectra and simulation of leaf spectra along a phylogeny. Ancestral leaf attributes evolve along a phylogenetic tree under a given evolutionary regime, generating the current leaf attributes that underlie spectra. From the evolved leaf attributes, radiative transfer models (RTMs) such as PROSPECT estimate spectra that carry the signature of the phylogeny

Fitting evolutionary models directly to spectra can be useful for identifying promising associations between phylogenetic history and plant spectral signatures. However, this approach is largely devoid of mechanism and does not allow us to verify that our inferences are biologically meaningful.

Another approach to integrate phylogenies and leaf spectra is to explicitly model the evolution of structural and chemical traits that underlie the spectrum. This approach matches more closely the reality of biology by acknowledging that any signal of evolution found in the spectra is an emerging property of the evolutionary dynamics of leaf traits (see Sect. 7.5.3). This idea can be implemented by coupling the models of trait evolution described in the previous section with leaf radiative transfer models (Fig. 7.4) that predict spectral profiles from a small set of leaf attributes (see Martin, Chap. 5; Ustin and Jacquemoud, Chap. 14).

This framework can be used in several ways. For example, we can simulate what leaf spectra would look like given a certain evolutionary model and phylogenetic tree (Sect. 7.4.1). Alternatively, given a phylogeny and a spectral data set, we can infer what ancestral spectra or ancestral traits were like if we assume a certain model of evolution. Finally, given a spectrum from an unknown plant, we could estimate how that plant is related to other plants (Sect. 7.4.3).

7.4.1 Simulating Leaf Spectra Under Different Evolutionary Regimes

A model that describes the evolution of leaf spectra mediated by the evolution of leaf traits enables us to simulate spectral data in a phylogenetically explicit way. This allows us to forecast how different evolutionary scenarios would affect the shape and diversity of spectral profiles we observe. For example, Fig. 7.5 shows how the different scenarios for the evolution of leaf structure—the number of layers parameter (N)

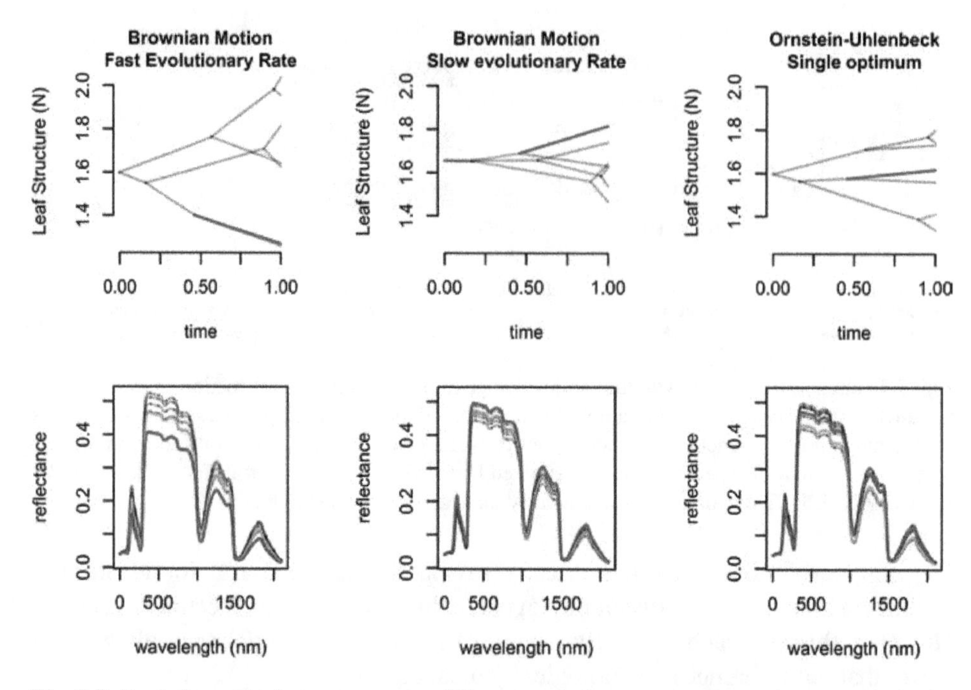

Fig. 7.5 Evolution of leaf structure under different evolutionary scenarios and consequences for leaf spectra. Top row depicts evolution according to an unbounded Brownian motion model at two different rates and according to an Ornstein–Uhlenbeck process. The bottom row shows spectra estimated with the PROSPECT5 model, where all leaf attributes evolved under the same model except for leaf structure, which evolved under the three scenarios outlined above

in PROSPECT5—result in different amounts of trait variability. A fast Brownian rate (top left, Fig. 7.5) results in higher trait variation than a slow Brownian rate (top center, Fig. 7.5). Evolution under an Ornstein–Uhlenbeck model also results in less variation than the fast Brownian model even though their rates of evolution are the same (top right, Fig. 7.5). The trait values shaped by evolution have a noticeable effect on the spectral profiles of those lineages (bottom panels, Fig. 7.5).

7.4.2 Making Evolutionary Inferences from Leaf Spectra

Integrating spectra and phylogenies raises the exciting prospect of leveraging spectra to estimate aspects of the evolutionary process and test hypotheses.

Some questions may be about evolutionary patterns in the spectra themselves. Those include investigations about phylogenetic signal or rates of evolution across the spectrum. For example, Cavender-Bares et al. (2016) and McManus et al. (2016) investigated how much phylogenetic signal is present in leaf spectra. Meireles et al. (in review) estimated how rates of evolution varied across the leaf spectrum of seed plants. Now, because we are interested in biology, evolutionary inference made at the spectral level will often need to be interpreted a posteriori.

Interpreting results correctly may pose some challenges, however. Who guarantees that the high rates of evolution in a particular spectral band really means that a certain trait is evolving at a fast pace? A potentially better approach is to infer traits from spectra first using either statistical (e.g., partial least squares regression) or RTM inversions (e.g., PROSPECT) and then study the evolution of those traits (see Serbin and Townsend, Chap. 3).

We can test hypotheses about how evolution affects leaf spectra because we can calculate the likelihood of spectral data being generated by different models of evolution, which can be compared to each other using a goodness of fit metric such as Akaike information criterion (AIC; Burnham and Anderson 2002). We foresee numerous interesting hypotheses being tested using this type of approach, especially related to evolutionary rates and convergent evolution.

Here is a hypothetical but realistic example: We could hypothesize that plant lineages that shift from sunny to shade habitats see an increase in their leaf chlorophyll content from 20 to 60 ug/cm^2, that is, they have a new chlorophyll content optimum, and that should be reflected in their spectra (Fig. 7.6). We used the predictive approach established in the previous subsection to simulate leaf spectra under that evolutionary scenario, which highlights the disparity in reflectance in the visible spectrum between sun and understory plants. We can then fit various models of evolution to the spectra (including one- and two-rate Brownian motion as well as a one-optimum Ornstein–Uhlenbeck and a two-optimum Ornstein–Uhlenbeck, the model under which the data were simulated), calculate their AIC, and compare models using AIC weights (Burnham and Anderson 2002), as shown in Fig. 7.6. In this simulated scenario, we find that indeed the best fit comes from having two different optima in the spectra correlating with chlorophyll content. However, in real data, we might find that there is a difference, but only in bands correlating with lignin content in leaves (which could reflect different herbivore or structural pressure); that there is a difference in optimum but that understory plants are much more constrained toward their optimum than plants from sunnier habitats; or that there is a change but it happens over longer time periods than we expect.

7.4.3 *Leaf Spectra, Biodiversity Detection, and Evolution*

As other chapters discuss, one approach to assessing biodiversity from plant spectra is to use spectral indices that correlate with species richness (Gamon et al., Chap. 16). Another approach is to use classification models (Clark et al. 2005; Asner and Martin 2011; Serbin and Townsend, Chap. 3). Using an empirical example within a single lineage (the oaks, genus *Quercus*), there is enough information in the spectra of leaves to significantly differentiate populations within a single species (*Quercus oleoides*) and assign them to the correct population most of the time. Different species can be correctly classified with even greater accuracy, and the four major oak clades in the example can be identified with very high accuracy (Cavender-Bares et al. 2016).

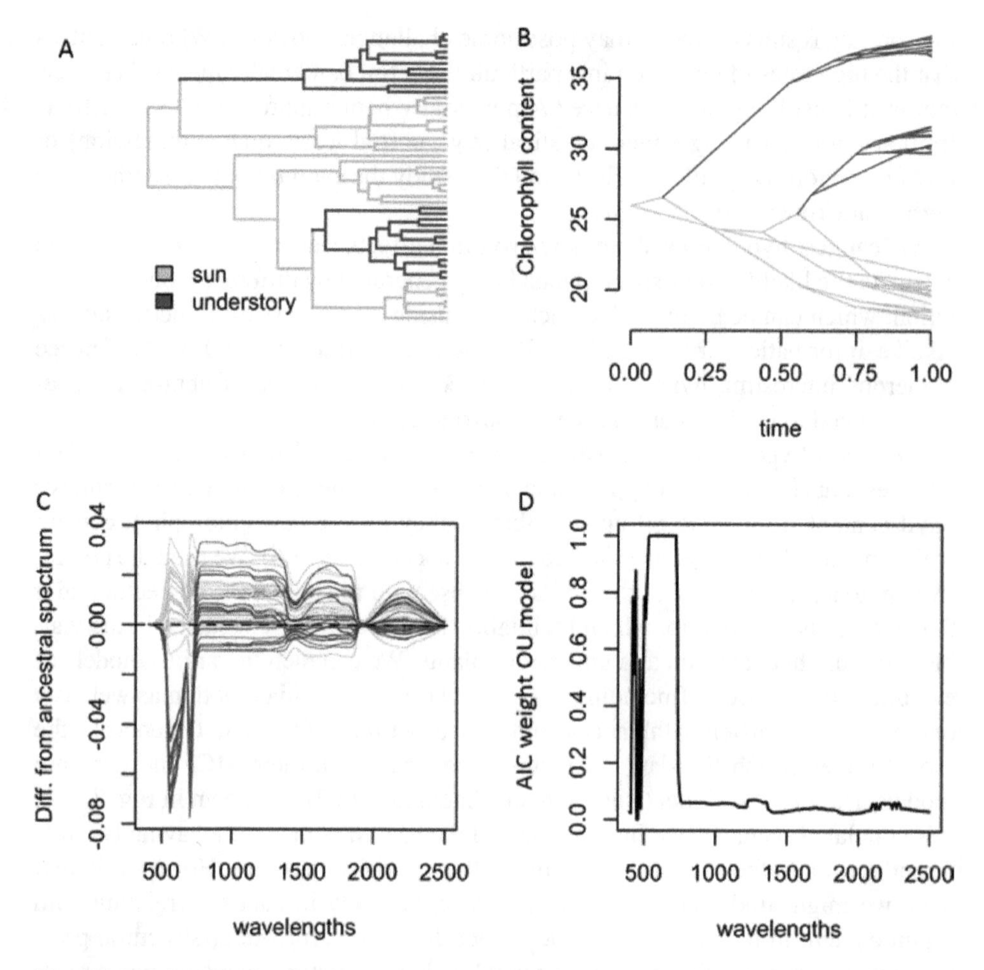

Fig. 7.6 Simulation of the evolution of chlorophyll content under a multiple optima Ornstein–Uhlenbeck model. (**a, b**) Macroevolutionary shifts from sun exposed to understory habitats (**a**) result in chlorophyll content being pulled toward different optima in different lineages (**b**). (**c**) Differences between the evolved spectra and the ancestral spectrum highlight the effect of chlorophyll evolution on the visible region of the spectrum. (**d**) We can use AIC to calculate how well various models of evolution, including the true multiple optima Ornstein–Uhlenbeck model, describe evolution across the spectrum. AIC weights suggest that the multiple optima Ornstein–Uhlenbeck model is preferred in the visible regions and nowhere else, which matches how the data were simulated

Evolutionarily-explicit diversity detection approaches could have enormous potential even when species cannot be identified. Biodiversity encompasses, among other things, which branches of the tree of life are found in an area how much evolutionary history that represents. Because plant spectral profiles can carry information about evolutionary history, they can be leveraged to assess the diversity of lineages instead of (or in addition to) the diversity in species or function. There are key conceptual advantages of taking this approach.

First, we can estimate lineage diversity at different phylogenetic scales when species-level detection performs poorly. As suggested in Fig. 7.7, leaf spectral

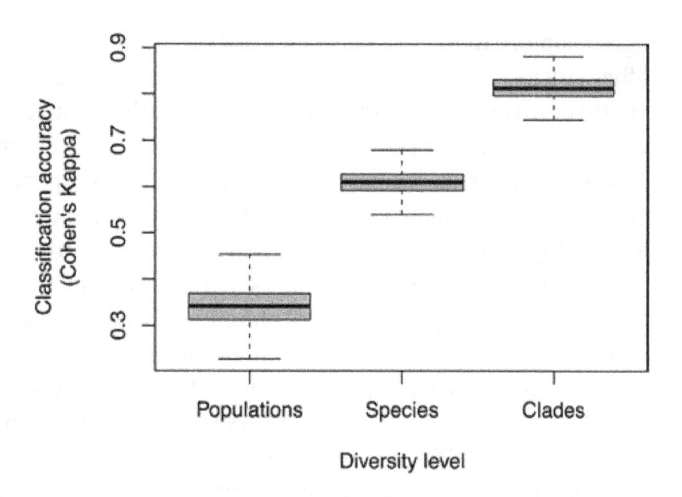

Fig. 7.7 Classification accuracy for different diversity levels of Quercus: (1) populations with Quercus oleoides, (2) 33 oak species, and (3) 4 clades of the genus Quercus. Accuracy was estimated from 300 independent PLS-DA iterations and summarized using Cohen's kappa. (Redrawn from Cavender-Bares et al. (2016))

information can more accurately identify broad oak clades (kappa 0.81) than species (kappa 0.61) and than population within a species (kappa 0.34). It is thus possible that classification models can detect broad clades more accurately than they can detect very young clades or species.

Second, we know that species definitions change over time and that many species in hyperdiverse ecosystems are still unknown to science. How can we classify species that we do not yet know about? Using evolutionary models to estimate where an unknown spectrum belongs on the tree of life reduces the need for labeling, because it can reveal the taxa that the unknown sample is related to. This provides a means to estimate the phylogenetic diversity of a site even without species identities. The models of evolution described above should allow us to calculate the probability that the unknown spectrum belongs to different parts of the tree assuming that we know the correct evolutionary model and its parameter values. Developing the framework to achieve this would require filling in many gaps and detecting species at appropriate spatial resolutions. It also would require trusting many assumptions that go into evolutionary models, because we know that as we go deeper in phylogenetic time and evolutionary history, these models become increasingly complex (see Sect. 7.5.2).

7.4.4 *Diversity Detection at Large Scales: Challenges and Ways Forward*

The fact that spectra are tightly coupled with evolutionary history helps explain why hyperspectral data can be used for accurate taxonomic classification. It also provides a basis for using remotely sensed hyperspectral data for biodiversity composition monitoring.

Using RS hyperspectral data for biodiversity detection requires moving from the leaf level to the whole canopy level (Serbin and Townsend, Chap. 3; Martin, Chap. 5; Gamon et al., Chap. 16). We expect that canopy spectra, like leaf spectra, will show tight coupling to phylogenetic information. Branching architecture, leaf angles, and other structural traits of plants that contribute to spectral signals at the canopy level are themselves evolved traits and are potentially phylogenetically conserved. To the extent that remotely sensed hyperspectral data can capture the spectral profiles of individual canopies (Gamon et al., Chap. 16), hyperspectral data should be capable of detecting and identifying species (Morsdorf et al., Chap. 4) and lineages, following the logic presented above for leaves. Such an effort would require assembling vast libraries of spectral information across the plant tree of life for a given region of interest and comparing spectra obtained through remote sensing to those libraries.

Developing accurate classification models as the number of species and clades grow can be challenging. For example, Fig. 7.8 shows randomly assembled communities with different species diversity levels, where species spectra were simulated using PROSPECT5. As the number of species grows, the ability of a PLS-DA classification model to correctly classify species decreases.

This classification problem can be simplified by circumscribing the possible species pool. This could be done by estimating the potential pool of species or clades in a region based on other biodiversity monitoring and prediction approaches, including herbarium records, plant inventories or other types of in-situ data collection, and habitat suitability predictions (Pinto-Ledezma and Cavender-Bares, Chap. 9). Combining classification methods using hyperspectral data with prediction of species pools at regional scales across the globe could allow global plant compo-

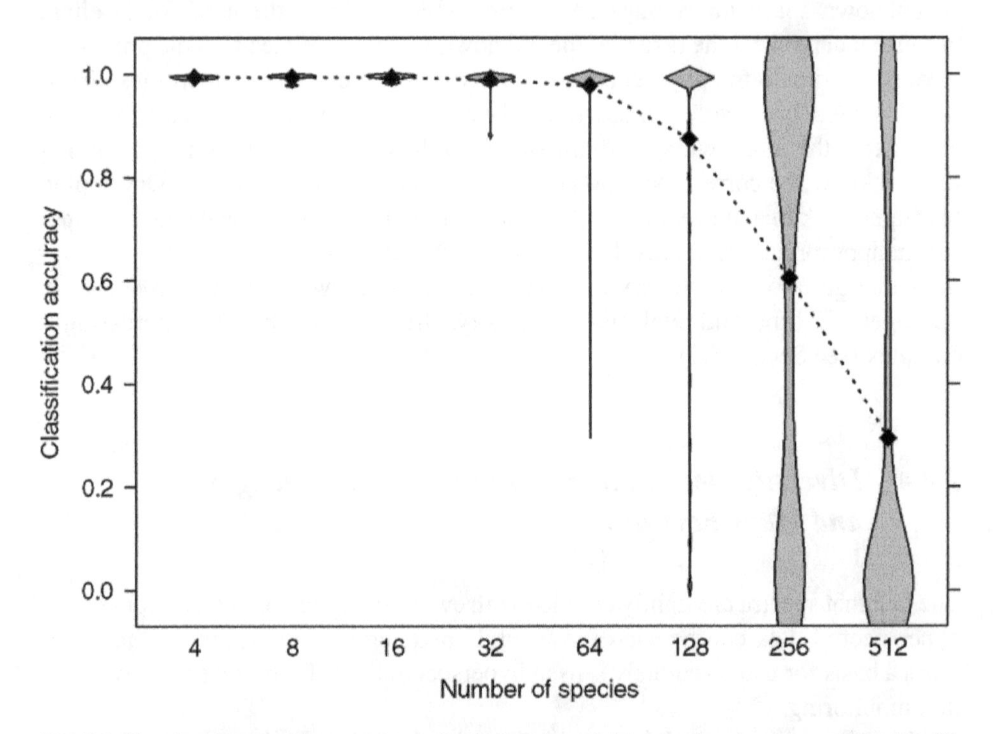

Fig. 7.8 Species classification accuracy of a PLS-DA model in simulated communities with different species richness. Spectra for each species were simulated using PROSPECT5, and communities with different diversity levels were randomly assembled

sition information to be detected when the spatial resolution of the data is sufficient to capture individuals.

7.5 Cautionary Notes

The integration of leaf spectra and phylogenies can provide breakthroughs in how we detect biodiversity, explain how spectral variation between species and lineages comes to be, and make inferences about the evolution of leaves. We should, nevertheless, be aware of the limitations inherent in making inferences about the deep past, be mindful of the sampling requirements and statistical assumptions of our analyses, and be careful to interpret our findings in a biologically meaningful way.

7.5.1 Is the Sampling Adequate for Making Evolutionary Inferences?

Inferences about the evolutionary process or that rely heavily on it—such as the degree of phylogenetic signal or the pace of evolution—are dependent on how well a lineage has been sampled. Evolutionary biologists usually target a particular lineage and strive to include in their analysis as many close relatives as possible regardless of their location. Ecologists, on the other hand, tend to focus on a specific geographic area of interest and end up sampling whatever species are there. This likely results in very severe undersampling of the total phylogenetic diversity represented by the particular species pool. For example, the 20 seed plants at a study site belong to a clade that has about 300,000 species, harbors incredible morphological and physiological diversity, and goes back 350 million years. Evolutionary analyses of undersampled will very likely yield poor estimates of the evolutionary parameters: The species in that area can tolerate a subset of all the climate conditions other seed plants can handle, for example.

In addition, ecological processes themselves can lead to bias in estimates of evolutionary parameters. For example, extremely arid conditions may act as an environmental filter that curbs colonization by species with low leaf water content, reducing the amount of variability in leaf succulence. As a consequence, estimates of the rate of evolution of leaf succulence based on species found in that hyper-arid community may be artificially low.

These caveats should be kept in mind when analyzing spectra in an evolutionary context. Finding that certain spectral regions have high phylogenetic signal in a large forest plot does not necessarily mean that those regions are truly phylogenetically conserved.

7.5.2 The More of the Tree of Life That Is Sampled, the More Complex Models Will (or Should) Be

Most of the models of evolution and phylogenetic signal statistics we saw here are actually rather simple. For example, a Brownian motion model has two parameters, the trait value at the root (mean) and the rate of evolution (variance). The single-rate Brownian motion model may reasonably describe the evolution of leaf water content in dogwoods (*Cornus*), but it would probably do a terrible job if you were analyzing all flowering plants because of the sheer heterogeneity and diversity that they possess (Felsenstein 2008; O'Meara 2012; Cornwell et al. 2014).

There is a trade-off: the most realistic model would have a different set of parameters at every time point on every branch but would have far more parameters to estimate than the data could support; a simple model of one set of parameters across all the time periods and species examined is clearly unrealistic. Most applications have used the simplest approach, but there are ways to allow for more complex models. Some of them test a priori hypotheses about heterogeneity in models of evolution: Biologists propose particular models linking sets of parameters on different parts of the tree (e.g., gymnosperms and angiosperms having different rates of evolution), and then the methods select between the possible models (Butler and King 2004; O'Meara et al. 2006). There are also methods that can automatically search across possible mappings to find the ones that fit best (Uyeda and Harmon 2014). In the case of multiple characters, such as reflectance at different wavelengths of light, there is also the question of whether different characters are evolving under the same or different models, and there are models to test that, as well (Adams and Otárola-Castillo 2013).

Early attempts to analyze spectra in an evolutionary context (Cavender-Bares et al. 2016; McManus et al. 2016; Meireles et al. in review) have used models that are maximally simple for each character (a single model applying for all taxa and times) and are nearly maximally complex between characters (each trait evolves independently of all others on the same common tree). Those approaches are computationally cheap but are at odds with our understanding of biology (i.e., models of evolution do vary among lineages) and physics (i.e., spectral bands do covary). Other ways of segregating complexity, such as models that incorporate heterogeneity among lineages and account for the covariance among spectral bands, remain potentially more fruitful ways of examining the diversity in leaf spectra.

7.5.3 Spectra Do not Evolve, Leaves Do!*

{*except when they do}

One could estimate the pace of evolution of the beaks of Darwin's finches from their photographs. But the photographs didn't evolve. Leaf spectra do capture many different aspects of the complex phenotype, and, we have seen in this chapter, each band of a spectrum can be analyzed as a trait in an evolutionary model. This does not necessarily mean that spectra themselves are traits nor that they themselves evolve. For example, there is no reason for evolution to favor lower reflectance at 660 nm. However, there may be biological reasons for natural selection to favor higher amounts of chlorophyll a in a leaf, which happens to absorb light at 660 nm. Terminology such as "evolution of spectra" or "spectral niches" may be efficient communication shortcuts but can also cause confusion. They may make it all too easy to lose sight of the biological mechanisms behind the observed phenomena.

Advances in analyzing spectral data in light of evolution will require keeping mechanisms in mind. That said, phylogenetic inference on spectra can be used as a discovery tool. Consistently finding high rates evolution in a spectral region not associated with a known function should trigger further investigation. Moreover, mechanistic thinking may end up proving us wrong and show that spectra in fact evolve (at least some regions). For example, increased leaf reflectance that prevents leaf overheating could be favored by evolution. In such a situation, high reflectance would result from "real" traits—such as bright hairs, cuticles, and waxes—but one can argue that there is biological meaning in the evolution of reflectance itself in this case.

7.5.4 Ignore Phylogeny at Your Peril

Phylogeny adds complexity to an analysis but has benefits in new insights (estimating ancestral leaf spectra, helping to go from observations to traits, and more). However, it can be tempting to analyze data on multiple species without accounting for shared evolutionary history. The problem with methods that ignore the phylogeny, such as partial least squares regression, is that they assume that species are independent data points. They are not! There is thus the risk of "overcounting" some parts of the tree of life: for example, if one wants to develop a model for all plants, and one has five oak species, a ginkgo, a pine, and a magnolia, the final model will essentially be an oak model with some deviations. However, the five oaks have shared much of their evolutionary history and so do not represent five independent instances of evolution. Phylogenies can be included into such analyses, and their importance appropriately scaled (in some cases, they will not affect results, but this is only knowable once the tree is used), and make results far more robust.

7.6 Moving Forward

The integration of leaf spectra and phylogenies using evolutionary models is still in its infancy. Phylogenetic models have the potential to unlock what drives evolution of the traits leading to different spectra. Spectra may have the potential, combined with phylogenies, to help identify species from afar, and even contain phylogenetic information themselves.

Acknowledgments We are grateful to Shan Kothari, Laura Williams, and Jesús Pinto-Ledezma for their feedback on an early version of this manuscript. We are also grateful to the National Institute for Mathematical and Biological Synthesis (NIMBioS) for funding the "Remotely Sensing Biodiversity" working group.

References

Adams DC, Otárola-Castillo E (2013) Geomorph: an R package for the collection and analysis of geometric morphometric shape data. Methods Ecol Evol 4(4):393–399

Antonelli A, Hettling H, Condamine FL, Vos K, Henrik Nilsson R, Sanderson MJ, Sauquet H et al (2016) Toward a self-updating platform for estimating rates of speciation and migration, ages, and relationships of taxa. Syst Biol 66(2):152–166. Oxford University Press: syw066

Asner G, Martin R (2011) Canopy phylogenetic, chemical and spectral assembly in a lowland Amazonian forest. New Phytol 189:999–1012

Blomberg SP, Garland T, Ives AR (2003) Testing for phylogenetic signal in comparative data: behavioral traits are more labile. Evolution 57(4):717–745

Burnham KP, Anderson DR (2002) Model selection and multimodel inference: a practical information-theoretic. Springer, New York, NY

Butler MA, King AA (2004) Phylogenetic comparative analysis: a modeling approach for adaptive evolution. Am Nat 164(6):683–695. The University of Chicago Press

Cavender-Bares J, Kozak KH, Fine PV, Kembel SW (2009) The merging of community ecology and phylogenetic biology. Ecol Lett 12:693–715

Cavender-Bares J, Meireles JE, Couture JJ, Kaproth MA, Kingdon CC, Singh A, Serbin SP et al (2016) Associations of leaf spectra with genetic and phylogenetic variation in oaks: prospects for remote detection of biodiversity. Remote Sens 8(3):221

Clark M, Roberts D, Clark D (2005) Hyperspectral discrimination of tropical rain 10 forest tree species at leaf to crown scales. Remote Sens Environ 96:375–398

Cornwell WK, Westoby M, Falster DS, FitzJohn RG, O'Meara BC, Pennell MW, McGlinn DJ et al (2014) Functional distinctiveness of major plant lineages. Edited by Amy Austin. J Ecol 102(2):345–356

Felsenstein J (1985) Phylogenies and the comparative method. Am Nat 125(1):1–15

Felsenstein J (2008) Comparative methods with sampling error and within-species variation: contrasts revisited and revised. Am Nat 171(6):713–725

Hinchliff CE, Smith SA, Allman JF, Gordon Burleigh J, Chaudhary R, Coghill LM, Crandall KA et al (2015) Synthesis of phylogeny and taxonomy into a comprehensive tree of life. Proc Natl Acad Sci U S A 112(41):12764–12769

Ives AR, Midford PE, Garland T (2007) Within-species variation and measurement error in phylogenetic comparative methods. Syst Biol 56(2):252–270

McManus K, Asner GP, Martin RE, Dexter KG, John Kress W, Field C (2016) Phylogenetic structure of foliar spectral traits in tropical forest canopies. Remote Sens 8(3):196

O'Meara BC (2012) Evolutionary inferences from phylogenies: a review of methods. Ann Rev Ecol Evol 43:267–285

O'Meara BC, Ané C, Sanderson MJ, Wainwright PC (2006) Testing for different rates of continuous trait evolution using likelihood. Evolution 60(5):922–933

Pagel M (1999) Inferring the historical patterns of biological evolution. Nature 401(6756):877–884

Pearse WD, Purvis A (2013) PhyloGenerator: an automated phylogeny generation tool for ecologists. Edited by Emmanuel Paradis. Methods Ecol Evol 4(7):692–698

Uyeda JC, Harmon LJ (2014) A novel Bayesian method for inferring and interpreting the dynamics of adaptive landscapes from phylogenetic comparative data. Syst Biol 63(6):902–918

Webb CO, Ackerly DD, McPeek MA, Donoghue MJ (2002) Phylogenies and Community Ecology. Annu Rev Ecol Syst 33:475–505

Webb CO, Donoghue MJ (2005) Phylomatic: tree assembly for applied phylogenetics. Mol Ecol Notes 5(1):181–183

8

Remote Sensing of Belowground Processes

Michael Madritch, Jeannine Cavender-Bares, Sarah E. Hobbie, and Philip A. Townsend

8.1 Framework

Remote sensing (RS) of belowground processes via aboveground ecosystem properties and plant foliar traits depends upon (1) the ability to quantify ecosystem productivity and relevant plant attributes—including plant chemical composition and diversity—and (2) tight linkages between above- and belowground systems. These linkages can occur through the effects of aboveground inputs into belowground systems and/or through relationships between above- and belowground attributes and, in turn, between belowground relationships between plant roots and microbial communities and processes (i.e., fine-root turnover, mycorrhizal associations). The increasing ability of remotely sensed information to accurately measure productivity, ecologically important plant traits (Serbin and Townsend, Chap. 3, this volume; Wang et al. 2019), and plant taxonomic, functional, and phylogenetic diversity (Wang et al. 2019; Schweiger et al. 2018; Gholizadeh et al. 2019) creates new opportunities to observe terrestrial ecosystems. While the focus of RS tools is generally on aboveground vegetation characteristics, the tight linkage between above- and belowground systems through productivity and foliar chemistry means that many belowground processes can be inferred from remotely sensed information. Here, we focus on how the productivity and composition of foliar traits in plant communities influence belowground processes such as decomposition and nutrient cycling. We specifically consider foliar traits that are increasingly measurable via

M. Madritch (✉)
Department of Biology, Appalachian State University, Boone, NC, USA
e-mail: madritchmd@appstate.edu

J. Cavender-Bares · S. E. Hobbie
Department of Ecology, Evolution and Behavior, University of Minnesota, Saint Paul, MN, USA

P. A. Townsend
Department of Forest and Wildlife Ecology, University of Wisconsin, Madison, WI, USA

airborne RS. Using two case studies, one in a clonal aspen (*Populus tremuloides*) forest system and one in a manipulated grassland biodiversity experiment, we demonstrate that plant foliar traits and vegetation cover, as measured via plant spectra (Wang et al. 2019), can provide critical information predictive of belowground processes.

8.2 How Are Belowground Processes and Microbial Communities Influenced by Aboveground Properties?

Belowground processes—including decomposition and nutrient cycling, which are mediated by microbial biomass, composition, and diversity—are heavily influenced by both the amount and chemistry of aboveground inputs. Quantifying the amount and quality of foliar components is a major aspect of trait-based ecology, which seeks to use functional traits, rather than taxonomic classification, to determine organisms' contributions to communities and ecosystems. Trait-based ecology has inherent strengths, including the ability to consider biological variation across both phylogenetic and spatial scales (Funk et al. 2017). While there is a range of accepted trait-based approaches in plant sciences (Funk et al. 2017), the emergence of the leaf economic spectrum (Wright et al. 2004) and the whole plant economic spectrum (Reich 2014) has clearly demonstrated that plant traits are important to ecosystem processes across multiple biological and spatial scales. Further, employing a trait-based approach to explore the relationships among plant function, biodiversity, and belowground processes allows us to take advantage of recent advances in RS to accurately measure plant traits across large spatial scales.

A. *Decomposition and Nutrient Cycling*—The productivity, composition, and diversity of aboveground communities influence belowground processes, in part through decomposition of leaf litter (Gartner and Cardon 2004; Hättenschwiler et al. 2005), root litter (Bardgett et al. 2014; Laliberté 2017), and root exudates (Hobbie 2015; Cline et al. 2018) and also through effects on soil organic matter (SOM) properties (Mueller et al. 2015) and soil physical structure (Gould et al. 2016). Several seminal reviews outlining the importance of biodiversity to ecosystem function (BEF) have focused specifically on the afterlife effects of litter diversity on decomposition (Hättenschwiler et al. 2005; Gessner et al. 2010).
B. *Microbial Community Composition*—Variation in the quantity and quality of organic inputs into belowground systems drives variation in belowground microbial communities and functioning (de Vries et al. 2012). Differences in aboveground communities are mirrored by those in belowground communities (Wardle et al. 2004; De Deyn and van der Putten 2005; Kardol and Wardle 2010). Across multiple spatial and taxonomic scales, variation in belowground microbial communities is driven by variation in plant traits associated with the leaf economic spectrum (de Vries et al. 2012). In general, fungi dominate decomposition of complex, low-quality substrates, while bacteria favor labile, high-quality substrates (Fig. 8.1, Bossuyt et al. 2001; Lauber et al. 2008).

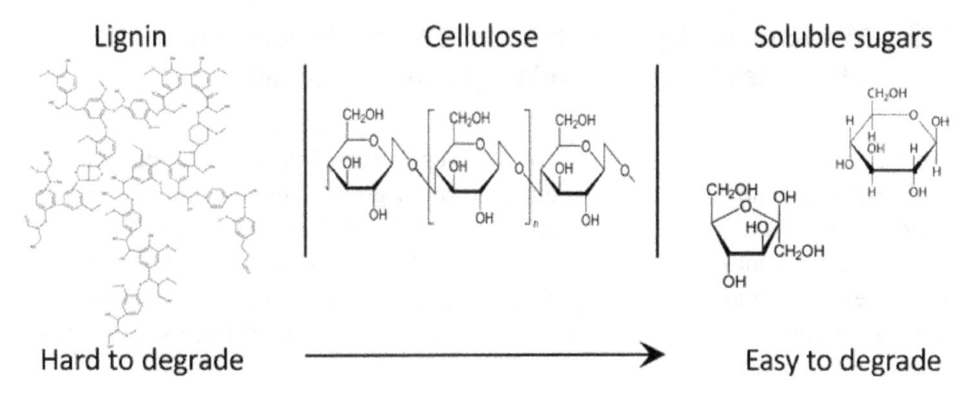

Fig. 8.1 Complex, recalcitrant compounds are typically degraded by fungi, while soluble, labile substrates are catabolized by bacteria

Microorganisms release extracellular enzymes, which degrade organic molecules outside of their cells, and likely differ among groups of microorganisms (Schneider et al. 2012). As a consequence, microbial composition and diversity are expected to influence decomposition and nutrient cycling. Most litter decomposition appears to be driven by fungal members, with *Ascomycota* dominating early degradation of cellulose and hemicellulose, followed by colonization by lignin-degrading *Basidiomycota* (Osono 2007; Schneider et al. 2012). Although lignin decomposition is dominated by fungal groups, some bacteria also degrade lignin (Kirby 2006; López-Mondéjar et al. 2016). Bacteria not directly involved with litter decomposition target the low molecular weight carbohydrates provided by fungal-derived extracellular enzymes (Allison 2005). The degradation of aromatic polyphenolics is largely limited to fungal member of the *Basidiomycota* phylum (Floudas et al. 2012). The wide structural variation among tannins (see section on carbon (polyphenols) results in a wide range of effects on specific microbial members (Kraus et al. 2003).

A challenge in predicting belowground processes such as decomposition and nutrient cycling from the diversity and quality of leaf litter inputs is that such an approach must also consider the diversity and function of belowground microbial communities. Belowground mycorrhizal communities can increase net primary production (NPP) and drive variation in plant communities (Wardle et al. 2004). Given the influence of plant traits on belowground processes, biodiversity may drive variation in decomposition through top-down (microbially driven) rather than bottom-up (substrate driven) forces (Srivasta et al. 2009). Several reviews have addressed the importance of belowground community diversity to ecosystem processes (e.g., Hättenschwiler et al. 2005; Gessner et al. 2010; Phillips et al. 2013; Bardgett and van der Putten 2014). Belowground diversity can influence aboveground factors such as NPP (Wardle et al. 2004; Eisenhauer et al. 2018) that then have important feedbacks to belowground processes. Decomposition is driven by a combination of both the microbial community and the quality and quantity of litter that those communities receive (e.g., Keiser et al. 2013; García-Palacios et al. 2016).

8.3 Mechanisms by Which Aboveground Vegetation Attributes Influence Belowground Processes

Aboveground community composition and vegetation chemistry are tightly linked with belowground communities through belowground inputs and subsequent decomposition and nutrient uptake (Hobbie 1992; Wardle et al. 2004). Plant biomass, structure, and chemical composition are all important drivers of belowground processes to such an extent that plant traits may be the dominant control on litter decomposition, outweighing the influence of climate even over large spatial scales (Cornwell et al. 2008).

8.3.1 Total Aboveground Inputs

Standing aboveground biomass and NPP are among the most important attributes of vegetation that impact belowground systems (Chapin et al. 2002) and are widely measured via RS techniques with increasing accuracy (Kokaly et al. 2009; Serbin et al., this issue). Belowground respiration is tightly linked with aboveground productivity (Högberg et al. 2001), and leaf litter can provide roughly half of organic inputs into some belowground systems (Coleman and Crossley 1996). The amount of aboveground biomass can be critical to litter decomposition (Lohbeck et al. 2015) and microbial community function and diversity (Fierer et al. 2009; Cline et al. 2018), and its influence may surpass the effects of plant quality, as measured by plant chemistry and functional traits (Lohbeck et al. 2015).

Plant traits related to biomass, such as leaf area index (LAI), are also linked to belowground processes, with belowground carbon (C) turnover peaking at intermediate LAI levels (Berryman et al. 2016; others). Importantly, LAI can be measured with RS products over large spatial scales (Serbin et al. 2014; Lausch et al., Chap. 13 this volume, Morsdorf et al. Chap. 4). While there have been few explicit links of remotely sensed LAI to soil respiration (but see Huang et al. 2015), the conceptual link has been recognized for decades (Landsberg and Waring 1997). Other remotely sensed variables tightly coupled with biomass, including vegetation cover (Wang et al. 2019), also predict soil respiration (Fig. 8.4).

The effects of biomass on belowground processes have been recognized by ecologists employing RS to estimate belowground C stocks (e.g., Bellassen et al. 2011). Across large scales, aboveground biomass is generally correlated with belowground root biomass (Cairns et al. 1997). While aboveground biomass is commonly measured, the calculation of belowground biomass is less common and is often limited to estimates of shoot biomass as a simple proportion of aboveground biomass (Mokany et al. 2006). Nonetheless, the belowground estimates based on aboveground measurements can be useful for estimating above- and belowground C stores via RS products over large spatial scales (Saatchi et al. 2011). Allocation of C to belowground systems varies among systems, with annual grassland systems differing from forested biomes in their allocation patterns of NPP (Litton et al. 2007). There are also large differences in above- and belowground linkages according to site fertility. In fertile sites the majority of NPP returned to the soil as labile fecal matter, whereas in infertile systems most NPP returned as recalcitrant plant litter (Wardle et al. 2004).

8.3.2 Chemical Composition of Vegetation

Beyond variation in total organic inputs to soil, variation in plant chemical composition is critical to belowground ecosystem processes. The physiological traits that comprise the plant economic spectrum developed by Wright et al. (2004) have important afterlife affects for belowground systems (Cornwell et al. 2008; Freschet et al. 2012; see review by Bardgett 2017). Variation in litter chemical quality can produce marked, long-term effects on litter decomposition rates and nutrient cycling in underlying soils, and litter quality has long been identified as key factor in determining decomposition rates (Tenney and Waksman 1929). Litter chemistry generally mirrors canopy chemistry (Hättenschwiler et al. 2008), making canopy chemistry a viable metric to estimate litter chemistry and subsequent belowground decomposition and nutrient cycling patterns. Aside from aboveground biomass, leaf nitrogen (N) and lignin content are often the dominant plant traits that drive variation in belowground process, particularly leaf litter decomposition (Aber and Mellilo 1982; Cadisch and Giller 1997), and both of these traits are readily derived from spectroscopy at multiple scales (Wessman et al. 1988; Serbin et al. 2014; Schweiger et al. 2018; Wang et al. 2019). RS of additional leaf traits important to belowground processes, such as plant secondary chemistry, is also increasingly measured via RS techniques (Kokaly et al. 2009; Asner et al. 2014; Serbin et al., this issue).

Nitrogen Foliar N is often the most important leaf trait driving variation in decomposition across biomes (Diaz et al. 2004; Cornwell et al. 2008; Handa et al. 2014). In some biomes leaf N is the only known leaf trait associated with leaf decomposition among wide ranges of species (Jo et al. 2016). Because canopy N has a tight correlation with plant carbon capture through photosynthesis, aboveground biomass, and belowground processes such as decomposition and N cycling rates, it is among the most common canopy traits measured via RS platforms (Martin and Aber 1997; Wessmen et al. 1998; Kokaly and Clark 1999; Martin et al. 1998, 2008; Ollinger et al. 2002; Townsend et al. 2003; Kokaly et al. 2009; Vitousek et al. 2009; Ollinger et al. 2013).

Leaf N is directly linked to plant productivity because most plant N is associated with metabolically active proteins, including RuBisCo. Leaf N content is driven by a trade-off between the benefits of increased photosynthetic potential and the costs associated with acquiring N along with the increased risk of herbivory (Diaz et al. 2016). In addition, leaf N can be indicative of plant growth strategies

(Wardle et al. 2004). Most short-term decomposition studies indicate that leaf N increases leaf litter decay (Cornwell et al. 2008). However, as decomposition progresses, leaf N may negatively affect the latter stages of decomposition, possibly due to interactions with lignified substrates (Berg 2014; discussed in brief below).

Carbon quality (lignin) The second most abundant natural polymer following cellulose is lignin, a complex phenolic polymer that wraps in and out of the structural polysaccharides in cell walls (Cadisch and Giller 1997). Due to its central roles in both aboveground biomass and belowground decomposition, lignin has been targeted as an important plant trait for RS techniques (Wessman et al. 1988; Serbin et al. 2014; Serbin and Townsend, Chap. 3). While lignin is a polyphenolic compound comprised of linked phenols (Horner et al. 1988), it is considered separate from other polyphenols because lignin is a primary structural component, whereas other polyphenols are a subset of secondary metabolites not directly involved with plant growth. The structure role of lignin and its low solubility also merit distinction from other polyphenolics when considering belowground processes (Hättenschwiler and Vitousek 2000). Lignin concentrations are negatively correlated with decomposition rates (Meentemeyer 1978; Melillo et al. 1982; Horner et al. 1988). The recalcitrant nature of lignin is due, in part, to its irregular structure and low energy yield, which largely limits its degradation to white-rot fungus members of *Basidiomycota* (Chapin et al. 2002).

The interaction of N and lignin during decomposition is not straightforward because N limits the early stages of decomposition, whereas lignin limits the latter stages of decomposition (Burns et al. 2013). Newly senesced leaves are composed largely of polysaccharides of holocellulose and lignin. High N availability will stimulate holocellulose decomposition in the early stages of decomposition but will then retard lignin decomposition in later stages of decomposition leading to lignified soil organic matter (SOM), potentially due to white-rot fungi favoring low N conditions (Berg 2014). The degradation of lignin is often a rate-limiting step during the later stages of decomposition because it protects cell wall polysaccharides physically and chemically (Talbot et al. 2012). Despite the changing roles that leaf N and lignin have over the course of decomposition, litter quality metrics such as C:N and lignin: N can explain variation in decomposition, with decomposition rates increasing with N in the early stages, but decreasing with N in the later stages, and decreasing with lignin (Fanin and Bertrand 2016).

While lignin almost universally retards decomposition, there is a large amount of variation within lignin compounds based on the proportion of specific monomers that varies across major plant groups (Thevenot et al. 2010). Angiosperm lignin tends to degrade more quickly than does gymnosperm lignin due to the specific identities of constituting moieties of lignin in each species (Higuchi 2006). The compact nature of gymnosperm lignin subunits is thought to protect them from enzymatic degradation (Hatakka and Hammel 2010). Functional measurements of lignin are often made via either acid digestion or thioglycolic acid methods that can then be used to calibrate spectroscopic methods (Brinkmann et al. 2002; Schweiger et al. 2018).

Carbon quality (polyphenols) In some ecosystems non-lignin carbon compounds (e.g., phenolics) explain more variation in decomposition than does either N or lignin (Hättenschwiler et al. 2011). Phenolics are the most widely distributed class of secondary plant metabolites and interact strongly with several aspects of nutrient cycling (Hättenschwiler and Vitousek 2000). Simple phenolics can prime (Fontaine et al. 2007), while large complex polyphenolics can retard (Coq et al. 2010) decomposition. Carbon quality—including the chemical composition of polyphenolics—can be more important to litter decomposition than is litter nutrient concentration (Hättenschwiler and Jørgensen 2010). Plant polyphenolics can be accurately measured via near-infrared spectroscopy (NIRS; Rupert-Nason et al. 2013), and by airborne imaging spectroscopy (Kokaly et al. 2009; Asner et al. 2014; Madritch et al. 2014; Serbin and Townsend, Chap. 3).

Though typically considered primarily for their aboveground defensive properties, phenolics in plant residues (leaf litter and roots) can have large influences on decomposition. Simple phenolics can increase soil respiration by providing a simple carbon source for microorganisms (Horner et al. 1988; Schimel et al. 1996; Madritch et al. 2007). Tannins are defined, in part, by their ability to bind to proteins (Bate-Smith 1975). The attributes of nonstructural polyphenolics that make them effective plant pathogen defenses also affect nonpathogenic fungi and microbes once litter enters the detrital food web; tannins do not discriminate between enzymes of plant pathogenic fungi or decomposing fungi. If tannins bind covalently with proteins to form polyphenolic-protein complexes, they become highly recalcitrant, and only basidiomycetes with polyphenol oxidase and earthworms can take advantage of these complex N sources (Hättenschwiler and Vitousek 2000). The inhibitory role of tannins on soil enzymes varies with specific tannin structure, which varies widely among species (Triebwasser et al. 2012). Tannins also have a limited ability to bind with carbohydrates and cellulose to form recalcitrant complexes (Horner et al. 1988; Kraus et al. 2003). The ability of polyphenolics to complex with proteins and other biochemicals is the primary method by which they influence soil respiration, litter decomposition, and soil N fluxes.

In addition to their influence on decomposition, nonstructural polyphenolics (which do not include lignin) influence N cycling by binding to and promoting retention of N-rich compounds including ammonium, amino acids, and proteins (Hättenschwiler and Vitousek 2000). Ayres (1997) suggested that condensed tannins may be more important to N cycling than to herbivore defense, since condensed tannins frequently have no anti-herbivory activity. Hättenschwiler et al. (2011) also proposed that polyphenolics, and tannins in particular, may be an important N conservation and recovery strategy for some species. This appears to be the case in *Populus tremuloides* systems, where high-tannin genotypes recovered more N than did low-tannin genotypes, especially when under severe herbivory (Madritch and Lindroth 2015). The high reactivity and branching structure of reactive hydroxyl sites also allow polyphenolics to complex with clay particles in soil and thereby influence several micronutrients in addition to N (Schnitzer et al. 1984).

Variation in plant phenolics is driven by several interacting factors. In general, polyphenolic concentrations in foliage are highest during the summer months (Feeny 1970). Summer coincides with both the onset of herbivory and the highest levels of photosynthetic activity. Herbivory-induced polyphenolic production is a well-documented aspect of plant-insect interactions (Herms and Mattson 1992; Baldwin 1994). The composition and quantity of phenolics vary among taxa at small and large phylogenetic scales. At large phylogenetic scales, condensed tannins are common in woody plants but almost absent in herbaceous species (Haslam 1989). At narrow phylogenetic scales, the concentration of polyphenolics is also under genetic control, and often there is considerable variation within the same species that can have important influences on belowground processes including litter decomposition and nutrient cycling (Lindroth et al. 2002; Schweitzer et al. 2005; Madritch et al. 2006, 2007).

8.3.3 Plant Diversity

Plant diversity, which can be accurately remotely sensed at some spatial scales (Wang et al. 2019; Gholizadeh et al. 2019), can influence belowground processes through its effects on productivity as well as on chemical diversity (Meier and Bowman 2008). Belowground diversity may be intrinsically linked to aboveground diversity because high plant diversity may provide a high diversity of litter quality and quantity to belowground systems that subsequently result in a high diversity of decomposers (Hooper et al. 2000). The specific relationship between aboveground plant communities and belowground microbial communities is context, system, and scale dependent (De Deyn and van der Putten 2005; Wu et al. 2011; Cline et al. 2018). For instance, Chen et al. (2018) found that plant diversity is coupled with soil beta diversity but not soil alpha diversity in grassland systems. Nonetheless, if aboveground diversity is indeed linked to belowground diversity, then aboveground estimates of plant diversity and plant traits could provide robust estimates of belowground processes.

Early work that focused on the influence of aboveground species diversity on litter decomposition yielded idiosyncratic results (Gartner and Cardon 2004; Hättenschwiler et al. 2005), with some studies reporting no effect of plant species diversity (e.g., Naeem et al. 1999; Wardle et al. 1999; Wardle et al. 2000; Knops et al. 2001), some reporting unpredictable results (Wardle and Nicholson 1996), and some reporting positive effects of plant species diversity on litter decomposition (Hector et al. 2000). Similar to aboveground processes, BEF studies that link aboveground diversity with belowground processes initially focused on aboveground species diversity (Scherer-Lorenzen et al. 2007; Ball et al. 2008; Gessner et al. 2010). The idiosyncratic relationship between species diversity and belowground processes led others to identify aboveground functional diversity and composition as more important to belowground processes than species diversity (Dawud et al. 2017).

Foliar chemistry is relevant to biodiversity and ecosystem functioning studies because plant chemistry varies widely among and within species and can influence belowground microbial communities and biogeochemical cycles (Cadisch and Giller 1997; Hättenschwiler and Vitousek 2000). It follows that variation in foliar traits important to decomposition (e.g., tannin concentration) will affect belowground microbial communities and the basic biogeochemical cycles that sustain forested ecosystems. Some studies have supported a chemical diversity approach toward elucidating the belowground effects of aboveground diversity (Hoorens et al. 2003; Smith and Bradford 2003). Epps et al. (2007) demonstrated that accounting for chemical variation was more informative regarding decomposition than was species diversity. While the usefulness of trait-based dissimilarity approaches remains somewhat equivocal (Frainer et al. 2015), there is increasing support for such trait-based approaches in explaining variation in leaf litter decomposition (Fortunel et al. 2009; Finerty et al. 2016; Jewell et al. 2017; Fujii et al. 2017). Handa et al. (2014) found that variation in leaf litter decomposition across widely different biomes was largely driven by commonly measured leaf traits such as N, lignin, and tannin content. At large scales, species traits rather than species diversity per se appears to at least partially drive variation in decomposition and belowground nutrient cycling.

In experimental systems, plant communities with high biodiversity result in high above- and belowground productivity (Tilman et al. 2001). The additional biomass that an ecosystem produces in diverse assemblages over what is expected from monocultures is called "overyielding" and has been documented in both grassland and forest experiments (Grossman et al. 2018; Weisser et al. 2017). The additional productivity results from several mechanisms acting simultaneously in more diverse communities, such as reduced pathogen attack, reduced seed limitation, and increased trait differences leading to "complementarity" in resource uptake (Weisser et al. 2017). Complementarity in resource use, particularly light harvesting, results in more efficient use of limiting resources and greater productivity (Williams et al. 2017). Similar patterns of greater productivity with higher diversity are observed in forest plots globally (Liang et al. 2016) although such patterns are scale dependent, and do not necessarily hold at large spatial extents (Chisholm et al. 2013). In naturally assembled grasslands, the relationship may not necessarily hold consistently (Adler et al. 2011). An open question, then, is the extent to which diversity and productivity are linked at large spatial scales in ecosystems globally. This is a question that can reasonably be addressed with remotely sensed measures of biodiversity and ecosystem productivity if scaling issues are appropriately considered (Gamon et al., Chap. 16). Plant diversity influences the quality of inputs and may allow for niche partitioning among functionally different microbes and may also influence productivity, the source of inputs of organic matter available to microbes, and microbial diversity. Through these linkages, foliar diversity has the potential to influence microbial diversity and function and hence belowground processes (Cline et al. 2018). The extent to which diversity and productivity, measured aboveground, can predict belowground microbial and soil processes is a question that is ready to be tackled at a range of scales across continents.

8.4 Case Studies

8.4.1 Remote Sensing of Belowground Processes via Canopy Chemistry Measurements

Plants act as aboveground signals for belowground systems. As such, RS of plant spectra can provide information about belowground systems. Plant spectra can provide a wealth of biological information important to plant physiology and community and ecosystem processes across multiple spatial scales (Cavender-Bares et al. 2017). Some researchers have used direct spectral measurements (e.g., NIRS) for direct measurements of soil characteristics (reviewed by Stenberg et al. 2010; Bellon-Maurel and McBratney 2011; Soriano-Disla et al. 2014), and there are limited examples of remotely sensed spectroscopic measurements of soils (reviewed by Ustin et al. 2004; Cecillon et al. 2009). Here we focus on remotely sensed spectral measurements of plant communities as a surrogate for belowground processes. The optical surrogacy hypothesis (sensu Gamon 2008) argues that plant spectra can serve as a surrogate for important belowground processes.

Direct spectral measurements have been used to assess belowground processes for decades. For instance, direct NIRS of leaf litter can be used to predict decomposition rates in a variety of systems (Gillon et al. 1993; Gillon et al. 1999; Shepherd et al. 2005; Fortunel et al. 2009; Parsons et al. 2011). RS of canopy traits to predict belowground processes is becoming increasingly useful. Spectroscopic measurement of $\delta^{15}N$ is of particular interest for ecosystem processes (Serbin et al. 2014) because stable N isotopes can provide important information regarding ecosystem N cycling (Robinson 2001; Hobbie and Hobbie 2006). RS of forest disturbance (e.g., fire severity) and subsequent belowground processes is relatively common (e.g., Holden et al. 2016). Sabetta et al. (2006) used hyperspectral imaging to predict leaf litter decomposition across four forest communities. Fisher et al. (2016) were able to distinguish between arbuscular and ectomycorrhizal tree-mycorrhizal associations using spectral information gleaned from Landsat data. While the above examples focus on remotely sensed spectral information, remotely sensed forest structural information developed from lidar data can also provide information about belowground systems, as Thers et al. (2017) were able to use remotely sensed lidar data to estimate belowground fungal diversity. The growing number of examples that employ remotely sensed data to provide information about belowground systems points to the potential of plant spectra to be used as surrogates for ecosystem processes.

8.4.2 Forest Systems: Aspen Clones Example

An example of optical surrogacy in practice is illustrated by work completed in trembling aspen (*Populus tremuloides*) systems across the Western and Midwestern USA. Trembling aspen is the most widespread native tree species in North America (Mitton and Grant 1996) and is an ecologically important foundation species across its native range (Lindroth and St. Clair 2013). Aspen is facing large and rapid declines in intraspecific biodiversity because concentrated patches of aspen are currently experiencing high mortality rates in North America (Frey et al. 2004; Worrall et al. 2008). This phenomenon, commonly referred to as sudden aspen decline (SAD), leads to the death of apparently healthy aspen stands in 3–6 years (Shields and Bockheim 1981; Frey et al. 2004). These natural history traits, combined with the ecological and economic significance of the species, make trembling aspen an ideal system to employ RS techniques to estimate genetic diversity and the consequences thereof for belowground processes.

Aspen typically reproduces clonally, often creating a patchwork of clones with many ramets (Fig. 8.2). Aspen clones vary widely in canopy chemistry traits that are important to belowground processes such as litter decomposition (Madritch et al. 2006). Several studies have highlighted the importance of plant genetic diversity to ecosystem processes (Madritch and Hunter 2002, 2003; Schweitzer et al. 2005; Crutsinger et al. 2006; Madritch et al. 2006, 2007) and community composition (Wimp et al. 2004, 2005; Johnson and Agrawal 2005). These recent advances demonstrate that genetic diversity affects fundamental ecosystem processes by influencing both above- and belowground communities (Hughes et al. 2008). The natural history traits of aspen, its clonal nature, genetically mediated variation in canopy chemistry, and the concomitant wide range of variation in foliar traits make it an ideal model system for RS of biodiversity.

Madritch et al. (2014) described how remotely sensed spectroscopic data from NASA's AVIRIS platform can be used to describe aboveground genetic and chemical variation in aspen forests across subcontinental spatial scales. This work built upon past work that demonstrated the ability of imaging spectroscopy to detect both aboveground chemistry (Townsend et al. 2003) and biodiversity (Clark et al. 2005) and employed imaging spectroscopy to discriminate intraspecific, genetic variation in aboveground chemistry and diversity. Because of the tight linkages between aboveground and belowground systems and because of the large variation in secondary chemistries important to belowground processes in aspen, this project also demonstrated the ability to predict belowground process via RS of forest canopy chemistry. Figure 8.3 illustrates both the

Fig. 8.2 Aerial photo showing color differentiation of genetically distinct aspen clones. Genotypes can be detected rapidly via remote sensing techniques

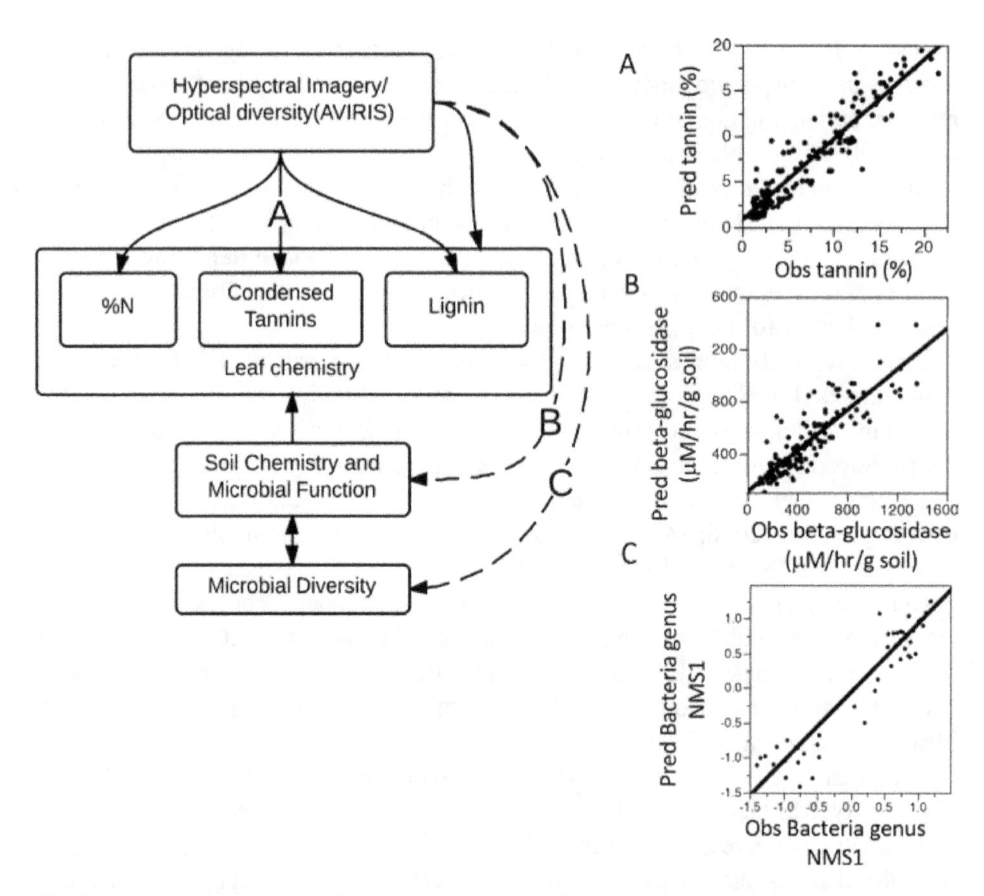

Fig. 8.3 Imaging spectroscopy links to several layers of ecological processes in aspen forests. (**a**) Partial least squares (PLS) prediction (pred) for condensed tannin concentration from AVIRIS data compared to observed (obs) tannin. (**b**) PLS prediction (pred) for soil b-glucosidase activity compared to observed (obs) b-glucosidase. (**c**) PLS prediction from AVIRIS spectra for bacterial diversity compared to observed bacterial diversity, where bacterial diversity is the first axis of an NMDS ordination of amplicon sequencing of rDNA (525f and 806r primers). (Tannin and soil enzyme data are from Madritch et al. (2014))

direct linkages between RS and canopy chemistry (A) and the subsequent indirect linkages to belowground function (B) and the microbial community (C). The indirect linkages represent the optical surrogacy hypothesis. Belowground attributes are not measured directly via RS, but rather RS of the forest canopy was able to provide detailed information regarding belowground process.

8.4.3 Experiment Prairie Grassland System: Cedar Creek Example

Vegetation differences between prairie and forested ecosystems have important consequences for above- and belowground linkages. Detrital inputs in forests are dominated by leaf litter, whereas they are dominated by root exudates and turnover in prairie systems that are frequently burned. We employed a parallel application of spectroscopic imagery to assess above- and belowground diversity and functioning at the grassland biodiversity experiment located at Cedar Creek Ecosystem Science Reserve (Tilman et al. 2001). Rather than a monospecific forest canopy, the grassland experiment consisted of replicated diversity treatments ranging from 1 to 16 perennial grassland species in 9 m × 9 m plots. This work had more technical challenges associated with it compared to the aspen forest project due to the inherent complexity of a mixed species system and the small spatial scale of the experimental plots.

The relationship between plant diversity and aboveground biomass in the Cedar Creek BioDIV experiment is well documented (Tilman et al. 2001, 2006). Schweiger et al. (2018) further demonstrate that both plant diversity and function are measurable via remotely sensed spectra within the experiment and that spectral diversity predicted productivity. Wang et al. (2019) used AVIRIS imagery to map functional traits across the experiment. Remotely sensed productivity and functional trait composition can thus be tested for linkages with belowground processes. In this system, the quantity of inputs had a large impact on fungal composition and diversity (Cline et al. 2018). Productivity, measured as annual aboveground biomass, given that it is annually burned, can be accurately detected as remotely sensed vegetation cover (Fig. 8.4a; Wang et al. 2019, following the method of Serbin et al. 2015). Remotely sensed vegetation cover, in turn, predicted fungal diversity, measured as operational taxonomic unit (OTU) richness (Fig. 8.4b), and cumulative soil respiration (Fig. 8.4c). In addition to the total organic matter inputs to the soil, chemical composition also influenced belowground microbial communities. For example, remotely sensed %N (Wang et al. 2019) was positively correlated with soil microbial biomass (Cavender-Bares et al., unpublished manuscript).

8.4.4 Challenges and Future Directions

Employing plant spectra to predict belowground processes has both caveats and advantages over traditional belowground sampling. One important caveat is that any prediction of belowground processes requires a solid understanding of the linkages between above- and belowground processes in any given system. Examples in the literature that link remotely sensed attributes of aboveground systems with belowground systems remain scarce, in part, because of the historic separation of the two disciplines. It is unclear how well remotely sensed plant attributes will predict microbial and soil processes across ecological systems. In the above forest example, aspen forests were generally uniform in canopy coverage. It was also a single-species system where leaf structure remained consistent across the study area, despite the large spatial sampling scheme. Consequently, most of the variation in aspen spectral signal was likely due to variation in canopy chemistry and biomass rather than leaf structure. Lastly, in this temperate forest system, leaf litter accounts for a large fraction of inputs into belowground systems, compared to systems such as Cedar Creek that are burned frequently and where fine-root turnover dominates belowground inputs.

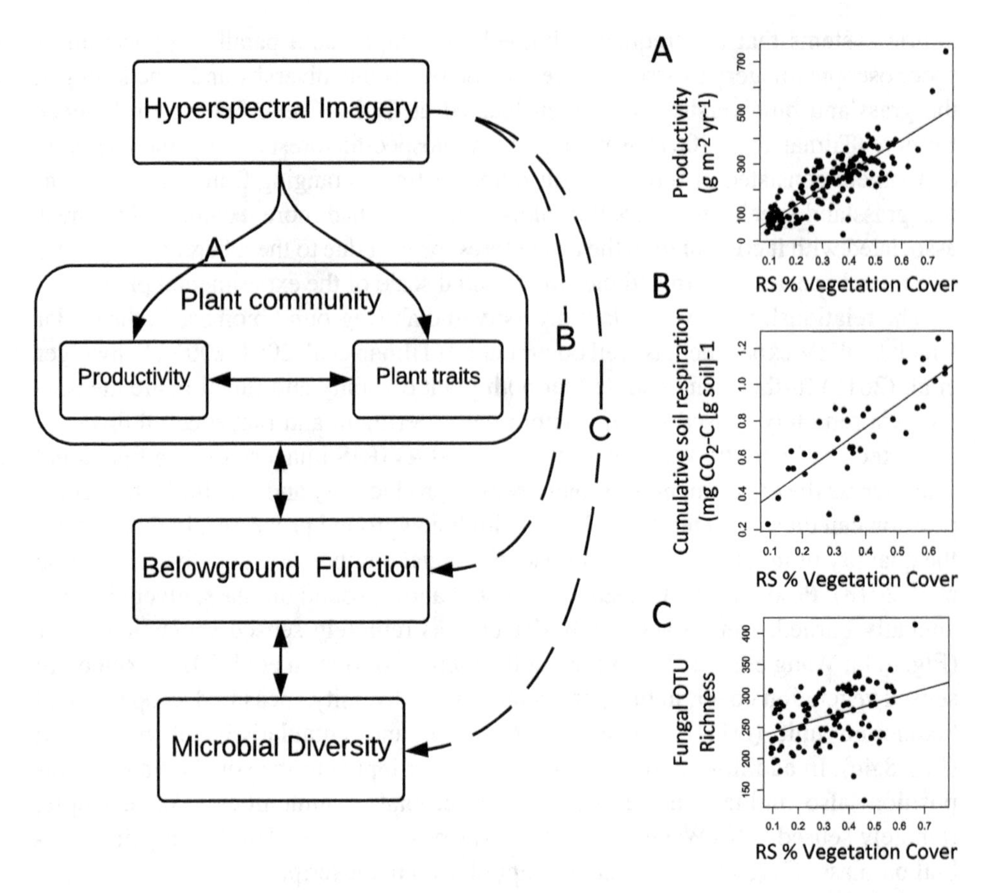

Fig. 8.4 Hyperspectral imagery links above- and belowground processes in a prairie ecosystem. (**a**) Remotely sensed vegetation cover significantly predicted aboveground plant productivity, $R^2 = 0.695$ (**a**); cumulative soil respiration (mg CO_2-C [g soil]$^{-1}$), $R^2 = 0.63$ (**b**); and fungal diversity, measured as OTU richness, $R^2 = 0.144$ (**c**). (Soil respiration and fungal diversity data are from Cline et al. (2018))

RS of aboveground properties poses further challenges that include separating the spectral signals important to canopy chemistry from those of physical properties of the forest canopy (Townsend et al. 2013). Larger challenges lie in the lack of accessibility of RS data and processing techniques to the broader ecological research community.

Several issues of scale present challenges to the application of RS to belowground systems. Large knowledge gaps remain in connecting the small spatial scale observations of traditional field studies with the large spatial scale observations of airborne or satellite RS platforms (Asner et al. 2015; Gamon et al., Chap. 16). In addition, there is a large mismatch in the spatial heterogeneity between above- and belowground systems, with belowground systems being notoriously heterogeneous across small spatial scales (Bardgett and van der Putten 2014). The majority of variation in belowground processes may be due to small, local-scale factors rather than large-scale factors such as climate (Bradford et al. 2016). In addition to issues associated with spatial scale, there are large spans in the scales of biodiversity and time. Speciose aboveground systems may contain upward of 600 species ha^{-1} (Lee et al. 2002), whereas soils contain many thousands of microbial "species" per gram of soil, with large numbers of endemics (Schloss and Handelsman 2006). Linking function to diversity remains a challenge in both systems and particularly in belowground systems where the functional role of the vast majority of species is unknown (Krause et al. 2014). Likewise, large differences in temporal scales exist between above- and belowground systems, with leaf responses to sunlight occurring on the order of seconds (Lambers et al. 1998), while the turnover of soil organic matter can take years to centuries (Bardgett and van der Putten 2014). Variation in temporal scales across systems is particularly important given that the importance of biodiversity to ecosystem processes increases with temporal scale (Cardinale et al. 2012; Reich et al. 2012).

Irrespective of RS, there are shortcomings associated with belowground measurement. For example, belowground measurements that use enzyme activity potentials as indicators of microbial function are widespread, but they are known to have numerous limitations (Nannipieri et al. 2018). Likewise, microbial diversity estimates based upon amplicon sequences of bacterial 16s rDNA have their own methodological and interpretive limitations (Schöler et al. 2017). Nonetheless, both enzyme activities and amplicon sequencing techniques provide useful information about belowground systems and are used widely enough to be compared across studies as long as protocols are consistent.

Advantages of using remotely sensed spectral properties of aboveground vegetation to predict belowground processes lie within the data-rich nature of imaging spectroscopy and the consequent ability to measure many more traits of the canopy than would otherwise be feasible with traditional benchtop methods. In Madritch et al. (2014), only four canopy traits were considered using traditional wet chemistry techniques (leaf tannin, N, C, lignin). These canopy foliar traits were expectedly well correlated with belowground processes. However, plant spectra themselves were better correlated with belowground processes than were plant leaf traits (Madritch et al. 2014). This strong relationship between plant spectra and belowground processing existed because the plant spectra provided quantitative information about many plant traits that were not measured via wet chemistry techniques. Potentially dozens of leaf traits important to belowground processes could be conveyed by plant spectra. The ability of plant spectra to capture many foliar attributes quickly and accurately is a large reason why plant spectra are useful for predicting belowground processes. In addition, identifying which regions of plant spectra are most variable and correlated with belowground process allows researchers to use spectra to identify plant traits important to soil processes. In short, the potential for RS products to link above- and belowground systems is promising but faces considerable obstacles.

References

Aber JD, Mellilo JM (1982) Nitrogen immobilization in decaying hardwood leaf litter as a function of initial nitrogen and lignin content. Can J Bot 60:2263–2269

Adler PB, Seabloom EW, Borer ET, Hillebrand H, Hautier Y, Hector A, Harpole WS, O'Halloran LR, Grace JB, Anderson TM, Bakker JD, Biederman LA, Brown CS, Buckley YM, Calabrese LB, Chu C-J, Cleland EE, Collins SL, Cottingham KL, Crawley MJ, Damschen EI, Davies KF, DeCrappeo NM, P a F, Firn J, Frater P, Gasarch EI, Gruner DS, Hagenah N, Hille Ris Lambers J, Humphries H, Jin VL, Kay AD, Kirkman KP, J a K, Knops JMH, La Pierre KJ, Lambrinos JG, Li W, MacDougall AS, McCulley RL, B a M, Mitchell CE, Moore JL, Morgan JW, Mortensen B, Orrock JL, Prober SM, D a P, Risch AC, Schuetz M, Smith MD, Stevens CJ, Sullivan LL, Wang G, Wragg PD, Wright JP, Yang LH (2011) Productivity is a poor predictor of plant species richness. Science 333:1750–1753

Allison SD (2005) Cheaters, diffusion, and nutrients constrain decomposition by microbial enzymes in spatially structured environments. Ecol Lett 8:626–635

Asner GP, Martin RE, Anderson CB, Knapp DE, (2015) Quantifying forest canopy traits: imaging spectroscopy versus field survey. Remote Sens Environ 158, 15–27

Asner GP, Martin RE, Carranza-Jiménez L, Sinca F, Tupayachi R, Anderson CB, Martinez P (2014) Functional and biological diversity of foliar spectra in tree canopies throughout the Andes to Amazon region. New Phytol 204:127–139

Ayres M, Clausen T, MacLean SF, Redman AM, Reichardt PB (1997) Diversity of structure and antiherbivore activity in condensed tannins. Ecology 78:1696–1712

Baldwin I (1994) Chemical changes rapidly induced by folivory. In: Bernays EA (ed) Insect-plant interactions. CRC Press, Tuscon, pp 1–23

Ball BA, Hunter MD, Kominoski JS, Swan CM, Bradford MA (2008) Consequences of non-random species loss for decomposition dynamics: experimental evidence for additive and non-additive effects. J Ecol 96:303–313

Bardgett RD (2017) Plant trait-based approaches for interrogating belowground function. Biol Environ Proc Royal Irish Acad 117:1–13

Bardgett RD, Van Der Putten WH (2014) Belowground biodiversity and ecosystem functioning. Nature 515:505–511

Bardgett RD, Mommer L, De Vries FT (2014) Going underground: root traits as drivers of ecosystem processes. Trends Ecol Evol 29:692–699

Bate-Smith E (1975) Phytochemistry of proanthocyanidins. Phytochemistry 14:1107–1113

Bellassen V, Delbart N, Le Maire G, Luyssaert S, Ciais P, Viovy N (2011) Potential knowledge gain in large-scale simulations of forest carbon fluxes from remotely sensed biomass and height. For Ecol Manag 261:515–530

Bellon-Maurel V, McBratney A (2011) Near-infrared (NIR) and mid-infrared (MIR) spectroscopic techniques for assessing the amount of carbon stock in soils - critical review and research perspectives. Soil Biol Biochem 43:1398–1410

Berg B (2014) Decomposition patterns for foliar litter – a theory for influencing factors. Soil Biol Biochem 78:222–232

Berryman E, Ryan MG, Bradford JB, Hawbaker TJ, Birdsey R (2016) Total belowground carbon flux in subalpine forests is related to leaf area index, soil nitrogen, and tree height. Ecosphere 7:e01418

Bossuyt H, Denef K, Six J, Frey SD, Merckx R, Paustian K (2001) Influence of microbial populations and residue quality on aggregate stability. Appl Soil Ecol 16:195–208

Bradford MA, Berg B, Maynard DS, Wieder WR, Wood SA (2016) Understanding the dominant controls on litter decomposition. J Ecol 104:229–238

Brinkmann K, Blaschke L, Polle A (2002) Comparison of different methods for lignin determination as a basis for calibration of near-infrared reflectance spectroscopy and implications of lignoproteins. J Chem Ecol 28:2483–2501

Burns RG, DeForest JL, Marxsen J, Sinsabaugh RL, Stromberger ME, Wallenstein MD, Weintraub MN, Zoppini A (2013) Soil enzymes in a changing environment: current knowledge and future directions. Soil Biol Biochem 58:216–234

Cadisch G, Giller K (1997) Driven by nature: plant litter quality and decomposition. CAB International, Wallingford

Cairns MA, Brown S, Helmer EH, Baumgardner GA (1997) Root biomass allocation in the world's upland forests. Oecologia 111:1–11

Cardinale BJ, Duffy JE, Gonzalez A, Hooper DU, Perrings C, Venail P, Narwani A, Mace GM, Tilman D, Wardle DA, Kinzig AP, Daily GC, Loreau M, Grace JB, Lariguaderie A, Srivasta DS, Naeem S (2012) Biodiversity loss and its impact on humanity. Nature 486:59–67

Cavender-Bares J, Gamon JA, Hobbie SE, Madritch MD, Meireles JE, Schweiger AK, Townsend PA (2017) Harnessing plant spectra to integrate the biodiversity sciences across biological and spatial scales. Am J Bot 104:966–969

Cecillon L, Barthes BG, Gomez C, Ertlen D, Genot V, Hedde M, Stevens A, Brun JJ (2009) Assessment and monitoring of soil quality using near infrared reflectance spectroscopy (NIRS). Eur J Soil Sci 60, 770–784

Chapin F, Matson P, Mooney H (2002) Principles of terrestrial ecosystem ecology. Springer, New York

Chen W, Xu R, Wu Y, Chen J, Zhang Y, Hu T, Yuan X, Zhou L, Tan T, Fan J (2018) Plant diversity is coupled with beta not alpha diversity of soil fungal communities following N enrichment in a semi-arid grassland. Soil Biol Biochem 116:388–398

Chisholm RA, Muller-Landau HC, Abdul Rahman K, Bebber DP, Bin Y, Bohlman SA, Bourg NA, Brinks J, Bunyavejchewin S, Butt N, Cao H, Cao M, Cárdenas D, Chang L-W, Chiang J-M, Chuyong G, Condit R, Dattaraja HS, Davies S, Duque A, Fletcher C, Gunatilleke N, Gunatilleke S, Hao Z, Harrison RD, Howe R, Hsieh C-F, Hubbell SP, Itoh A, Kenfack D, Kiratiprayoon S, Larson AJ, Lian J, Lin D, Liu H, Lutz JA, Ma K, Malhi Y, McMahon S, McShea W, Meegaskumbura M, Mohd. Razman S, Morecroft MD, Nytch CJ, Oliveira A, Parker GG, Pulla S, Punchi-Manage R, Romero-Saltos H, Sang W, Schurman J, Su S-H, Sukumar R, Sun I-F, Suresh HS, Tan S, Thomas D, Thomas S, Thompson J, Valencia R, Wolf A, Yap S, Ye W, Yuan Z, Zimmerman JK (2013) Scale-dependent relationships between tree species richness and ecosystem function in forests. J Ecol 101:1214–1224

Clark ML, Roberts DA, Clark DB (2005) Hyperspectral discrimination of tropical rain forest tree species at leaf to crown scales. Remote Sens Environ 96:375–398

Cline LC, Hobbie SE, Madritch MD, Buyarski CR, Tilman D, Cavender-Bares JM (2018) Resource availability underlies the plant-fungal diversity relationship in a grassland ecosystem. Ecology 99:204–216

Coleman DC, Crossley DA (1996) Fundamentals of soil ecology. Academic Press, San Diego

Coq S, Souquet J-M, Meudec E, Cheynier V, Hättenschwiler S (2010) Interspecific variation in leaf litter tannins drives decomposition in a tropical rain forest of French Guiana. Ecology 91:2080–2091

Cornwell WK, Cornelissen JHC, Amatangelo K, Dorrepaal E, Eviner VT, Godoy O, Hobbie SEHB, Kurokawa H, Pérez-Harguindeguy N, Quested HM, Santiago LS, Wardle DA, Wright IJ, Aerts R, Allison SD, van Bodegom P, Brovkin V, Chatain A, Callaghan TV, Diaz S, Garnier E, Gurvich DE, Kazakou E, Klein JA, Read J, Reich RB, Soudzilovskaia JA, Vaieretti MW, Westoby M (2008) Plant species traits are the predominant control on litier decomposition rates within biomes worldwide. Ecol Lett 11:1065–1071

Crutsinger GM, Collins MD, Fordyce JA, Gompert Z, Nice CC, Sanders NJ (2006) Plant genotypic diversity predicts community structure and governs an ecosystem process. Science 313:966–968

Dawud SM, Raulund-Rasmussen K, Ratcliffe S, Domisch T, Finér L, Joly FX, Hättenschwiler S, Vesterdal L (2017) Tree species functional group is a more important driver of soil properties than tree species diversity across major European forest types. Funct Ecol 31:1153–1162

De Deyn GB, van der Putten WH (2005) Linking aboveground and belowground diversity. Trends Ecol Evol 20:625–633

De Vries FT, Manning P, Tallowin JRB, Mortimer SR, Pilgrim ES, Harrison KA, Hobbs PJ, Quirk H, Shipley B, Cornelissen JHC, Kattge J, Bardgett RD (2012) Abiotic drivers and plant traits explain landscape-scale patterns in soil microbial communities. Ecol Lett 15:1230–1239

Díaz S, Hodgson JG, Thompson K, Cabido M, Cornelissen JHC, Jalili A, Montserrat-Martí G, Grime JP, Zarrinkamar F, Asri Y, Band SR, Basconcelo S, Castro- Díez P, Funes G, Hamzehee B, Khoshnevi M, Pérez-Harguindeguy N, Pérez-Rontomé MC, Shirvany FA, Vendramini F, Yazdani S, Abbas-Azimi R, Bogaard A, Boustani S, Charles M, Dehghan M, De Torres-Espuny L, Falczuk V, Guerrero-Campo J, Hynd A, Jones G, Kowsary E, Kazemi-Saeed F, Maestro-Martínez M, Romo-Díez A, Shaw S, Siavash B, Villar-Salvador P, Zak MR (2004) The plant traits that drive ecosystems: Evidence from three continents. Journal of Vegetation Science 15:295–304

Díaz S, Kattge J, Cornelissen JHC, Wright IJ, Lavorel S, Dray S, Reu B, Kleyer M, Wirth C, Prentice IC, Garnier E, Bönisch G, Westoby M, Poorter H, Reich PB, Moles AT, Dickie J, Gillison AN, Zanne AE, Pierce S, Shipley B, Kirkup D, Casanoves F, Joswig JS, Günther A, Falczuk V (2016) The global spectrum of plant form and function. Nature 529:167–171

Eisenhauer N, Vogel A, Jensen B, Scheu S (2018) Decomposer diversity increases biomass production and shifts aboveground-belowground biomass allocation of common wheat. Sci Rep 8:17894

Epps KY, Comerford NB, Reeves JB III, Cropper WP Jr, Araujo QR (2007) Chemical diversity—highlighting a species richness and ecosystem function disconnect. Oikos 116:1831–1840

Fanin N, Bertrand I (2016) Aboveground litter quality is a better predictor than belowground microbial communities when estimating carbon mineralization along a land-use gradient. Soil Biol Biochem 94:48–60

Feeny P (1970) Seasonal changes in oak leaf tannins an nutrients as a cause of spring feeding by winter moth caterpillars. Ecology 51:565–579

Fierer N, Strickland MS, Liptzin D, Bradford MA, Cleveland CC (2009) Global patterns in belowground communities. Ecol Lett 12:1238–1249

Finerty GE, de Bello F, Bílá K, Berg MP, Dias ATC, Pezzatti GB, Moretti M (2016) Exotic or not, leaf trait dissimilarity modulates the effect of dominant species on mixed litter decomposition. J Ecol 104:1400–1409

Fisher JB, Sweeney S, Brzostek ER, Evans TP, Johnson DJ, Myers JA, Bourg NA, Wolf AT, Howe RW, Phillips RP (2016) Tree-mycorrhizal associations detected remotely from canopy spectral properties. Glob Chang Biol 22:2596–2607

Floudas D, Binder M, Riley R, Barry K, Blanchette RA, Henrissat B, Martínez AT, Otillar R, Spatafora JW, Yadav JS, Aerts A, Benoit I, Boyd A, Carlson A, Copeland A, Coutinho PM, de Vries RP, Ferreira P, Findley K, Foster B, Gaskell J, Glotzer D, Górecki P, Heitman J, Hesse C, Hori C, Igarashi K, Jurgens JA, Kallen N, Kersten P, Kohler A, Kües U, Kumar TKA, Kuo A, LaButti K, Larrondo LF, Lindquist E, Ling A, Lombard V, Lucas S, Lundell T, Martin R, McLaughlin DJ, Morgenstern I, Morin E, Murat C, Nagy LG, Nolan M, Ohm RA, Patyshakuliyeva A, Rokas A, Ruiz-Dueñas FJ, Sabat G, Salamov A, Samejima M, Schmutz J, Slot JC, St. John F, Stenlid J, Sun H, Sun S, Syed K, Tsang A, Wiebenga A, Young D, Pisabarro A, Eastwood DC, Martin F, Cullen D, Grigoriev IV, Hibbett DS (2012) The Paleozoic origin of enzymatic lignin decomposition reconstructed from 31 fungal genomes. Science 336:1715–1719

Fontaine S, Barot S, Barré P, Bdioui N, Mary B, Rumpel C (2007) Stability of organic carbon in deep soil layers controlled by fresh carbon supply. Nature 450:277–280

Fortunel C, Garnier E, Joffre R, Kazakou E, Quested H, Grigulis K, Lavorel S, Ansquer P, Castro H, Cruz P, Doležal J, Eriksson O, Freitas H, Golodets C, Jouany C, Kigel J, Kleyer M, Lehsten V, Lepš J, Meier T, Pakeman R, Papadimitriou M, Papanastasis VP, Quétier F, Robson M, Sternberg M, Theau J-P, Thébault A, Zarovali M (2009) Leaf traits capture the effects of land use changes and climate on litter decomposability of grasslands across Europe. Ecology 90:598–611

Frainer A, Moretti MS, Xu W, Gessner MO (2015) No evidence for leaf-trait dissimilarity effects on litter decomposition, fungal decomposers, and nutrient dynamics. Ecology 96:550–561

Freschet GT, Aerts R, Cornelissen JHC (2012) A plant economics spectrum of litter decomposability. Funct Ecol 16:56–65
Frey BR, Lieffers VJ, Hogg EH, Landhausser SM (2004) Predicting landscape patterns of aspen dieback: mechanisms and knowledge gaps. Can J For Res 34:1379–1390
Fujii S, Mori AS, Koide D, Makoto K, Matsuoka S, Osono T, Isbell F (2017) Disentangling relationships between plant diversity and decomposition processes under forest restoration. J Appl Ecol 54:80–90
Funk JL, Larson JE, Ames GM, Butterfield BJ, Cavender-Bares J, Firn J, Laughlin DC, Sutton-Grier AE, Williams L, Wright J (2017) Revisiting the Holy Grail: using plant functional traits to understand ecological processes. Biol Rev 92:1156–1173
Gamon J (2008) Tropical remote sensing: opportunities and challenges. In: Kalacska M, Sanchez-Azofeifa GA (eds) Hyperspectral remote sensing of tropical and subtropical forests. CRC Press, New York, pp 297–305
García-Palacios P, Prieto I, Ourcival JM, Hättenschwiler S (2016) Disentangling the litter quality and soil microbial contribution to leaf and fine root litter decomposition responses to reduced rainfall. Ecosystems 19:490–503
Gartner TB, Cardon ZG (2004) Decomposition dynamics in mixed-species leaf litter. Oikos 104:230–246
Gessner MO, Swan CM, Dang CK, McKie BG, Bardgett RD, Wall DH, Hättenschwiler S (2010) Diversity meets decomposition. Trends Ecol Evol 25:372–380
Gholizadeh H, Gamon JA, Townsend PA, Zygielbaum AI, Helzer CJ, Hmimina GY, Yu R, Moore RM, Schweiger AK, Cavender-Bares J (2019) Detecting prairie biodiversity with airborne remote sensing. Remote Sens Environ 221:38–49
Gillon D, Joffre R, Dardenne P (1993) Predicting the stage of decay of decomposing leaves by near-infrared reflectance spectroscopy. Can J For Res-Revue Canadienne de Recherche Forestiere 23:2552–2559
Gillon D, Houssard C, Joffre R (1999) Using near- infrared reflectance spectroscopy to predict carbon, nitrogen and phosphorus content in heterogeneous plant material. Oecologia 118:173–182
Gould IJ, Quinton JN, Weigelt A, De Deyn GB, Bardgett RD (2016) Plant diversity and root traits benefit physical properties key to soil function in grasslands. Ecol Lett 19:1140–1149
Grossman JJ, Vanhellemont M, Barsoum N, Bauhus J, Bruelheide H, Castagneyrol B, Cavender-bares J, Eisenhauer N, Ferlian O, Gravel D, Hector A, Jactel H, Kreft H, Mereu S, Messier C, Muys B, Nock C, Paquette A, Parker J, Perring MP, Ponette Q, Reich PB, Schuldt A, Staab M, Weih M, Clara D, Scherer-lorenzen M, Verheyen K (2018) Synthesis and future research directions linking tree diversity to growth, survival, and damage in a global network of tree diversity experiments ☆. Environ Exp Bot 152:68–89
Handa IT, Aerts R, Berendse F, Berg MP, Bruder A, Butenschoen O, Chauvet E, Gessner MO, Jabiol J, Makkonen M, McKie BG, Malmqvist B, Peeters ETHM, Scheu S, Schmid B, Van Ruijven J, Vos VCA, Hättenschwiler S (2014) Consequences of biodiversity loss for litter decomposition across biomes. Nature 509:218–221
Haslam E (1989) Plant polyphenols: vegetable tannins revisited. Cambridge University Press, Cambridge
Hatakka A, Hammel KE (2010) Fungal biodegradation of lignocelluloses. In: Osiewacz HD (ed) The Mycota: a comprehensive treatise on fungi as experimental systems for basic and applied research. Springer, New York, pp 319–334
Hättenschwiler S, Jørgensen HB (2010) Carbon quality rather than stoichiometry controls litter decomposition in a tropical rain forest. J Ecol 98:754–763
Hättenschwiler S, Vitousek P (2000) The role of polyphenols in terrestrial ecosystem nutrient cycling. TREE 15:238–243
Hättenschwiler S, Tiunov AV, Scheu S (2005) Biodiversity and litter decomposition in terrestrial ecosystems. Ann Rev Ecol Evol Syst 36:191–218
Hättenschwiler S, Aeschlimann B, Coûteaux M-M, Roy J, Bonal D (2008) High variation in foliage and leaf litter chemistry among 45 tree species of a neotropical rainforest community. New Phytol 179:165–75

Hättenschwiler S, Coq S, Barantal S, Handa IT (2011) Leaf traits and decomposition in tropical rainforests: revisiting some commonly held views and towards a new hypothesis. New Phytol 189:950–965

Hector A, Beale AJ, Minns A, Otway SJ, Lawton JH (2000) Consequences of the reduction of plant diversity for litter decomposition: effects through litter quality and microenvironment. Oikos 90:357–371

Herms DA, Mattson WJ (1992) The dilemma of plants: to grow or defend. Q Rev Biol 67:283–335

Higuchi T (2006) Look back over the studies of lignin biochemistry. J Wood Sci 52:2–8

Hobbie SE (1992) Effects of plant species on nutrient cycling. Trends Ecol Evol 7:336–339

Hobbie JE, Hobbie EA (2006) 15N in symbiotic fungi and plants estimates nitrogen and carbon flux rates in Arctic tundra. Ecology 87:816–822

Hobbie SE (2015) Plant species effects on nutrient cycling: revisiting litter feedbacks. Trends Ecol Evol 30:357–63

Högberg P, Nordgren A, Buchmann N, Taylor AFS, Ekblad A, Högberg MN, Nyberg G, Ottosson-Löfvenius M, Read DJ (2001) Large-scale forest girdling shows that current photosynthesis drives soil respiration. Nature 411:789–792

Holden SR, Rogers BM, Treseder KK, Randerson JT (2016) Fire severity influences the response of soil microbes to a boreal forest fire. Environ Res Lett 11:035004

Hooper DU, Bignell DE, Brown VK, Brussard L, Dangerfield JM, Wall DH, Wardle DA, Coleman DC, Giller KE, Lavelle P, van der Putten WH, de Ruiter PC, Rusek J, Silver WL, Tiedje JM, Wolters V (2000) Interactions between aboveground and belowground biodiversity in terrestrial ecosystems: patterns, mechanisms, and feedbacks. Bioscience 50:1049–1061

Hoorens B, Aerts R, Stroetenga M (2003) Does initial litter chemistry explain litter mixture effects on decomposition? Oecologia 137:578–586

Horner JD, Gosz JR, Cates RG (1988) The role of carbon-based plant secondary metabolites in decomposition in terrestrial ecosystems. Am Nat 132:869–883

Huang N, Gu L, Black TA, Wang L, Niu Z (2015) Remote sensing-based estimation of annual soil respiration at two contrasting forest sites 2306–2325

Hughes AR, Inouye BD, Johnson MTJ, Underwood N, Vellend M (2008) Ecological consequences of genetic diversity. Ecol Lett 11:609–623

Jewell MD, Shipley B, Low-Décarie E, Tobner CM, Paquette A, Messier C, Reich PB (2017) Partitioning the effect of composition and diversity of tree communities on leaf litter decomposition and soil respiration. Oikos 126:959–971

Jo I, Fridley JD, Frank DA (2016) More of the same? In situ leaf and root decomposition rates do not vary between 80 native and nonnative deciduous forest species. New Phytol 209:115–122

Johnson MTJ, Agrawal AA (2005) Plant genotype and environment interact to shape a diverse arthropod community on evening primrose (*Oenothera biennis*). Ecology 86:874–885

Kardol P, Wardle DA (2010) How understanding aboveground-belowground linkages can assist restoration ecology. Trends Ecol Evol 25:670–679

Keiser AD, Knoepp JD, Bradford MA (2013) Microbial communities may modify how litter quality affects potential decomposition rates as tree species migrate. Plant Soil 372:167–176

Kirby R (2006) Actinomycetes and lignin degradation. Adv Appl Microbiol 58:125–168

Knops JMH, Wedin D, Tilman D (2001) Biodiversity and decomposition in experimental grassland ecosystems. Oecologia 126:429–433

Kokaly RF, Clark RN (1999) Spectroscopic determination of leaf biochemistry using band-depth analysis of absorption features and stepwise multiple linear regression. Remote Sens Environ 67:267–287

Kokaly RF, Asner GP, Ollinger SV, Martin ME, Wessman CA (2009) Characterizing canopy biochemistry from imaging spectroscopy and its application to ecosystem studies. Remote Sens Environ 113:S78–S91

Kraus TEC, Dahlgren RA, Zasoski RJ (2003) Tannins in nutrient dynamics of forest ecosystems—a review. Plant Soil 256:41–66

Krause S et al (2014) Trait-based approaches for understanding microbial biodiversity and ecosystem functioning. Front Microbiol 5:251

Laliberté E (2017) Tansley insight - Below-ground frontiers in trait-based plant ecology. New Phytol 213:1597–1603
Lambers H, Chapin FS, Pons TL (1998) Plant physiological ecology. Springer-Verlag New York Inc., New York
Landsberg JJ, Waring RH (1997) A generalised model of forest productivity using simplified concepts of radiation-use efficiency, carbon balance and partitioning. For Ecol Manag 95:209–228
Lauber CL, Strickland MS, Bradford MA, Fierer N (2008) The influence of soil properties on the structure of bacterial and fungal communities across land-use types. Soil Biol Biochem 40:2407–2415
Lee HS, Davies SJ, Lafrankie J, Ashton PS (2002) Floristic and structural diversity of mixed dipterocarp forest in Lambir Hill National Park, Sarawak, Malaysia. J Trop For Sci 14:379–400
Liang J, Crowther TW, Picard N, Wiser S, Zhou M, Alberti G, Schulze E, Mcguire AD, Bozzato F, Pretzsch H, Paquette A, Hérault B, Scherer-lorenzen M, Barrett CB, Glick HB, Hengeveld GM, Nabuurs G, Pfautsch S, Viana H, Vibrans AC, Ammer C, Schall P, Verbyla D, Tchebakova N, Fischer M, Watson JV, Chen HYH, Lei X, Schelhaas M, Lu H, Gianelle D, Parfenova EI, Salas C, Lee E, Lee B, Kim HS, Bruelheide H, Coomes DA, Piotto D, Sunderland T, Schmid B, Gourlet-fleury S, Sonké B, Tavani R, Zhu J, Brandl S, Baraloto C, Frizzera L, Ba R, Oleksyn J, Peri PL, Gonmadje C, Marthy W, Brien TO, Martin EH, Marshall AR, Rovero F, Bitariho R, Niklaus PA, Alvarez-loayza P, Chamuya N, Valencia R, Mortier F, Wortel V, Engone-obiang NL, Ferreira LV (2016) Positive biodiversity-productivity relationship predominant in global forests. Science 354:aaf8957
Lindroth RL, St. Clair SB (2013) Adaptations of quaking aspen (*Populus tremuloides* Michx.) for defense against herbivores. For Ecol Manag 299:14–21
Lindroth RL, Osier TL, Barnhill HR, Wood SA (2002) Effects of genotype and nutrient availability on phytochemistry of trembling aspen (*Populus tremuloides* Michx.) during leaf senescence. Biochem Syst Ecol 30:297–307
Litton CM, Raich JW, Ryan MG (2007) Carbon allocation in forest ecosystems. Glob Chang Biol 13:2089–2109
Lohbeck M, Poorter L, Martínez-Ramos M, Bongers F (2015) Biomass is the main driver of changes in ecosystem process rates during tropical forest succession. Ecology 96:1242–1252
López-Mondéjar R, Zühlke D, Becher D, Riedel K, Baldrian P (2016) Cellulose and hemicellulose decomposition by forest soil bacteria proceeds by the action of structurally variable enzymatic systems. Sci Rep 6:25279
Madritch MD, Hunter MD (2002) Phenotypic diversity influences ecosystem functioning in an oak sandhills community. Ecology 83:2084–2090
Madritch MD, Hunter MD. 2003. Intraspecific litter diversity and nitrogen deposition affect nutrient dynamics and soil respiration. Oecologia 136:124–8.
Madritch MD, Kingdon CC, Singh A, Mock KE, Lindroth RL, Townsend PA, B PTRS, Madritch MD, Kingdon CC, Singh A, Mock KE, Lindroth RL, Townsend PA (2014) Imaging spectroscopy links aspen genotype with below-ground processes at landscape scales Imaging spectroscopy links aspen genotype with belowground processes at landscape scales
Madritch MD, Lindroth RL (2015) Condensed tannins increase nitrogen recovery by trees following insect defoliation. New Phytol 208:410–420
Madritch MD, Donaldson JR, Lindroth RL (2006) Genetic identity of *Populus tremuloides* litter influences decomposition and nutrient release in a mixed forest stand. Ecosystems 9:528–537
Madritch MD, Jordan LM, Lindroth RL (2007) Interactive effects of condensed tannin and cellulose additions on soil respiration. Can J For Res 37:2063–2067
Martin ME, Aber JD (1997) High spectral resolution remote sensing of forest canopy lignin, nitrogen, and ecosystem processes. Ecol Appl 7:431–443
Martin ME, Newman SD, Aber JD, Congalton RG (1998) Determining forest species composition using high spectral remote sensing data. Remote Sens Environ 65:249–254

Martin ME, Plourde LC, Ollinger SV, Smith M-L, McNeil BE (2008) A generalizable method for remote sensing of canopy nitrogen across a wide range of forest ecosystems. Remote Sens Environ 112:3511–3519

Meentemeyer V (1978) Macroclimate and lignin control of litter decomposition rates. Ecology 59(3):465–472

Meier CL, Bowman WD (2008) Links between plant litter chemistry, species diversity, and below-ground ecosystem function. Proc Natl Acad Sci 105:19780–19785

Melillo JM, Aber JD, Muratone JF (1982) Nitrogen and lignin control of hardwood leaf litter decomposition dynamics. Ecology 63:621–626

Mitton JB, Grant MC (1996) Genetic variation and the natural history of quaking aspen. Bioscience 46:25–31

Mokany K, Raison RJ, Prokushkin AS (2006) Critical analysis of root:shoot ratios in terrestrial biomes. Glob Change Biol 12:84–96

Mueller KE, Hobbie SE, Chorover J, Reich PB, Eisenhauer N, Castellano MG, Chadwick OA, Dobies T, Hale CM, Jagodziński AM, Kalucka I, Kieliszewska-Rokicka B, Modrzyński J, Rożen A, Skorupski M, Sobczyk Ł, Stasińska M, Trocha LK, Weiner J, Wierzbicka A, Oleksyn J (2015) Effects of litter traits, soil biota, and soil chemistry on soil carbon stocks at a common garden with 14 tree species. Biogeochemistry 123:313–327

Naeem S, Tjossem SF, Byers D, Bristow C, Li S (1999) Plant neighborhood diversity and production. Ecoscience 6:355–365

Nannipieri P, Trasar-Cepeda C, Dick RP (2018) Soil enzyme activity: a brief history and biochemistry as a basis for appropriate interpretations and meta-analysis. Biol Fertil Soils. (2018 54:11–19

Ollinger SV, Smith ML, Martin ME, Hallett RA, Goodale CL, Aber JD (2002) Regional variation in foliar chemistry and N cycling among forests of diverse history and composition. Ecology 83:339–355

Ollinger SV, Reich PB, Frolking S, Lepine LC, Hollinger DY, Richardson AD (2013) Nitrogen cycling, forest canopy reflectance, and emergent properties of ecosystems. Proc Natl Acad Sci U S A 110:E2437

Osono T (2007) Ecology of ligninolytic fungi associated with leaf litter decomposition. Ecol Res 22:955–974

Parsons SA, Lawler IR, Congdon RA, Williams SE (2011) Rainforest litter quality and chemical controls on leaf decomposition with near-infrared spectrometry. J Plant Nutr Soil Sci 174:710–720

Phillips RP, Brzostek E, Midgley MG (2013) The mycorrhizal-associated nutrient economy: a new framework for predicting carbon-nutrient couplings in temperate forests. New Phytol 199:41–51

Reich PB, Tilman D, Isbell F, Mueller K, Hobbie SE, Flynn DFB, Eisenhauer N (2012) Impacts of biodiversity loss escalate through time as redundancy fades. Science 336:589–592

Reich PB (2014) The world-wide 'fast–slow' plant economics spectrum: a traits manifesto. J Ecol 102:275–301

Robinson D (2001) δ15N as an integrator of the nitrogen cycle. Trends Ecol Evol 16:153–162

Rubert-Nason KF, Holeski LM, Couture JJ, Gusse A, Undersander DJ, Lindroth RL (2013) Rapid phytochemical analysis of birch (Betula) and poplar (Populus) foliage by near-infrared reflectance spectroscopy. Anal Bioanal Chem 405:1333–44

Sabetta L, Zaccarelli N, Mancinelli G, Mandrone S, Salvatori R, Costantini ML, Zurlini G, Rossi L (2006) Mapping litter decomposition by remote-detected indicators. Ann Geophys 49:219–226

Saatchi SS, Harris NL, Brown S, Lefsky M, Mitchard ETA, Salas W (2011) Benchmark map of forest carbon stocks in tropical regions across three continents. 108

Scherer-Lorenzen M, Bonilla JL, Potvin C (2007) Tree species richness affects litter production and decomposition rates in a tropical bio-diversity experiment. Oikos 116:2108–2124

Schimel J, Van Cleve K, Cates R, Clausen T, Reichardt P (1996) Effects of balsam poplar (*Populus balsamifera*) tannins and low molecular weight phenolics on microbial activity in taiga flood-plain soil: implications for changes in N cycling during succession. Can J Bot 74:84–90

Schloss PD, Handelsman J (2006) Toward a census of bacteria in soil. PLoS Comput Biol 2(7):e92

Schneider T, Keiblinger KM, Schmid E, Sterflinger-Gleixner K, Ellersdorfer G, Roschitzki B, Richter A, Eberl L, Zechmeister-Boltenstern S, Riedel K (2012) Who is who in litter decomposition? Metaproteomics reveals major microbial players and their biogeochemical functions. ISME J 6:1749–1762

Schnitzer M, Barr M, Hartenstein R (1984) Kinetics and characteristics of humic acids produced from simple phenols. Soil Biol Biochem 16:371–376

Schöler A, Jacquiod S, Vestergaard G, Schulz S, Schloter M (2017) Analysis of soil microbial communities based on amplicon sequencing of marker genes. 485–489

Schweiger AK, Cavender-Bares J, Townsend PA, Hobbie SE, Madritch MD, Wang R, Tilman D, Gamon JA (2018) Plant spectral diversity integrates functional and phylogenetic components of biodiversity and predicts ecosystem function. Nat Ecol Evol 2:976. https://doi.org/10.1038/s41559-018-0551-1

Schweitzer JA, Bailey JK, Hart SC, Wimp GM, Chapman SK, Whitham TG (2005) The interaction of plant genotype and herbivory deccelerate leaf litter decomposition and alter nutrient dynamics. Oikos 110:133–145

Serbin SP, Singh A, McNeil BE, Kingdon CC, Townsend PA (2014) Spectroscopic determination of leaf morphological and biochemical traits for northern temperate and boreal tree species. Ecol Appl 24:1651–1669

Serbin SP, Singh A, Desai AR, Dubois SG, Jablonski AD, Kingdon CC, Kruger EL, Townsend PA (2015) Remotely estimating photosynthetic capacity, and its response to temperature, in vegetation canopies using imaging spectroscopy. Remote Sens Environ 167, 78–87

Shepherd KD, Vanlauwe B, Gachengo CN, Palm CA (2005) Decomposition and mineralization of organic residues predicted using near infrared spectroscopy. Plant and Soil 277:315–333.

Shields WJ, Bockheim JG (1981) Deterioration of trembling aspen clones in the Great-Lakes region. Can J For Res 11:530–537

Smith VC, Bradford MA (2003) Do non-additive effects on decomposition in litter-mix experiments result from differences in resource quality between litters? Oikos 102:235–242

Soriano-Disla JM, Janik LJ, Viscarra Rossel RA, Macdonald LM, McLaughlin MJ (2014) The performance of visible, near-, and mid-infrared reflectance spectroscopy for prediction of soil physical, chemical, and biological properties. Appl Spectrosc Rev 49:139–186

Srivastava DS, Cardinale BJ, Downing AL, Duffy JE, Jouseau C, Sankaran M, Wright JP (2009) Diversity has stronger top-down than bottom-up effects on decomposition. Ecology 90:1073–1083

Stenberg B, Rossel RAV, Mouazen AM, Wetterlind J (2010) Visible and near infrared spectroscopy in soil science. In: Sparks DL (ed) Advances in agronomy, vol 107, pp 163–215

Talbot JM, Yelle DJ, Nowick J, Treseder KK (2012) Litter decay rates are determined by lignin chemistry. Biogeochemistry 108:279–295

Tenney FG, Waksman SA (1929) Composition of natural organic materials and their decomposition in the soil. IV. The nature and rapidity of decomposition of the various organic complexes in different plant materials, under aerobic conditions. Soil Sci 28:55–84

Thers H, Brunbjerg AK, Læssøe T, Ejrnæs R, Bøcher PK, Svenning J (2017) Lidar-derived variables as a proxy for fungal species richness and composition in temperate Northern Europe. Remote Sens Environ 200:102–113

Thevenot M, Dignac MF, Rumpel C (2010) Fate of lignins in soils: a review. Soil Biol Biochem 42:1200–1211

Tilman D, Reich PB, Knops J, Wedin D, Mielke T, Lehman C (2001) Diversity and productivity in a long-term grassland experiment. Science 294:843–845

Tilman D, Reich PB, Knops J (2006) Biodiversity and ecosystem stability in a decade-long grassland experiment. Nature 441:629–632

Townsend PA, Foster JR, Chastain RA, Currie WS (2003) Application of imaging spectroscopy to mapping canopy nitrogen in the forests of the central Appalachian Mountains using Hyperion and AVIRIS. IEEE Trans Geosci Remote Sens 41:1347–1354

Townsend PA, Serbin SP, Kruger EL, Gamon JA (2013) Disentangling the contribution of biological and physical properties of leaves and canopies in imaging spectroscopy data. Proc Natl Acad Sci 110:E1074–E1074

Triebwasser DJ, Tharayil N, Preston CM, Gerard PD (2012) The susceptibility of soil enzymes to inhibition by leaf litter tannins is dependent on the tannin chemistry, enzyme class and vegetation history. New Phytol 196:1122–1132

Ustin SL, Roberts DA, Gamon JA, Asner GP, Green RO (2004) Using imaging spectroscopy to study ecosystem processes and properties. Bioscience 54:523

Vitousek P, Asner GP, Chadwick OA, Hotchkiss S (2009) Landscape-level variation in forest structure and biogeochemistry across a substrate age gradient in Hawaii. Ecology 90:3074–3086

Wang Z, Townsend PA, Schweiger AK, Couture JJ, Singh A, Hobbie SE, Cavender-Bares J (2019) Mapping foliar functional traits and their uncertainties across three years in a grassland experiment. Remote Sens Environ 221:405–416

Wardle DA (1999) Is "sampling effect" a problem for experiments investigating biodiversity-ecosystem function relationships? Oikos 87:403–407

Wardle DA, Nicholson KS (1996) Synergistic effects of grassland plant species on soil microbial biomass and activity: implications for ecosystem-level effects of enriched plant diversity. Funct Ecol 10:410–416

Wardle DA, Bonner KI, Barker GM (2000) Stability of ecosystem properties in response to above-ground functional group richness and composition. Oikos 89:11–23

Wardle DA, Bardgett RD, Klironomos JN, Setälä H, van der Putten WH, Wall DH (2004) Ecological linkages between aboveground and belowground biota. Science 304:1629–1633

Weisser WW, Roscher C, Meyer ST, Ebeling A, Luo G, Allan E, Beßler H, Barnard RL, Buchmann N, Buscot F, Engels C, Fischer C, Fischer M, Gessler A, Gleixner G, Halle S, Hildebrandt A, Hillebrand H, de Kroon H, Lange M, Leimer S, Le Roux X, Milcu A, Mommer L, Niklaus PA, Oelmann Y, Proulx R, Roy J, Scherber C, Scherer-Lorenzen M, Scheu S, Tscharntke T, Wachendorf M, Wagg C, Weigelt A, Wilcke W, Wirth C, Schulze E-D, Schmid B, Eisenhauer N (2017) Biodiversity effects on ecosystem functioning in a 15-year grassland experiment: patterns, mechanisms, and open questions. Basic Appl Ecol 23:1–73

Wessman CA, Aber JD, Peterson DL, Melillo JM (1988) Remote sensing of canopy chemistry and nitrogen cycling in temparate forest ecosystems. Nature 335:154–156

Williams LJ, Paquette A, Cavender-Bares J, Messier C, Reich PB (2017) Spatial complementarity in tree crowns explains overyielding in species mixtures. Nat Ecol Evol 1:0063

Wimp GM, Young WP, Woolbright SA, Martinsen GD, Keim P, Whitham TG (2004) Conserving plant genetic diversity for dependent animal communities. Ecol Lett 7:776–780

Wimp GM, Martinsen GD, Floate KD, Bangert RK, Whitham TG (2005) Plant genetic determinants of arthropod community structure and diversity. Evolution 59:61–69

Worrall JJ, Egeland L, Eager T, Mask RA, Johnson EW, Kemp PA, Shepperd WD (2008) Rapid mortality of *Populus tremuloides* in southwestern Colorado, USA. For Ecol Manag 255:686–696

Wright IJ, Reich PB, Westoby M, Ackerly DD, Baruch Z, Bongers F, Cavender- Bares J, Chapin T, Cornelissen JHC, Diemer M, Flexas J, Garnier E, Groom PK, Gulias J, Hikosaka K, Lamont BB, Lee T, Lee W, Lusk C, Midgley JJ, Navas M-L, Niinemets Ü, Oleksyn J, Osada N, Poorter H, Poot P, Prior L, Pyankov VI, Roumet C, Thomas SC, Tjoelker MG, Veneklaas EJ, Villar R (2004) The worldwide leaf economics spectrum. Nature 428:821–827

Wu T, Ayres E, Bardgett RD, Wall DH, Garey JR (2011) Molecular study of worldwide distribution and diversity of soil animals. PNAS 108:17720–17725

9

Prediction of Species Distribution using Satellite Remote Sensing

Jesús N. Pinto-Ledezma and Jeannine Cavender-Bares

9.1 Introduction

What drives species distributions? This is one of the most fundamental questions in ecology, evolution, and biogeography, and it drew the attention of early naturalists (Gaston 2009; Guisan et al. 2017). Although the question is classic and its answers sometime seem obvious—for example, Alfred Russel Wallace recognized the effect of geographical and environmental features on species distributional ranges (Wallace 1860)—the answers are highly complex as a consequence of historical evolutionary and biogeographic processes and the spatial and temporal dynamics of abiotic and biotic factors (Soberón and Peterson 2005; Soberón 2007; Colwell and Rangel 2009).

Here we explore the potential of satellite remote sensing (S-RS) products to quantify species-environment relationships that predict species distributions. We propose several new metrics that take advantage of the high temporal resolution in Moderate Resolution Imaging Spectroradiometer (MODIS) leaf area index (LAI) and MODIS normalized difference vegetation index (NDVI) data products. Evaluating the potential of remotely sensed data in environmental niche modeling (ENM) and species distribution modeling (SDM) is an important step toward the long-term goal of improving our ability to monitor and predict changes in biodiversity globally. To achieve this, we first modeled the environmental/ecological niches for the American live oak species (*Quercus* section *Virentes*) using environmental variables derived from (1) interpolated climate surfaces data (i.e., WorldClim) and

J. N. Pinto-Ledezma (✉) · J. Cavender-Bares
Department of Ecology, Evolution and Behavior, University of Minnesota, Saint Paul, MN, USA
e-mail: jpintole@umn.edu

(2) S-RS products. Live oaks are a small lineage, descended from a common ancestor (for a discussion on phylogenetics, see Meireles et al., Chap. 7) that includes seven species that vary in geographic range size and climatic breadth and are distributed in both temperate and tropical climates from the southeastern United States, Mesoamerica, and the Caribbean (Cavender-Bares et al. 2015). Their variation in range size and climatic distributions, their distributions in both *highly* studied and *understudied* regions of the globe, and the second author's expert knowledge of their distributions make them an interesting case study for comparing SDM/ENMs that rely on classic data sources to those that use remotely sensed data sources, which have more consistent data accuracy and resolution. We used the live oaks as a test clade to evaluate the relationships among the modeled niches estimated from both sources of environmental data.

Given that the interpolated climate surfaces from WorldClim (Hijmans et al. 2005) are the most widely used data set for the study of species-environment relationships, we compare the performance of SDMs based on S-RS products to those based on WorldClim data. If there is a tight relationship between models from the two sources, this would indicate that the resultant models from S-RS products have similar performance to the resultant models from the WorldClim climatic predictors. Remotely sensed data products may provide an advantage in predicting species distributions in regions where climatic data is sparsely sampled. Although WorldClim provides interpolated climate surfaces for land areas across the world at multiple spatial resolutions, from 30 arc seconds (~1 km) to 10 arcmin (~18.5 km) (Hijmans et al. 2005), the spatial distribution of the base information (i.e., weather or climatic stations) used for interpolations is unevenly distributed across the world (Fig. 9.2c). This is not a small issue given the uncertainty associated with interpolated climatic variables when modeling species-environment relationships, especially in many tropical countries, where weather stations are frequently few and far apart (Soria-Auza et al. 2010). Given that tropical regions are precisely the regions where most species occur (Fig. 9.2c), finding alternative means to predict species is important for efforts to monitor and manage biodiversity globally. S-RS products, which provide quasi-global coverage of land and sea surfaces at high temporal and spatial resolution, represent promising alternatives that may be particularly important in the world's most biologically diverse regions. Our aim here is to provide an understanding of the potential of S-RS products to quantify species ecological niches and estimate species distributions rather than to develop a definitive ecological and geographical profile for the live oaks themselves. If the consistent accuracy and high spatial resolution of S-RS products can actually improve estimates of species distributions, they will represent an advance in our ability to predict where species are likely persist under changing environments. Ultimately, such predictions can be combined with other remote sensing (RS) means of detecting species and biodiversity (Meireles et al., Chap. 7; Bolch et al., Chap. 12; Record et al., Chap. 10) to enable global-scale biodiversity change detection.

9.2 Theoretical Background

9.2.1 The BAM Diagram

One way to explore the ubiquitous relationship between the spatial and temporal dynamics of abiotic and biotic factors is through the BAM framework (Fig. 9.1, Soberón and Peterson 2005; Soberón 2007), which formally describes the individual and joint effects of biotic factors (B; e.g., species interactions), abiotic factors (A; e.g., environmental conditions or abiotically suitable area), and movement (M; e.g., species dispersal capacity) in determining species distributions in a geographical space (G; e.g., the study region). Notice that all factors in the BAM framework are placed within a spatial context. Within the geographical space (G), three overlapping circles are shown, each of which represents suitable conditions for a given species. The intersection between all factors "B∩A∩M" represents the occupied distributional area (G0) or the "realized" or occupied niche. The intersection between biotic and abiotic factors "B∩A" represents the invadable distributional area (G1) or areas that can be colonized because suitable biotic and abiotic conditions and both present. The intersection between abiotic and movement factors "A∩M" represents the area where the species cannot be found. Finally, the union between the occupied and invadable areas "G0∪G1" represents the geographic potential distribution area (GP) or biotically reduced niche (see Soberón 2007; Peterson et al. 2011 for detailed explanations).

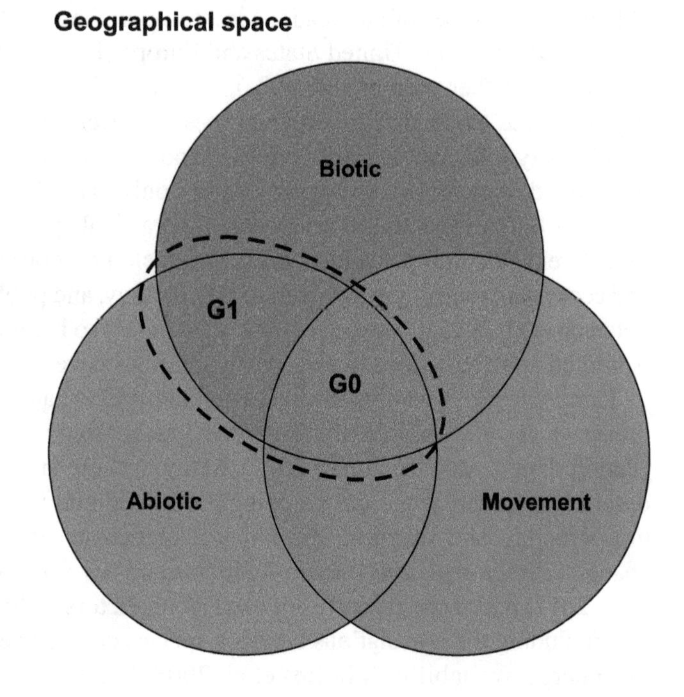

Fig. 9.1 The BAM diagram, where B biotic, A abiotically suitable area, and M movement or migration, illustrates the relationship among the three major determinants of species distributions. G1, the invadable distributional area, and G0, the occupied distributional area, represent the outcomes from the intersection between the major determinants

9.2.2 *Where Are We Now?*

Although the BAM framework was developed to understand and quantify species-environment relationships (Soberón 2007; see also Soberón and Peterson 2005), the concept and investigation of species-environment relationships are long-standing, dating back to Wallace (Wallace 1860) and early ecologists (Grinnell 1904, 1917; Elton 1927; Holdridge 1947; Hutchinson 1957). These early naturalists originally established the theoretical principles to analyze and describe biogeographical distributions in relation to environmental patterns (Colwell and Rangel 2009). Interestingly, despite the large body of theoretical advances and empirical applications, the quantification of ecological niches and estimation of species distributions is still a challenging task (but see Sanín and Anderson 2018; Smith et al. 2018) and one of the most active areas in macroecological and biodiversity research (Franklin 2010; Peterson et al. 2011; Anderson 2013; Guisan et al. 2017).

In fact, since the first algorithm for modeling species-environment relationship was presented (BIOCLIM, Nix 1986), the number of publications has increased dramatically (Lobo et al. 2010; Booth et al. 2013). A simple search in Google Scholar for the terms "ecological niche model" and "species distribution model" (last accessed on December 30, 2018) returned 2,950 and 6,400 citations, respectively, for 1990–2018 (Fig. 9.2a). Interestingly, the number of publications on these topics increased markedly in the past 10 years (Fig. 9.2, see also Lobo et al. 2010) and continues to grow, particularly in studies that emphasize the application of ENMs and SDMs to environmental assessment, forecasting, and hindcasting species distributions (Anderson 2013; Elith and Franklin 2013; Guisan et al. 2017). Interestingly, although the number of publications increased in the last 10 years, most of the studies were performed in United States and Europe (Fig. 9.2b) in countries with a high density of weather stations (Fig. 9.2c), with much less emphasis on the most diverse regions of the globe. The increasing access to species occurrence data (e.g., Global Biodiversity Information Facility, GBIF) and environmental data (climatic and satellite derived) has created the opportunity not only to model species-environment relationships but to expand the theoretical and practical applications of ENM and SDM to different research programs and fields, including conservation biology, wildlife and ecosystem management, evolutionary biology, and public health (Franklin 2010; Peterson et al. 2011; Guisan et al. 2017), and to do so in remote regions where access is limited and predictions of species distributions have disproportionate importance.

Parallel to the development and evolution of ENM and SDM theory and applications, we have witnessed the growth of technological tools and S-RS products (Pettorelli et al. 2014a; Turner 2014). Many of these are particularly applicable for describing, quantifying, and mapping the spatial and temporal patterns of vegetation structure and function, the impacts of human activities, and environmental change (Turner et al. 2003; Pinto-Ledezma and Rivero 2014; Jetz et al. 2016; Cord et al. 2017) and more recently are used as predictors of broad patterns of biodiversity, including the associations between species co-occurrence patterns and ecosystem energy availability (Phillips et al. 2008; Pigot et al. 2016; Hobi et al. 2017). In addition, an unprecedented number of S-RS data and data products (S-RS) have

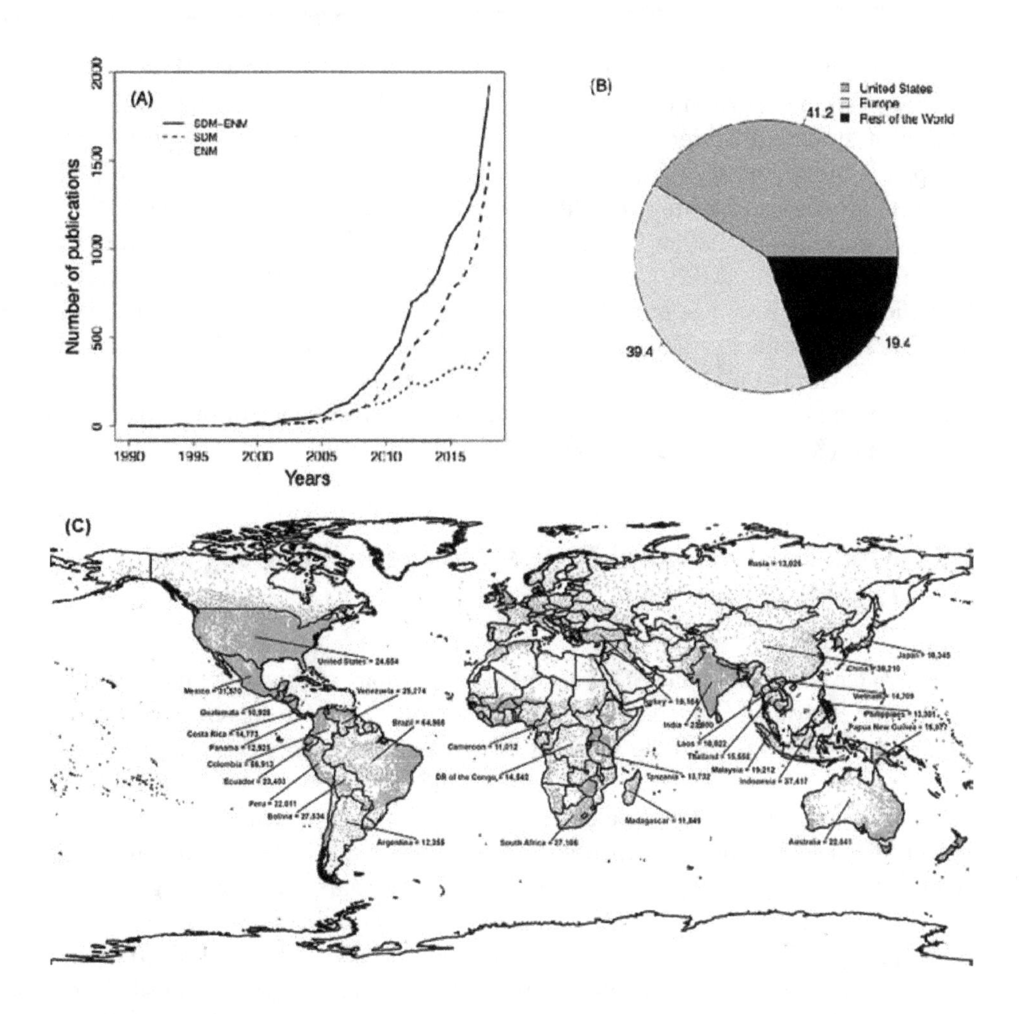

Fig. 9.2 (**a**) Number of publications containing the term "species distribution model" or "ecological niche model" between 1990 and 2018 (Google Scholar search December 31, 2018). The solid line represents the combination of both SDM and ENM, while dashed and dotted lines indicate the individual terms. (**b**) Percentage of ENM/SDM studies performed in the United States and Europe in relation to the total number of publications from (**a**). (**c**) Distribution of weather stations (green dots) used to create the interpolated climate surfaces (i.e., WorldClim) and the number of species for the 30 most diverse countries. The numbers correspond to the estimated number of species—vertebrates and vascular plants—for each country. Notice that for most countries the weather stations are sparse and have low coverage.

been made freely available (Turner et al. 2003; Hobi et al. 2017) with the potential to track the spatial variation in the chemical composition of vegetation (Wang et al. 2019; Serbin and Townsend, Chap. 3), physiology, structure, and function (Lausch et al., Chap. 13; Serbin and Townsend, Chap. 3; Myneni et al. 2002; Saatchi et al. 2008; Jetz et al. 2016).

Despite the potential of S-RS products for measuring and modeling biodiversity (Gillespie et al. 2008; Pettorelli et al. 2014a, b; Turner 2014; Cord et al. 2013;

Zimmermann et al. 2007), attention has only recently turned to using these data in studies of species-environment relationships (Cord et al. 2013; West et al. 2016), and most studies use bioclimatic data such as WorldClim (but see Paz et al., Chap. 11; Record et al., Chap. 10). Although early attempts indicated that S-RS products do not seem to improve the accuracy in estimating species distributions (Pearson et al. 2004; Thuiller 2004; Zimmermann et al. 2007), more recent publications (Kissling et al. 2012; Cord et al. 2013) suggest that despite these apparent limitations, S-RS products provide better spatial resolution that allow the discrimination of habitat characteristics not captured when bioclimatic data are used (Saatchi et al. 2008; Cord et al. 2013), and they can be used as surrogates of biotic and/or functional predictors such as LAI that increase the performance of individual species models (Kissling et al. 2012; Cord et al. 2013).

9.3 Modeling Ecological Niches and Predicting Geographic Distributions

Although the terms ENM and SDM are often used synonymously in the literature, the two are not equivalent (Anderson 2012; Soberón et al. 2017). A comprehensive discussion of this topic is beyond the scope of this chapter but is provided elsewhere (see Peterson et al. 2011; Anderson 2012; Soberón et al. 2017). A crucial step in differentiating the two terms is to establish a distinction between environmental space and geographical space (Hutchinson's duality; Colwell and Rangel 2009). On the one hand, environmental space corresponds to a suite of environmental conditions at a given time (e.g., climate, topography); on the other hand, geographical space is the extent of a particular region or study area (Soberón and Nakamura 2009; Peterson et al. 2011) and includes important historical context. Thus, when modeling species ecological niches, we are modeling the existing abiotically suitable conditions for the species or the biotically reduced niche (Peterson et al. 2011; see also Fig. 9.1). However, when modeling species distributions, the intent is to project objects into geographical space (Fig. 9.1), and, depending on the factors considered, it is possible to estimate the occupied distributional area or the invadable distributional area (Soberón and Nakamura 2009; Peterson et al. 2011; Anderson 2012; Soberón et al. 2017).

9.3.1 Methods

9.3.1.1 Oak Species Data Sets

Occurrence data were downloaded from iDigBio between 20 and 24 July 2018, including localities collected by the authors, and cleaned for accuracy. Any botanical garden localities were discarded. All points were visually examined, and any localities that were outside the known range of the species, or in unrealistic locations (e.g., water bodies), were discarded.

9.3.1.2 Environmental Data Sets

For comparative purposes we obtained environmental data from two sources: (1) environmental variables derived from WorldClim and (2) S-RS data products (Fig. 9.3). Environmental variables derived from climatic data were obtained from the 10 to 2.5 arcmin WorldClim (Hijmans et al. 2005; spatial resolution of ~18.5 and 4.5 km at the equator, respectively) for annual mean temperature (BIO1), temperature seasonality (BIO4), minimum temperature of coldest month (BIO6), mean

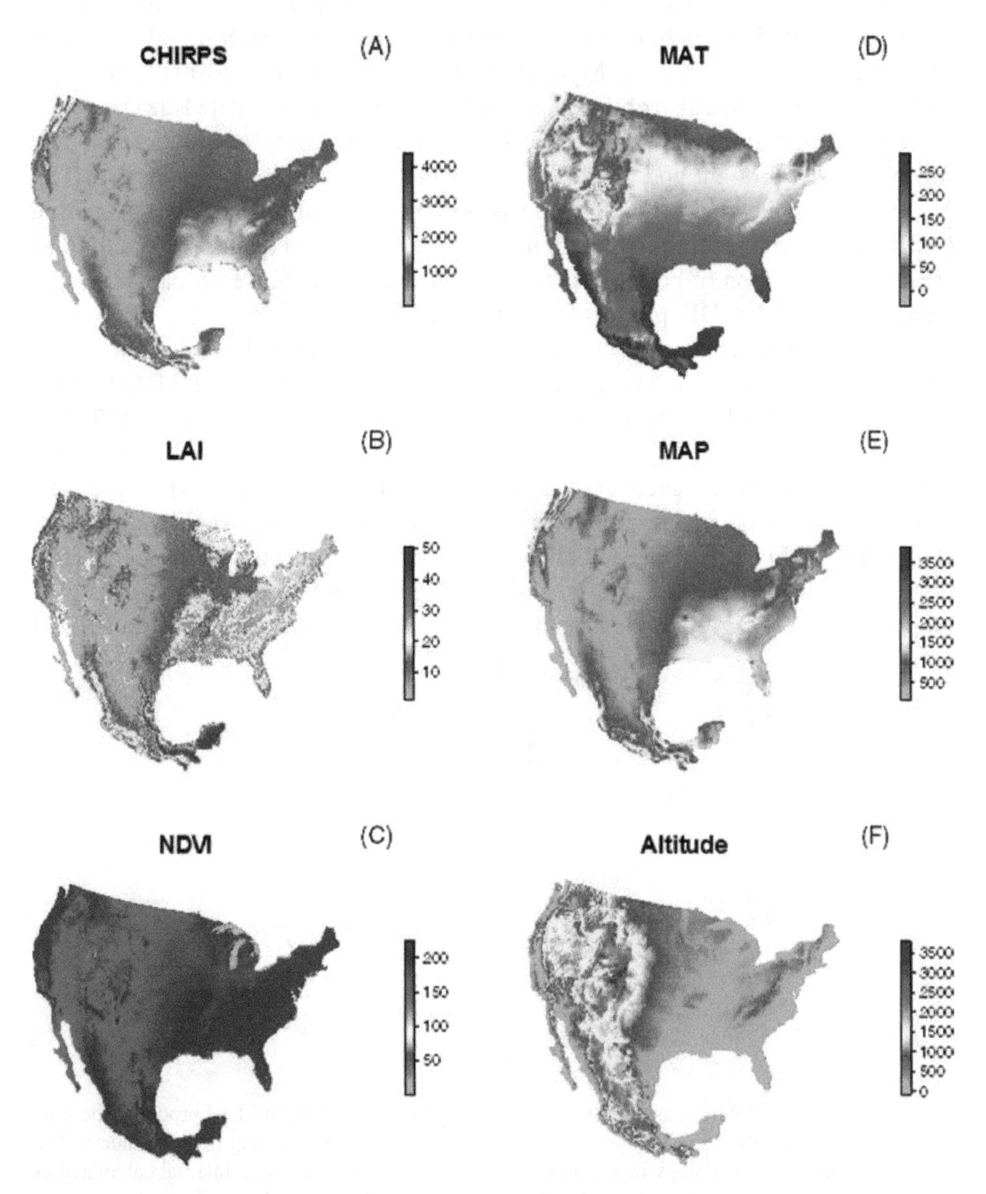

Fig. 9.3 A selection of S-RS products and climatic variables used in this study. The panels show (**a**) Climate Hazards group Infrared Precipitation with Stations (CHIRPS); (**b**) MODIS mean LAI; (**c**) MODIS mean NDVI; (**d**) mean annual temperature; (**e**) mean annual precipitation; and (**f**) altitude or mean elevation from Shuttle Radar Topography Mission (SRTM)

temperature of warmest quarter (BIO10), annual precipitation (BIO12), and precipitation seasonality (BIO15). These environmental variables were selected as critical for the distribution of oak species (Hipp et al. 2017) generally and were previously shown to be important in differentiating live oak (*Virentes*) species specifically (Cavender-Bares et al. 2011; Koehler et al. 2011; Cavender-Bares et al. 2015).

Environmental variables from S-RS products were obtained from MODIS over a 15-year period (2001–2015) from NASA using the interface EOSDIS Earthdata (https://earthdata.nasa.gov). Data include two MODIS Collection 5 land products: LAI (8-day temporal resolution) and NDVI (16-day temporal resolution). LAI and NDVI products (Fig. 9.3b, c) are derived from Terra/Aqua MOD15A2 and Terra MOD13A2, respectively (see Myneni et al. 2002 for a detailed explanation of MODIS products). We also obtained precipitation data from Climate Hazards group Infrared Precipitation with Stations (CHIRPS, Fig. 9.3a), an S-RS product designed for monitoring drought and global environmental land change (Funk et al. 2015). Notice that the original MODIS products present a spatial resolution of 1 km and CHIRPS, a spatial resolution of 3 arcmin or ~5.5 km at the equator. To standardize the spatial resolution of both MODIS products and CHIRPS, we upscaled the spatial resolution of MODIS products to that of CHIRPS.

Prior to following the ENM/SDM procedures (outlined below), we calculated five new metrics taking advantage of the high temporal resolution LAI and NDVI data by doing simple arithmetic calculations: LAI/NDVI cumulative, LAI/NDVI mean, LAI/NDVI max, LAI/NDVI min, and LAI/NDVI seasonality or the coefficient of variation (see Saatchi et al. 2008; Hobi et al. 2017 for details). These metrics represent the spatial variation in vegetation productivity over a year (Berry et al. 2007; Hobi et al. 2017) and allow detection of biodiversity changes, description of habitats of different species, and tracking of phenology within species geographical ranges (Fig. 9.4). We used these two S-RS products given LAI and NDVI

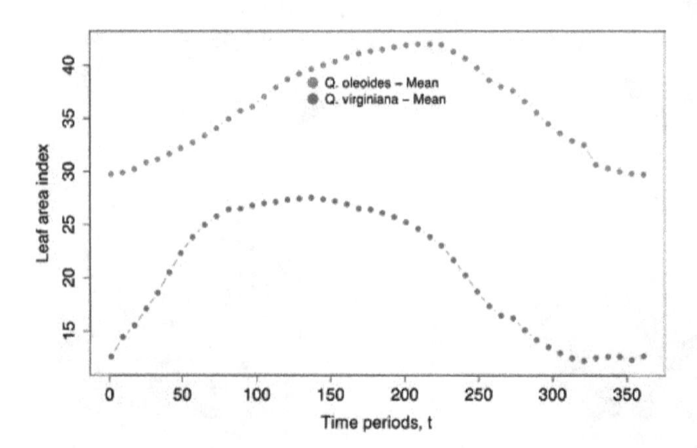

Fig. 9.4 Satellite remotely sensed vegetation phenology based on MODIS LAI product. The time periods (t) represent the 46 time intervals every 8 days within a year starting from 1 January. The curves for vegetation phenology represent the variation in LAI over a 1-year interval calculated as the mean LAI within a species geographical distribution at 8-day intervals averaged over 15 years (see Hobi et al. 2017 for details). Shown is seasonal variation in the temperate forest vegetation where *Q. virginiana* occurs in North America compared with seasonal variation in the tropical dry forest vegetation where *Q. oleoides* occurs in Mexico and Central America

provide information on net primary productivity, dynamics of the growing season, and vegetation seasonality, all potentially important variables for characterizing plant species ranges (Myneni et al. 2002; Saatchi et al. 2008). All data processing and metric calculations were performed in R v3.5 (R Core Team 2018) using customized scripts and core functions from the packages raster (Hijmans 2018), gdalUtils (Greenberg and Mattiuzzi 2018), and rgdal (Bivand et al. 2018). R scripts for data processing and metric calculations can be found at https://github.com/jesusNPL/RS-SDM_ENM.

9.3.1.3 Modeling Procedure

To model the ecological niche and distribution for oak species, we used an ensemble framework—prediction of a niche or a distributional area made by combining results of different modeling algorithms (Araújo and New 2007; Diniz-Filho et al. 2009). Within this framework we fit six species models and projected potential distributions for current environmental conditions for both environmental data sets (Table 9.1). The modeling algorithms included three statistical models (generalized linear models [GLM], generalized additive models [GAM], and adaptive regression splines [MARS]) and three machine learning models (MAXENT, support vector

Table 9.1 Combinations of environmental variables used for modeling live oak species-environment relationship under an ensemble framework

Source	Environmental predictors	Description
S-RS	CHIRPS	Climate Hazards group Infrared Precipitation with Stations
	LAI maximum	MODIS maximum leaf area index calculated over a year
	LAI mean	MODIS mean leaf area index calculated over a year
	LAI seasonality	MODIS seasonality of leaf area index calculated over a year
	LAI minimum	MODIS minimum leaf area index calculated over a year
	Altitude	Mean elevation from Shuttle Radar Topography Mission
S-RS2	CHIRPS	–
	LAI maximum	–
	LAI mean	–
	LAI seasonality	–
	LAI minimum	–
	NDVI maximum	MODIS maximum normalized difference vegetation index calculated over a year
	NDVI mean	MODIS mean normalized difference vegetation index calculated over a year
	NDVI seasonality	MODIS seasonality of normalized difference vegetation index calculated over a year
	NDVI minimum	MODIS minimum normalized difference vegetation index calculated over a year
	Altitude	–

(continued)

Table 9.1 continued

Source	Environmental predictors	Description
WorldClim	BIO 1	Mean annual temperature
	BIO 4	Temperature seasonality
	BIO 6	Minimum temperature of coldest month
	BIO 10	Mean temperature of warmest quarter
	BIO 12	Mean annual precipitation
	BIO 15	Precipitation seasonality
	Altitude	–

S-RS satellite remote sensing products. For comparative purposes, we used the same environmental variables from WorldClim at two spatial resolutions, 10 and 2.5 arcmin

machines [SVM], and Random Forest [RF]). A description for each algorithm is detailed in Franklin (2010) (see also Peterson et al. 2011). All algorithms were fit in R and used the packages dismo (Hijmans et al. 2017), kernlab (Karatzoglou et al. 2004), randomForest (Liaw and Wiener 2002), mgcv (Wood 2006), and earth (Milborrow 2016).

Within our ensemble framework, species' ecological niches are modeled using the six algorithms by fitting the occurrences of a single species and the predictors. The resulting six species models (one for each algorithm) are stacked into a single species model by averaging all models (Araújo and New 2007). We chose this approach because a major source of uncertainty in ENM/SDM arises from the algorithm used for modeling (Diniz-Filho et al. 2009; Qiao et al. 2015) and because the choice of the "best" modeling algorithm depends on the aims of the modeling applications (Peterson et al. 2011). Finally, using the stacked species models, we estimated macroecological patterns of species richness and the uncertainty associated with model parametrization. These patterns are less interesting in their own right for a small clade with only seven species, but they demonstrate an effective approach that can be applied to much larger groups of species.

We estimated live oak species richness by summing the projected potential species distributions; uncertainty was estimated as the variance attributable to the source of uncertainty (i.e., algorithms and their interactions) by performing a one-way analysis of variance (ANOVA) without replicates (Sokal and Rohlf 1995). The resulting uncertainty map shows regions with low and high uncertainty associated with the source of uncertainty (i.e., algorithm).

Statistical Analyses

To explore the performance of environmental data derived from RS for ENM/SDM compared to traditionally used environmental data from climatic variables (e.g., WorldClim), we evaluated the relationship between the modeled ecological niches from: (1) S-RS products; and (2) environmental variables from WorldClim. In doing so, we used correlation analyses corrected according Clifford's method to obtain the effective degrees of freedom for Pearson's coefficients while controlling for spatial autocorrelation (Clifford et al. 1989). Statistical analyses were performed in R using the package SpatialPack (Vallejos and Osorio 2014).

9.3.2 Results

Live oak models calibrated using different sources and combinations of environmental predictors (Table 9.1) within the ensemble framework generally provided similar suitability distributions (Fig. 9.5). Interestingly, increasing the number of predictors or increasing model complexity (S-SR2 in Table 9.1) affected model performance as measured by the Cohen's Kappa coefficient and AUC (area under the receiver operating characteristic curve) indices (Table 9.2), and thus affected the geographic predictions: Complex models tended to have higher statistical performance but to underestimate the distributions of live oak species when compared with simpler models (Fig. 9.5). Individual live oak species models made from S-RS products and WorldClim differed somewhat in their performances (see Table 9.2 and Fig. 9.5). Models from WorldClim tended to have slightly better statistical performance in inferring species distributions based on the AUC and Kappa criteria. However, these metrics do not capture differences in the precision and spatial resolution of the approaches. In several species, the WorldClim models predicted low precision locations compared to the S-RS data. In particular, the IUCN (International Union for Conservation of Nature) red-listed narrow endemic Brandegee Oak (*Quercus brandegeei*) in southern Baja California is very imprecisely predicted compared with the S-RS data. Using high-resolution interpolated climatic predictors did not improve the performance of individual models (WC25 in Table 9.2) and returned similar suitability predictions to those estimated under lower spatial resolution climatic predictors (Fig. 9.5). Although WorldClim models seems to have better statistical performance as shown in Table 9.2, we can at most discriminate the accuracy of interpolating continuous surface-derived models, only when we are inferring habitat suitability models (ENM) and not the projected species geographical distribution (SDM). Using S-RS data as predictors not only helps to identify the species habitat suitability but also incorporates local ecological conditions necessary to predict local species distributions and co-occurrence (Radeloff et al. 2019). This is because S-RS data have the potential to get at biological mechanisms, for example, through the detection of species phenological variation over space and time (Figs. 9.3c and 9.4).

When macroecological patterns of species richness and uncertainty maps were constructed, we observed similar patterns of species richness between maps made from the simpler combination of S-RS and WorldClim models (Fig. 9.6a, c, and d; see Table 9.1 for a description of the environmental combinations of S-RS and WorldClim). Notably, species richness estimation from the complex S-RS tends to restrict live oak assemblages to southeastern North America (Fig. 9.6b), which is the

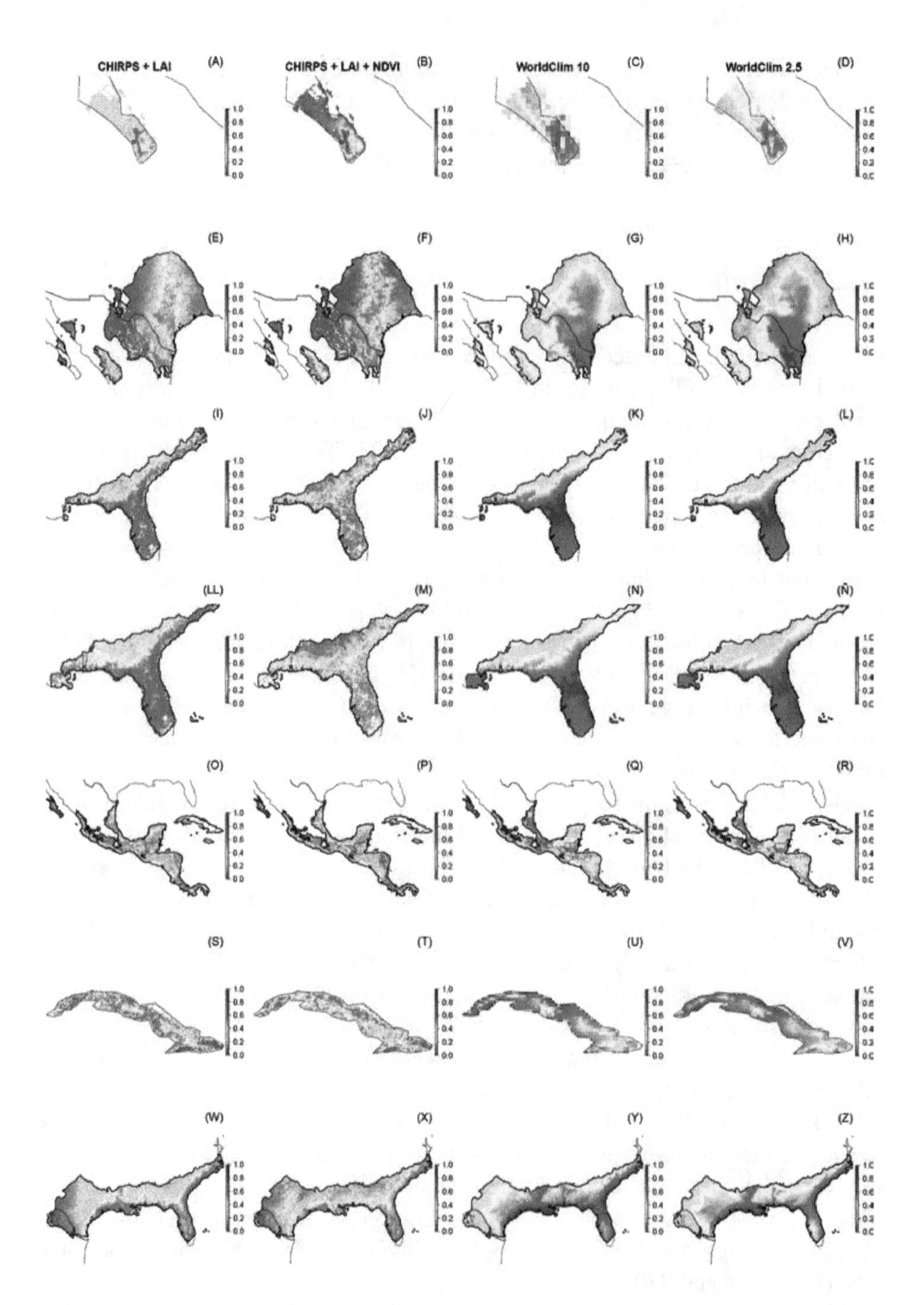

Fig. 9.5 Maps of predicted distributions for seven live oak species under three combinations of environmental variables. Legend colors represent values of suitability, where 1 and 0 represent maximum and minimum suitability, respectively. **(a–d)** *Quercus brandegeei*; **(e–h)** *Quercus fusiformis*; **(i–l)** *Quercus geminata*; (LL–Ñ) *Quercus minima*; **(o–r)** *Quercus oleoides*; **(s–v)** *Quercus sagraena*; and **(w–z)** *Quercus virginiana*. Notice that each species model was estimated using an ensemble framework such that each species model represents the average of six algorithms weighted by their AUC. For representation purposes we cropped the predicted distributions using the species geographical ranges obtained from BIEN database (Enquist et al. 2016)

Table 9.2 Model accuracy assessment for the three combinations of environmental predictors

Species	Environmental combinations	Threshold	AUC	Kappa
Quercus brandegeei	S-RS	0.5221	0.9901	0.7989
	S-RS2	0.4808	0.9983	0.9044
	WC10	0.5865	0.9992	0.9244
	WC25	0.5816	0.9996	0.9321
Quercus fusiformis	S-RS	0.4764	0.9224	0.6192
	S-RS2	0.5026	0.9432	0.7115
	WC10	0.4137	0.9205	0.6073
	WC25	0.4153	0.9020	0.5814
Quercus geminata	S-RS	0.4730	0.9470	0.6439
	S-RS2	0.4196	0.9737	0.7365
	WC10	0.4958	0.9853	0.7965
	WC25	0.5829	0.9621	0.7650
Quercus minima	S-RS	0.4311	0.9587	0.6241
	S-RS2	0.4601	0.9848	0.7311
	WC10	0.4985	0.9813	0.7429
	WC25	0.5232	0.9708	0.7158
Quercus oleoides	S-RS	0.4917	0.8408	0.5678
	S-RS2	0.4947	0.8660	0.6113
	WC10	0.5527	0.9320	0.7788
	WC25	0.4891	0.9128	0.7121
Quercus sagraena	S-RS	0.4975	0.9788	0.7185
	S-RS2	0.5196	0.9945	0.8311
	WC10	0.5423	0.9809	0.7522
	WC25	0.5758	0.9818	0.7666
Quercus virginiana	S-RS	0.3712	0.9088	0.6292
	S-RS2	0.3613	0.9396	0.7064
	WC10	0.4204	0.9424	0.7263
	WC25	0.4314	0.9231	0.7215

AUC area under the ROC curve, Kappa Cohen's Kappa coefficient, *S-RS* CHIRPS + LAI + Altitude, *S-RS2* CHIRPS + LAI + NDVI + Altitude, *WC10 and WC25* WorldClim + Altitude at spatial resolution of 10 and 2.5 arcmin, respectively

only region where three live oak species co-occur. Uncertainty maps (Fig. 9.6e–h) show that uncertainty values for the simpler S-RS models (Fig. 9.6e) are lower when compared with the other models and WorldClim models tend to show high uncertainty values across south-central North America (e.g., Texas).

The geographical relationship between the individual species models made for the four combinations of environmental predictors varied depending on the species evaluated (Table 9.3), although they show positive relationships in all cases. In general, spatial relationships between species models from S-RS products (i.e., S-RS vs S-RS2) and WorldClim (i.e., WC10 vs WC25) were strongly correlated (see also Paz et al., Chap. 11), while the relationship between models made from S-RS products

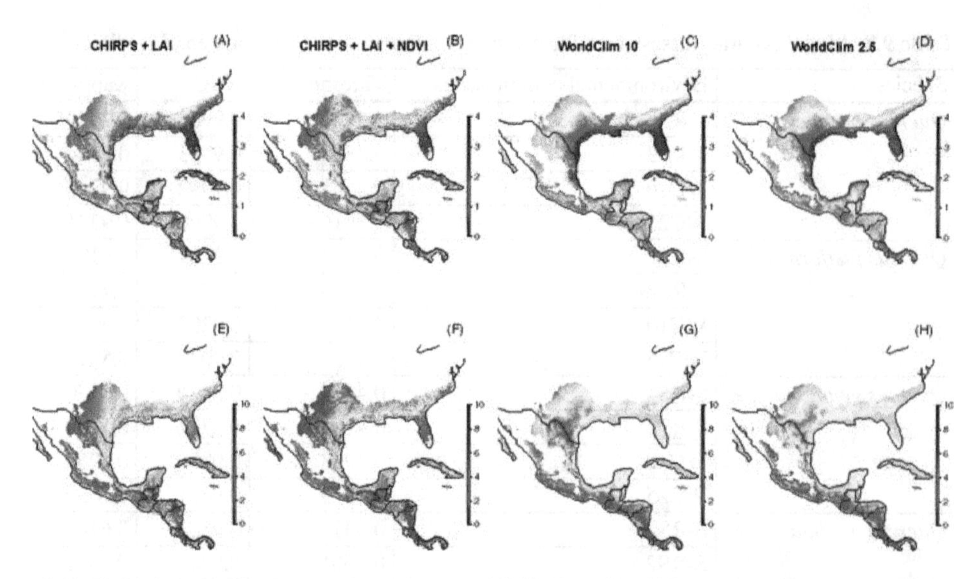

Fig. 9.6 Macroecological patterns of species richness (top panel) and uncertainty (bottom panel) for live oak species quantified under four combinations of environmental variables. WorldClim data were used at two spatial resolutions, 10 and 2.5 arcmin. See Table 9.1 for a description of the environmental combinations. Numbers on legends for the top panel represent the number of species within each pixel, where 4 means that four live oak are co-occurring in those pixels. Numbers on legends on the bottom panel represent the percentage of uncertainty or variance between algorithms, where higher values represent higher uncertainty

and WorldClim data varied slightly; the simpler models made from the S-RS data showed stronger spatial relationship to the WorldClim models (Table 9.3). Interestingly, the spatial relationships increased as a function of increasing the species potential distributions. For example, weaker spatial relationships ($r = 0.5384$ for S-RS/WC10 and $r = 0.6360$ for S-RS/WC25) were found for *Quercus brandegeei*, and stronger spatial relationships were found for dwarf live oak (*Quercus minima*; $r = 0.8150$ for S-RS/WC10 and $r = 0.8332$) and southern live oak (*Quercus virginiana*; $r = 0.8872$ for S-RS/WC10 and $r = 0.8523$ for S-RS/WC25) that are distributed across the southeastern United States (Fig. 9.5LL–Ñ and Fig. 9.5W–Z, respectively).

Finally, we found that macroecological patterns of species richness derived from the four sets of environmental predictors were strongly correlated (Table 9.4). Although the algorithms used for modeling have been emphasized as a major source of uncertainty (Diniz-Filho et al. 2009; Qiao et al. 2015), by applying the ensemble framework, we found that uncertainties due to the modeling algorithm were low (<10%) for the four sets of environmental predictors and showed similar distribution estimates (Fig. 9.6e–h). These results suggest that our results are not biased by applying a particular algorithm. Interestingly, we found low correlations between uncertainty predictions under S-RS and WC comparisons (Table 9.4), which potentially could suggest an associated error due to the predictors used to build the models.

Table 9.3 Spatial correlations between live oak ENMs estimated under three combinations of environmental variables

Species	Correlation	*r*	*F*	d.f.	*P*
Quercus brandegeei	RS/RS2	0.8441	40.8471	16.4851	0.0000
	RS/WC10	0.5384	5.7733	14.1398	0.0306
	RS/WC25	0.6360	10.3226	15.2004	0.0057
	RS2/WC10	0.6315	4.4031	6.6384	0.0762
	RS2/WC25	0.7043	7.0943	7.2068	0.0315
	WC10/WC25	0.8844	20.2487	5.6377	0.0048
Quercus fusiformis	RS/RS2	0.9105	102.5446	21.1581	0.0000
	RS/WC10	0.5735	8.1106	16.5511	0.0113
	RS/WC25	0.6117	8.7692	14.6656	0.0099
	RS2/WC10	0.5251	8.2619	21.6973	0.0089
	RS2/WC25	0.5652	8.9715	19.1088	0.0074
	WC10/WC25	0.9588	128.5055	11.2890	0.0000
Quercus geminata	RS/RS2	0.8764	62.1065	18.7489	0.0000
	RS/WC10	0.7099	14.3529	14.1270	0.0020
	RS/WC25	0.6951	12.1338	12.9824	0.0040
	RS2/WC10	0.6615	9.5259	12.2447	0.0092
	RS2/WC25	0.6589	8.6050	11.2139	0.0134
	WC10/WC25	0.9868	260.9438	7.0118	0.0000
Quercus minima	RS/RS2	0.8758	35.2577	10.7082	0.0001
	RS/WC10	0.8150	15.2887	7.7296	0.0048
	RS/WC25	0.8332	17.4921	7.7073	0.0033
	RS2/WC10	0.8051	13.1280	7.1263	0.0082
	RS2/WC25	0.8240	15.0107	7.0984	0.0059
	WC10/WC25	0.9878	195.5017	4.8702	0.0000
Quercus oleoides	RS/RS2	0.8945	930.5677	232.4530	0.0000
	RS/WC10	0.3946	29.0474	157.5251	0.0000
	RS/WC25	0.4130	35.6662	173.4187	0.0000
	RS2/WC10	0.3913	22.1329	122.4110	0.0000
	RS2/WC25	0.4059	26.5171	134.4395	0.0000
	WC10/WC25	0.9290	423.5265	67.2515	0.0000
Quercus sagraena	RS/RS2	0.7268	76.7276	68.5351	0.0000
	RS/WC10	0.6090	33.5887	56.9651	0.0000
	RS/WC25	0.6626	42.6603	54.5144	0.0000
	RS2/WC10	0.4627	17.5238	64.3187	0.0001
	RS2/WC25	0.5028	20.8935	61.7460	0.0000
	WC10/WC25	0.9333	83.3701	12.3397	0.0000
Quercus virginiana	RS/RS2	0.9064	47.3462	10.2891	0.0000
	RS/WC10	0.8872	42.5539	11.5129	0.0000
	RS/WC25	0.8523	44.9094	16.9107	0.0000
	RS2/WC10	0.7865	24.3299	14.9996	0.0002
	RS2/WC25	0.7748	32.6135	21.7171	0.0000
	WC10/WC25	0.9662	279.5733	19.9292	0.0000

S-RS CHIRPS + LAI + Altitude, *S-RS2* CHIRPS + LAI + NDVI + Altitude, WC10 and WC25 WorldClim + Altitude, at spatial resolution of 10 and 2.5 arc-min, respectively. *r* Pearson's correlation coefficient, *F* F-statistic, *d.f.* degrees of freedom, *P* associated p-value

Table 9.4 Spatial correlation between estimations of live oak species richness and uncertainty quantified under three combinations of environmental variables

Component	Correlation	*r*	*F*	d.f.	*P*
Species richness	RS/RS2	0.8991	98.5384	23.3534	0.0000
	RS/WC10	0.7420	27.0162	22.0477	0.0000
	RS/WC25	0.7224	25.6607	23.5094	0.0000
	RS2/WC10	0.7269	29.3608	26.2069	0.0000
	RS2/WC25	0.7120	28.6270	27.8452	0.0000
	WC10/WC25	0.9858	791.2799	22.9570	0.0000
Uncertainty	RS/RS2	0.8163	47.6775	23.8806	0.0000
	RS/WC10	0.3549	2.5399	17.6208	0.1288
	RS/WC25	0.4561	5.4356	20.6974	0.0299
	RS2/WC10	0.2315	1.2401	21.9056	0.2775
	RS2/WC25	0.3164	2.8755	25.8479	0.1019
	WC10/WC25	0.9425	136.3926	17.1534	0.0000

S-RS CHIRPS + LAI + Altitude, *S-RS2* CHIRPS + LAI + NDVI + Altitude, *WC10 and WC25* WorldClim + Altitude at spatial resolution of 10 and 2.5 arcmin, respectively

9.4 Perspectives

The field of ecological niche and species distribution modeling contributes significantly to our capacity to evaluate and describe the effect of geographical and environmental features on species distributions and has become in one of the most widely applied tools for the assessment of the impact of climate change and human activities on species and communities, biological invasions, epidemiology, and conservation biology (Peterson et al. 2011; Guisan et al. 2017). However, despite important advances in theory (Soberón 2007; Colwell and Rangel 2009; Soberón and Nakamura 2009; Peterson et al. 2011), methods, and algorithms (reviewed in Duarte et al. 2019; see also Warren et al. 2018) and practical applications (Guillera-Arroita et al. 2015; Cord et al. 2017; Sanín and Anderson 2018), most studies still rely on the use of interpolated climate data as environmental predictors (Saatchi et al. 2008; Waltari et al. 2014). In this article we compare the performance of environmental data derived from interpolated climate surfaces data (i.e., WorldClim) and S-RS products data (i.e., LAI and NDVI). Specifically, using live oaks as a case study, we show the advances and potential caveats in using S-RS data in describing and predicting species-environment relationships. Overall, our analyses show that S-RS products perform, as well as products from interpolated climate surfaces as environmental predictors (Tables 9.2, 9.3, and 9.4), and indeed present quite similar results for both species environmental suitability and macroecological patterns (Figs. 9.5 and 9.6), similar to Paz et al. (Chap. 11). However, they have the potential to provide more precise estimates of species distributions at higher spatial resolution.

In our example, we used different grain sizes for both data sets: WorldClim (10 and 2.5 arcmin or ~18.5 and ~ 4.5 km at the equator, respectively) and S-RS products (3 arcmin or ~5.5 km at the equator). Although changing grain size in the predictors might be anticipated to affect model performance, different lines of evidence indicate that model performance is not affected by grain resolution, but rather by species response to the environmental conditions in the study region (Guisan et al. 2007, see also Fig. 9.5). Our results show that enhancing spatial resolution of interpolated climatic data does not improve the spatial resolution at which species distributions can be accurately predicted. The quality of interpolated climate surfaces such as WorldClim, which depends on climatic stations as data sources, has been ignored as a source of uncertainty in studies of species-environment relationships—for example, Hijmans et al. (2005) used a variable number of weather stations for their interpolations, 47.554, 24.542, and 14.930 for precipitation, mean temperature, and maximum and minimum temperature, respectively—especially in the tropics (Fig. 9.2c), where weather stations are sparse (Hijmans et al. 2005; Soria-Auza et al. 2010). This source of uncertainty can be avoided using S-RS products, which have continuous (from daily to monthly) and quasi-global environmental information, including precipitation, temperature, and biophysical variables that represent different components of vegetation and ecosystems (Funk et al. 2015; Cord et al. 2017; Radeloff et al. 2019).

Fine and broad spatial and temporal scale data derived from S-RS, which have only been available in the last ~20 years (Turner 2014), can be used to improve the evaluation of species-environment relationships. A number of research avenues remain to be pursued to better understand the potential of S-RS data and their products in quantifying species ecological niches and estimating species distributions. For example, applying the same framework presented here to other species or clades (including vertebrates and invertebrates) or applying more complex frameworks (e.g., Peterson and Nakazawa 2008; Waltari et al. 2014) may shed light on the potential of S-RS products as predictors for the analysis of species-environment relationships. This is important because ENMs/SDMs are used as predictive models that can be extrapolated across space and time to forecast and monitor biodiversity under a changing global climate (Peterson and Nakazawa 2008; Warren 2012).

9.4.1 Should We Use S-RS Data for ENM/SDM?

Whether S-RS data should replace other environmental data in modeling niches and projecting species distribution depends on the modeling purposes (Peterson et al. 2011). In fact, modeling species niches and projecting distributions involves relating a set of species occurrences to relevant environmental predictors. In essence, ENM/SDM based only on climatic variables would tend to return broad predictions (Coudun et al. 2006, see also right panel in Fig. 9.5), particularly because climatic data are useful in describing macroecological patterns of species distributions and communities (Lin and Wiens 2017; Manzoor et al. 2018), while ENM/SDM based on S-RS data alone allows the discrimination of local features not captured by climatic information (Coudun et al. 2006; Saatchi et al. 2008; Radeloff et al. 2019).

Nevertheless, it seems that overall statistical model accuracy in this example is not improved (Pearson et al. 2004; Thuiller 2004; see also Table 9.2). Given that climatic and S-RS data provide information at different spatial and temporal scales, a promising option would be to use both sources of environmental predictors to model species distributions to achieve "the best of both worlds" (Saatchi et al. 2008; Pradervand et al. 2013).

More accurate predictions of species distributions are critical for the development of conservation and management actions if we are to meet the challenges posed by global change (Coudun et al. 2006; Zimmermann et al. 2007; Cord et al. 2013). Our point here is to facilitate and demonstrate the potential for the use of S-RS data for predicting species distributions and modeling environmental niches. The results we show here and those of others (e.g., Saatchi et al. 2008; Waltari et al. 2014) indicate that S-RS data provide a valuable complement to other environmental variables for ENM/SDM.

Another potential and important research direction is the use of S-RS products that have high temporal resolution, such as LAI (Fig. 9.3b), as biophysical variables that represent ecosystem functions (Cord et al. 2013, 2017). These products allow the exploration of dynamics of vegetation growth and seasonality in vegetation function, fundamental features that characterize vegetation form and function (Myneni et al. 2002; Hobi et al. 2017). Here, using metrics derived from MODIS LAI in combination with other S-RS products (Table 9.1), we show that, using relevant biophysical variables, it is possible to predict distributions similar to those predicted from climate data alone (Fig. 9.5, Tables 9.2 and 9.3). In fact, a recent study (Simões and Peterson 2018) found that including biotic predictors can improve ENMs even while increasing model complexity, such that the combination of abiotic and biotic predictors improves model performance (Simões and Peterson 2018). To confirm this conclusion, substantial effort would be needed, including new methodological and conceptual approaches, to disentangle the real contribution of S-RS products—spatial and temporal features of S-RS products that improve statistical model performance—as predictors of species distributions. Nonetheless, our results highlight an advance on the use of relevant predictors for modeling species-environment relationships.

In addition, recent macroecological studies have used these products to relate annual vegetation productivity to continental and global patterns of species richness (Pigot et al. 2016; Hobi et al. 2017; Coops et al. 2018), providing spatially explicit support for the use of satellite data products in predicting biodiversity. These advances point to an exciting avenue for the study of the distribution and assembly of biological communities (Ferrier and Guisan 2006). For example, S-RS products can be used for the development of stacked species distribution models (S-SDM, see Fig. 9.6) that can be integrated into novel biodiversity modeling frameworks, such as Spatially Explicit Species Assemblage Modelling (SESAM, Guisan and Rahbek 2011) or the Hierarchical Modelling of Species Communities (HMSC, Ovaskainen et al. 2017), aimed at predicting composition and distribution of species and communities (Mateo et al. 2017).

Finally, the example presented here is meant to spur further theoretical, methodological, and empirical research aimed at developing a Global Biodiversity Observatory (Geller et al., Chap. 20). Explicit incorporation of biotic information into species-environment modeling may turn our focus away from the use only of climatic information toward the "complete" evaluation of the drivers that determine the species distributions (Fig. 9.1).

9.4.2 Enabling Large-Scale Biodiversity Change Detection

Since the last millennium, rising human population and activity have been major drivers of environmental change on Earth, with consequences for the distribution and abundance of biodiversity and associated ecosystem functioning (Tilman 1997; Tylianakis et al. 2008). Thus, improving large-scale biodiversity change detection is crucial to the development of effective policies that advance conservation and management of species and communities.

Such efforts are critical to enhancing efforts to develop a Global Biodiversity Observatory (Geller et al., Chap. 20; Jetz et al. 2016). Research interest in using S-RS has increased in recent years given its high potential for monitoring global biodiversity and detecting change (Turner 2014; Jetz et al. 2016). For example, it is possible to identify shifts in vegetation structure or to monitor the dynamics of the growing season of an entire region, or within a specific species geographical range (Fig. 9.4) using time series S-RS products such as LAI—half of the total green leaf area per unit of horizontal ground surface area (Xiao et al. 2014)—which has a temporal resolution of 8 days. This is particularly important given that ENM/SDM theory assumes that species' niches are stable across time and space and that species and their environments are at pseudo-equilibrium, suggesting that species are occupying all suitable areas (Guisan and Thuiller 2005). However, the environment is dynamic and can change even at small scales; species ranges can thus expand and retract across time, varying within species lifetimes as well as over evolutionary timescales encompassing many generations. Long-term series of S-RS data products (i.e., spatial and temporal) supply remarkable opportunities for assessing and monitoring the state of the Earth's surface and, combining with species-environment relationship modeling, provide new frontiers for the prediction of species distributions and species monitoring across time and space (Randin et al. 2020). Indeed, using biophysical variables derived from high-resolution S-RS products (i.e., LAI) allows the identification of geographic areas where species actually occur (Fig. 9.7) and thus has the potential for enhancing the predictions of a set of species that could occur in an area—species pool—that is used for species assignments from direct RS detection using hyperspectral data (see simulation in Fig. 7.8, Section 9.4.2 in Meireles et al., Chap. 7).

In addition, enhancing predictive models of the species expected to be present in a given geographic region can be coupled with other means of detecting which species are present based on spectroscopic imaging (Serbin et al. 2015; Bolch et al.,

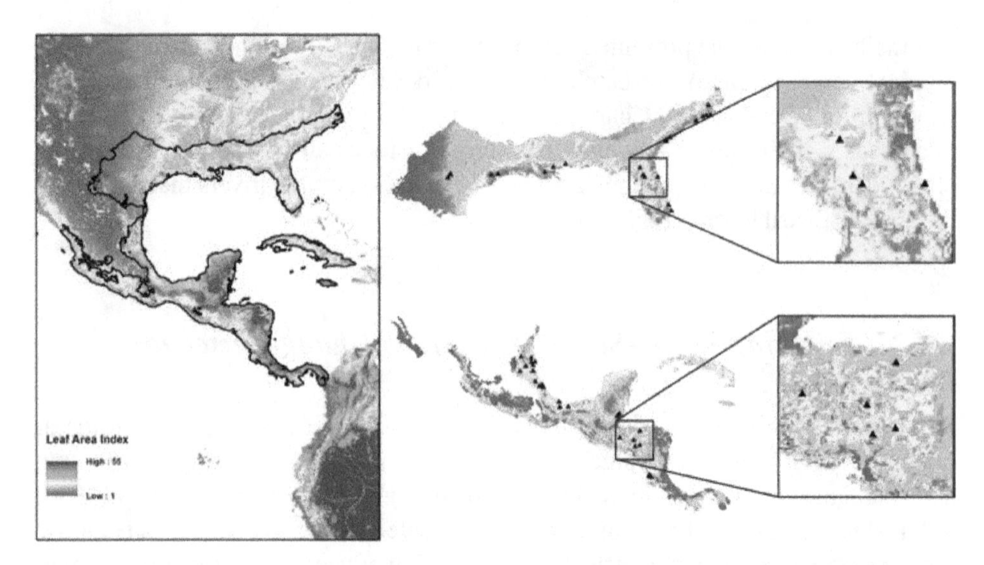

Fig. 9.7 Mean LAI estimated at 8-day intervals averaged over 15 years (left panel). The overlaid continuous lines correspond to the geographical ranges of *Q. virginiana* and *Q. oleoides* obtained from BIEN database. Predicted distributions for *Q. virginiana* (top right panel) and *Q. oleoides* (bottom right panel) are based on S-RS products, which include the temporal variation in LAI shown in A. The triangles over the maps represent occurrence points used for calibration (where the authors have collected specimens), and the boxes represent a zoom over a specific area of the predicted species distributions. Note that high values of the predicted distributions coincide with the occurrence points

Chap. 12; Meireles et al., Chap. 7), reducing the complexity of species identification algorithms. For example, imaging spectroscopy allows mapping of functional traits, by estimating vegetation traits for each pixel in an image (Wang et al. 2019; Asner et al. 2017; Martin, Chap. 5). Plant spectra obtained from imaging spectroscopy at different spatial resolutions can in turn be used to detect different aspects and traits—within- and between-species differences in morphology, foliar chemistry, life history strategies—of plant species (Ustin and Gamon 2010; Cavender-Bares et al. 2017; Schweiger et al. 2018) and the correct identification of different taxonomic levels from populations to species to clades (Cavender-Bares et al. 2016). Thus, the integration of spectral approaches with techniques for modeling species ecological niches has the potential to produce reliable information of species distributions and co-occurrence, filling current gaps about species-environment relationships at a range of spatial scales and levels of organization—from species to communities—increasing the accuracy of direct detection assignments, and enabling monitoring of changes in biodiversity, one of the premises for the sustainable management of the biosphere (Pinto-Ledezma and Rivero 2014; Fernández et al., Chap. 18).

Acknowledgments We thank Julián Velasco for his advice with ENM/SDM analyses. J.N.P-L. was supported by the University of Minnesota College of Biological Sciences' Grand Challenges in Biology Postdoctoral Program. We acknowledge the bioDISCOVERY community at the University of Zürich for important discussions and the NSF Research Coordination Network "Biodiversity Across Scales."

References

Anderson RP (2012) Harnessing the world's biodiversity data: promise and peril in ecological niche modeling of species distributions. Ann N Y Acad Sci 1260(1):66–80. https://doi.org/10.1111/j.1749-6632.2011.06440.x

Anderson RP (2013) A framework for using niche models to estimate impacts of climate change on species distributions. Ann N Y Acad Sci 1297(1):8–28. https://doi.org/10.1111/nyas.12264

Araújo MB, New M (2007) Ensemble forecasting of species distributions. Trends Ecol Evol 22(1):42–47. https://doi.org/10.1016/j.tree.2006.09.010

Asner GP, Martin RE, Knapp DE et al (2017) Airborne laser-guided imaging spectroscopy to map forest trait diversity and guide conservation. Science 355(6323):385–389. https://doi.org/10.1126/science.aaj1987

Berry S, Mackey B, Brown T (2007) Potential applications of remotely sensed vegetation greenness to habitat analysis and the conservation of dispersive fauna. Pacific Cons Bio 13(2):120. https://doi.org/10.1071/pc070120

Bivand R, Keitt T, Rowlingson B (2018) rgdal: bindings for the 'geospatial' data abstraction library. R package version 1.3-4. https://CRAN.R-project.org/package=rgdal.

Booth TH, Nix HA, Busby JR et al (2013) Bioclim: the first species studies. Divers Distrib 20(1):1–9. https://doi.org/10.1111/ddi.12144

Cavender-Bares J, Gonzalez-Rodriguez A, Pahlich A et al (2011) Phylogeography and climatic niche evolution in live oaks (Quercus series Virentes) from the tropics to the temperate zone. J Biogeogr 38(5):962–981. https://doi.org/10.1111/j.1365-2699.2010.02451.x

Cavender-Bares J, Gonzalez-Rodriguez A, Eaton DA et al (2015) Phylogeny and biogeography of the American live oaks (Quercus subsection Virentes): a genomic and population genetics approach. Mol Ecol 24(14):3668–3687. https://doi.org/10.1111/mec.13269

Cavender-Bares J, Meireles JE, Couture J et al (2016) Associations of leaf spectra with genetic and phylogenetic variation in oaks: prospects for remote detection of biodiversity. Remote Sens 8(3):221. https://doi.org/10.3390/rs8030221

Cavender-Bares J, Gamon JA, Hobbie SE et al (2017) Harnessing plant spectra to integrate the biodiversity sciences across biological and spatial scales. Am J Bot 104(7):966–969. https://doi.org/10.3732/ajb.1700061

Clifford P, Richardson S, Hemon D (1989) Assessing the significance of the correlation between two spatial processes. Biometrics 45(1):123–134. https://doi.org/10.2307/2532039

Colwell RK, Rangel TF (2009) Hutchinson's duality: the once and future niche. Proc Natl Acad Sci U S A 106(Supplement_2):19651–19658. https://doi.org/10.1073/pnas.0901650106

Coops NC, Kearney SP, Bolton DK et al (2018) Remotely-sensed productivity clusters capture global biodiversity patterns. Sci Rep 8(1):16261. https://doi.org/10.1038/s41598-018-34162-8

Cord AF, Meentemeyer RK, Leitão PJ et al (2013) Modelling species distributions with remote sensing data: bridging disciplinary perspectives. J Biogeogr 40(12):2226–2227. https://doi.org/10.1111/jbi.12199

Cord AF, Brauman KA, Chaplin-Kramer R et al (2017) Priorities to advance monitoring of ecosystem services using earth observation. Trends Ecol Evol 32(6):416–428. https://doi.org/10.1016/j.tree.2017.03.003

Coudun C, Gégout JC, Piedallu C et al (2006) Soil nutritional factors improve models of plant species distribution: an illustration with *Acer campestre* (L.) in France. J Biogeogr 33(10):1750–1763. https://doi.org/10.1111/j.1365-2699.2005.01443.x

Diniz-Filho JAF, Bini L, Rangel TF et al (2009) Partitioning and mapping uncertainties in ensembles of forecasts of species turnover under climate change. Ecography 32(6):897–906. https://doi.org/10.1111/j.1600-0587.2009.06196.x

Duarte A, Whitlock SL, Peterson JT (2019) Species distribution modeling. In: Encyclopedia of Ecology, pp 189–198. https://doi.org/10.1016/b978-0-12-409548-9.10572-x

Elith J, Franklin J (2013) Species distribution modeling. In: Encyclopedia of biodiversity, pp 692–705. https://doi.org/10.1016/b978-0-12-384719-5.00318-x

Elton C (1927) Animal Ecology. Sedgwick and Jackson, London
Enquist BJ, Condit R, Peet RK et al (2016) Cyberinfrastructure for an integrated botanical information network to investigate the ecological impacts of global climate change on plant biodiversity. PeerJ Preprints 4:e2615v2. https://doi.org/10.7287/peerj.preprints.2615v2
Ferrier S, Guisan A (2006) Spatial modelling of biodiversity at the community level. J Appl Ecol 43(3):393–404. https://doi.org/10.1111/j.1365-2664.2006.01149.x
Franklin J (2010) Mapping species distribution: spatial inference and prediction. Cambridge University Press, Cambridge
Funk C, Peterson P, Landsfeld M et al (2015) The climate hazards infrared precipitation with stations—a new environmental record for monitoring extremes. Sci Data 2:150066. https://doi.org/10.1038/sdata.2015.66
Gaston KJ (2009) Geographic range limits: achieving synthesis. Proc R Soc Lond B Biol Sci 276(1661):1395–1406. https://doi.org/10.1098/rspb.2008.1480
Gillespie TW, Foody GM, Rocchini D et al (2008) Measuring and modelling biodiversity from space. Prog Phys Geogr 32(2):203–221. https://doi.org/10.1177/0309133308093606
Greenberg JA, Mattiuzzi M (2018) gdalUtils: wrappers for the geospatial data abstraction library (GDAL) utilities. R package version 2.0.1.14. https://CRAN.R-project.org/package=gdalUtils.
Grinnell J (1904) The Origin and Distribution of the Chest-Nut-Backed Chickadee. The Auk, 21(3), 364–382. https://doi.org/10.2307/4070199
Grinnell J (1917) The Niche-Relationships of the California Thrasher. The Auk, 34(4), 427–433. https://doi.org/10.2307/4072271
Guillera-Arroita G, Lahoz-Monfort JJ, Elith J et al (2015) Is my species distribution model fit for purpose? Matching data and models to applications. Glob Ecol Biogeogr 24(3):276–292. https://doi.org/10.1111/geb.12268
Guisan A, Thuiller W (2005) Predicting species distribution: offering more than simple habitat models. Ecol Lett 8(9):993–1009. https://doi.org/10.1111/j.1461-0248.2005.00792.x
Guisan A, Graham CH, Elith J et al (2007) Sensitivity of predictive species distribution models to change in grain size. Divers Distrib 13(3):332–340. https://doi.org/10.1111/j.1472-4642.2007.00342.x
Guisan A, Rahbek C (2011) SESAM - a new framework integrating macroecological and species distribution models for predicting spatio-temporal patterns of species assemblages. Journal of Biogeography, 38(8), 1433–1444. https://doi.org/10.1111/j.1365-2699.2011.02550.x
Guisan A, Thuiller W, Zimmermann NE (2017) Habitat suitability and distribution models. Cambridge University Press, Cambridge. https://doi.org/10.1017/9781139028271
Hijmans RJ, Phillips S, Leathwick J et al (2017) dismo: species distribution modeling. R package version 1.1-4. https://CRAN.R-project.org/package=dismo.
Hijmans RJ (2018) raster: Geographic Data Analysis and Modeling. R package version 3.0-7. https://CRAN.R-project.org/package=raster
Hijmans RJ, Cameron SE, Parra JL et al (2005) Very high resolution interpolated climate surfaces for global land areas. Int J Climatol 25(15):1965–1978. https://doi.org/10.1002/joc.1276
Hipp AL, Manos PS, González-Rodríguez A et al (2017) Sympatric parallel diversification of major oak clades in the Americas and the origins of Mexican species diversity. New Phytol 217(1):439–452. https://doi.org/10.1111/nph.14773
Hobi ML, Dubinin M, Graham CH et al (2017) A comparison of dynamic habitat indices derived from different MODIS products as predictors of avian species richness. Remote Sens Environ 195:142–152. https://doi.org/10.1016/j.rse.2017.04.018
Holdridge LR (1947) Determination of world plant formations from simple climatic data. Science 105(2727):367–368. https://doi.org/10.1126/science.105.2727.367
Hutchinson GE (1957) Concluding Remarks. Cold Spring Harbor Symposia on Quantitative Biology, 22(0), 415–427. https://doi.org/10.1101/sqb.1957.022.01.03
Jetz W, Cavender-Bares J, Pavlick R et al (2016) Monitoring plant functional diversity from space. Nat Plants 2(3):16024. https://doi.org/10.1038/nplants.2016.24
Karatzoglou A, Smola A, Hornik K et al (2004) Kernlab – an S4 package for Kernel methods in R. J Stat Softw 11(9):1–20. http://www.jstatsoft.org/v11/i09/.

Kissling WD, Dormann CF, Groeneveld J et al (2012) Towards novel approaches to modelling biotic interactions in multispecies assemblages at large spatial extents. J Biogeogr 39(12): 2163–2178. https://doi.org/10.1111/j.1365-2699.2011.02663.x

Koehler K, Center A, Cavender-Bares J (2011) Evidence for a freezing tolerance-growth rate trade-off in the live oaks (Quercus series Virentes) across the tropical-temperate divide. New Phytol 193(3):730–744. https://doi.org/10.1111/j.1469-8137.2011.03992.x

Liaw A, Wiener M (2002) Classification and regression by randomForest. R News 2:18–22

Lin LH, Wiens JJ (2017) Comparing macroecological patterns across continents: evolution of climatic niche breadth in varanid lizards. Ecography 40(8):960–970. https://doi.org/10.1111/ecog.02343

Lobo JM, Jiménez-Valverde A, Hortal J (2010) The uncertain nature of absences and their importance in species distribution modelling. Ecography 33(1):103–114. https://doi.org/10.1111/j.1600-0587.2009.06039.x

Manzoor SA, Griffiths G, Lukac M (2018) Species distribution model transferability and model grain size – finer may not always be better. Sci Rep 8(1):7168. https://doi.org/10.1038/s41598-018-25437-1

Mateo RG, Mokany K, Guisan A (2017) Biodiversity models: what if unsaturation is the rule? Trends Ecol Evol 32(8):556–566. https://doi.org/10.1016/j.tree.2017.05.003

Milborrow S (2016) Earth: multivariate adaptive regression splines. R package version 4.4.4. Retrieved from https://CRAN.R-project.org/package=earth.

Myneni R, Hoffman S, Knyazikhin Y et al (2002) Global products of vegetation leaf area and fraction absorbed PAR from year one of MODIS data. Remote Sens Environ 83(1–2):214–231. https://doi.org/10.1016/s0034-4257(02)00074-3

Nix HA (1986) A biogeographic analysis of Australian elapid snakes. In: Longmore R (ed) Atlas of elapid snakes of Australia: Australian Flora and Fauna Series, vol 7. Bureau of Flora and Fauna, Canberra, pp 4–15

Ovaskainen O, Tikhonov G, Norberg A et al (2017) How to make more out of community data? A conceptual framework and its implementation as models and software. Ecol Lett 20(5):561–576. https://doi.org/10.1111/ele.12757

Pearson RG, Dawson TP, Liu C (2004) Modelling species distributions in Britain: a hierarchical integration of climate and land-cover data. Ecography 27(3):285–298. https://doi.org/10.1111/j.0906-7590.2004.03740.x

Peterson AT, Martínez-Meyer E, Soberón J et al (2011) Ecological niches and geographic distributions. Monographs in population biology, vol 49. Princeton University Press, Princeton

Peterson AT, Nakazawa Y (2008) Environmental data sets matter in ecological niche modelling: an example with *Solenopsis invicta* and *Solenopsis richteri*. Glob Ecol Biogeogr 17(1):135–144. https://doi.org/10.1111/j.1466-8238.2007.00347.x

Pettorelli N, Safi K, Turner W (2014a) Satellite remote sensing, biodiversity research and conservation of the future. Philos Trans R Soc Lond Ser B Biol Sci 369(1643):20130190–20130190. https://doi.org/10.1098/rstb.2013.0190

Pettorelli N, Laurance WF, O'Brien TG et al (2014b) Satellite remote sensing for applied ecologists: opportunities and challenges. J Appl Ecol 51(4):839–848. https://doi.org/10.1111/1365-2664.12261

Phillips LB, Hansen AJ, Flather CH (2008) Evaluating the species energy relationship with the newest measures of ecosystem energy: NDVI versus MODIS primary production. Remote Sens Environ 112(9):3538–3549. https://doi.org/10.1016/j.rse.2008.04.012

Pigot AL, Tobias JA, Jetz W (2016) Energetic constraints on species coexistence in birds. PLoS Biol 14(3):e1002407. https://doi.org/10.1371/journal.pbio.1002407

Pinto-Ledezma JN, Rivero ML (2014) Temporal patterns of deforestation and fragmentation in lowland Bolivia: implications for climate change. Clim Chang 127(1):43–54. https://doi.org/10.1007/s10584-013-0817-1

Pradervand JN, Dubuis A, Pellissier L et al (2013) Very high resolution environmental predictors in species distribution models. Prog Phys Geogr 38(1):79–96. https://doi.org/10.1177/0309133313512667

Qiao H, Soberón J, Peterson AT (2015) No silver bullets in correlative ecological niche modelling: insights from testing among many potential algorithms for niche estimation. Methods Ecol Evol 6(10):1126–1136. https://doi.org/10.1111/2041-210x.12397

R Core Team (2018) R: a language and environment for statistical computing. R Foundation for Statistical Computing, Vienna, Austria. https://www.R-project.org/

Radeloff VC, Dubinin M, Coops NC et al (2019) The dynamic habitat indices (DHIs) from MODIS and global biodiversity. Remote Sens Environ 222:204–214. https://doi.org/10.1016/j.rse.2018.12.009

Randin CF, Ashcroft MB, Bolliger J et al. (2020) Monitoring biodiversity in the Anthropocene using remote sensing in species distribution models. Remote Sens Environ 239:111626 https://doi.org/10.1016/j.rse.2019.111626

Smith AB, Godsoe W, Rodríguez-Sánchez F et al (2018) Niche estimation above and below the species level. Trends Ecol Evol 34:260. https://doi.org/10.1016/j.tree.2018.10.012

Saatchi S, Buermann W, ter Steege H et al (2008) Modeling distribution of Amazonian tree species and diversity using remote sensing measurements. Remote Sens Environ 112(5):2000–2017. https://doi.org/10.1016/j.rse.2008.01.008

Sanín C, Anderson RP (2018) A framework for simultaneous tests of abiotic, biotic, and historical drivers of species distributions: empirical tests for north American Wood warblers based on climate and pollen. Am Nat 192(2):E48–E61. https://doi.org/10.1086/697537

Serbin SP, Singh A, Desai AR et al (2015) Remotely estimating photosynthetic capacity, and its response to temperature, in vegetation canopies using imaging spectroscopy. Remote Sens Environ 167:78–87. https://doi.org/10.1016/j.rse.2015.05.024

Simões MVP, Peterson AT (2018) Importance of biotic predictors in estimation of potential invasive areas: the example of the tortoise beetle *Eurypedus nigrosignatus*, in Hispaniola. PeerJ 6:e6052. https://doi.org/10.7717/peerj.6052

Soberón J (2007) Grinnellian and Eltonian niches and geographic distributions of species. Ecol Lett 10(12):1115–1123. https://doi.org/10.1111/j.1461-0248.2007.01107.x

Soberón J, Nakamura M (2009) Niches and distributional areas: concepts, methods, and assumptions. Proc Natl Acad Sci U S A 106(Supplement_2):19644–19650. https://doi.org/10.1073/pnas.0901637106

Soberón J, Peterson AT (2005) Interpretation of models of fundamental ecological niches and species' distributional areas. Biodivers Inform 2. https://doi.org/10.17161/bi.v2i0.4

Soberón J, Osorio-Olvera L, Peterson AT (2017) Diferencias conceptuales entre modelación de nichos y modelación de áreas de distribución. Rev Mex Biodivers 88(2):437–441. https://doi.org/10.1016/j.rmb.2017.03.011

Sokal RR, Rohlf FJ (1995) Biometry: the principles and practice of statistics in biological research, 3rd edn. W. H. Freeman and Co., New York

Soria-Auza RW, Kessler M, Bach K et al (2010) Impact of the quality of climate models for modelling species occurrences in countries with poor climatic documentation: a case study from Bolivia. Ecol Model 221(8):1221–1229. https://doi.org/10.1016/j.ecolmodel.2010.01.004

Schweiger AK, Cavender-Bares J, Townsend PA et al (2018) Plant spectral diversity integrates functional and phylogenetic components of biodiversity and predicts ecosystem function. Nat Ecol Evol 2(6):976–982. https://doi.org/10.1038/s41559-018-0551-1

Thuiller W (2004) Patterns and uncertainties of species' range shifts under climate change. Glob Chang Biol 10(12):2020–2027. https://doi.org/10.1111/j.1365-2486.2004.00859.x

Tilman D (1997) The influence of functional diversity and composition on ecosystem processes. Science 277(5330):1300–1302. https://doi.org/10.1126/science.277.5330.1300

Turner W (2014) Sensing biodiversity. Science 346(6207):301–302. https://doi.org/10.1126/science.1256014

Turner W, Spector S, Gardiner N et al (2003) Remote sensing for biodiversity science and conservation. Trends Ecol Evol 18(6):306–314. https://doi.org/10.1016/s0169-5347(03)00070-3

Tylianakis JM, Didham RK, Bascompte J et al (2008) Global change and species interactions in terrestrial ecosystems. Ecol Lett 11(12):1351–1363. https://doi.org/10.1111/j.1461-0248.2008.01250.x

Ustin SL, Gamon JA (2010) Remote sensing of plant functional types. New Phytol 186(4):795–816. https://doi.org/10.1111/j.1469-8137.2010.03284.x
Vallejos R, Osorio F (2014) Effective sample size of spatial process models. Spat Stat 9:66–92. https://doi.org/10.1016/j.spasta.2014.03.003
Wallace AR (1860) On the zoological geography of the Malay archipelago. Zool J Linnean Soc 4(16):172–184. https://doi.org/10.1111/j.1096-3642.1860.tb00090.x
Waltari E, Schroeder R, McDonald K et al (2014) Bioclimatic variables derived from remote sensing: assessment and application for species distribution modelling. Methods Ecol Evol 5(10):1033–1042. https://doi.org/10.1111/2041-210x.12264
Wang Z, Townsend PA, Schweiger AK et al (2019) Mapping foliar functional traits and their uncertainties across three years in a grassland experiment. Remote Sens Environ 221:405–416. https://doi.org/10.1016/j.rse.2018.11.016
Warren DL (2012) In defense of "niche modeling". Trends Ecol Evol 27(9):497–500. https://doi.org/10.1016/j.tree.2012.03.010
Warren DL, Beaumont LJ, Dinnage R et al (2018) New methods for measuring ENM breadth and overlap in environmental space. Ecography 42:444. https://doi.org/10.1111/ecog.03900
West AM, Evangelista PH, Jarnevich CS et al (2016) Integrating remote sensing with species distribution models; mapping tamarisk invasions using the software for assisted habitat modeling (SAHM). J Vis Exp (116). https://doi.org/10.3791/54578
Wood SN (2006) Generalized additive models. Chapman and Hall/CRC, Boca Raton
Xiao Z, Liang S, Wang J et al (2014) Use of general regression neural networks for generating the GLASS leaf area index product from time-series MODIS surface reflectance. IEEE Trans Geosci Remote Sens 52(1):209–223. https://doi.org/10.1109/tgrs.2013.2237780
Zimmermann NE, Edwards TC, Moisen GG et al (2007) Remote sensing-based predictors improve distribution models of rare, early successional and broadleaf tree species in Utah. J Appl Ecol 44(5):1057–1067. https://doi.org/10.1111/j.1365-2664.2007.01348.x

10

Biodiversity-Geodiversity Relationships

Sydne Record, Kyla M. Dahlin, Phoebe L. Zarnetske, Quentin D. Read, Sparkle L. Malone, Keith D. Gaddis, John M. Grady, Jennifer Costanza, Martina L. Hobi, Andrew M. Latimer, Stephanie Pau, Adam M. Wilson, Scott V. Ollinger, Andrew O. Finley, and Erin Hestir

10.1 Conserving Nature's Stage

Biodiversity is essential for ecosystem functioning and ecosystem services (Chapin et al. 1997; Yachi and Loreau 1999). Yet rapid global change is altering biodiversity and endangering its vital functions, with human-caused habitat deterioration being the number one cause of biodiversity loss (Sala et al. 2000). In addition, climate change is directly affecting individual species abundances and distributions and indirectly affecting species via biotic interactions (Walther et al. 2002). When combined, these effects lead to novel ecological communities for which there are no modern analogs (Williams and Jackson 2007). Although species have continually experienced shifts in climate, the recent rate of temperature change is more rapid than in any other timeframe in the past 10,000 years (Marcott et al. 2013), and temperatures are expected to rise even faster in the near future (Smith et al. 2015). In light of these rapid global changes, a major challenge for biodiversity scientists is to generate robust statistical models that describe and predict biodiversity in space and

S. Record (✉)
Department of Biology, Bryn Mawr College, Bryn Mawr, PA, USA
e-mail: srecord@brynmawr.edu

K. M. Dahlin
Department of Geography, Environment, & Spatial Sciences, Michigan State University, East Lansing, MI, USA

Ecology, Evolutionary Biology, and Behavior Program, Michigan State University, East Lansing, MI, USA

P. L. Zarnetske · Q. D. Read
Ecology, Evolutionary Biology, and Behavior Program, Michigan State University, East Lansing, MI, USA

Department of Integrative Biology, Michigan State University, East Lansing, MI, USA time, from which changes in hot spots (highs) and cold spots (lows) of biodiversity may indicate shifts in ecosystem functions and services.

Contemporary strategies for addressing and managing biodiversity loss align with a metaphor developed by G. Evelyn Hutchinson in his book *The Ecological Theater and the Evolutionary Play* from Shakespeare's *As You Like It* (Hutchinson 1965). In Act II, Scene VII, of *As You Like It*, Shakespeare wrote, "All the world's a stage, and all the men and women merely players. They have their exits and their entrances." In Hutchinson's metaphor, the world's biota comprises the players, and the script is an evolutionary play. More recently, the metaphor has been extended to consider the Earth's abiotic setting as the stage (Beier et al. 2015).

Conservation efforts often emphasize management plans for the actors [e.g., Essential Biodiversity Variables (EBVs)] (Fernandez and Pereira, Chap. 18).

S. L. Malone
Department of Biological Sciences, Florida International University, Miami, FL, USA

K. D. Gaddis
National Aeronautics and Space Administration, Washington, D.C., USA

J. M. Grady
Department of Biology, Bryn Mawr College, Bryn Mawr, PA, USA

Ecology, Evolutionary Biology, and Behavior Program, Michigan State University, East Lansing, MI, USA

Department of Integrative Biology, Michigan State University, East Lansing, MI, USA

J. Costanza
Department of Forestry and Environmental Resources, NC State University, Research Triangle Park, NC, USA

M. L. Hobi
Swiss Federal Research Institute WSL, Birmensdorf, Switzerland

SILVIS Lab, Department of Forest and Wildlife Ecology, University of Wisconsin-Madison, Madison, WI, USA

A. M. Latimer
Department of Plant Sciences, UC Davis, Davis, CA, USA

S. Pau
Department of Geography, Florida State University, Tallahassee, FL, USA

A. M. Wilson
Geography Department, University at Buffalo, Buffalo, NY, USA

S. V. Ollinger
Department of Natural Resources and the Environment, University of New Hampshire, Durham, NH, USA

A. O. Finley
Department of Geography, Environment, & Spatial Sciences, Michigan State University, East Lansing, MI, USA

Department of Forestry, Michigan State University, East Lansing, MI, USA

E. Hestir
School of Engineering, University of California, Merced, CA, USA

For instance, the US Endangered Species Act and International Union for Conservation of Nature (IUCN) Red List focus on individual species (ESA 1973; IUCN 2001). However, an inherent challenge to managing species is that, during the course of a play, the actors move across the stage. Geo-referenced fossils from the paleoecological record provide evidence of how species' geographic ranges shifted in the past as Earth's climate fluctuated (Williams and Jackson 2007; Veloz et al. 2012). For instance, in terms of estimating EBVs (Fernandez and Pereira, Chap. 18), species distribution models (SDMs) are one of the most common tools for understanding how species ranges might shift over time and space (Elith and Leathwick 2009; Record and Charney 2016), but they are fraught with statistical (Record et al. 2013) and biological shortcomings (Belmaker et al. 2015; Charney et al. 2016; Evans et al. 2016) that hamper their ability to reliably inform management. Given the challenges of managing species whose ranges might be shifting in response to climate change (Veloz et al. 2012), there is interest in focusing conservation efforts on areas that are likely to support biodiversity and on the processes that generate it (Pressey et al. 2007; Anderson and Ferree 2010; Beier and Brost 2010). Indeed, The Nature Conservancy, one of the world's leading nonprofit conservation organizations, has adopted the rallying cry of "conserving nature's stage" (Beier et al. 2015). Conserving nature's stage entails identifying parcels of Earth that are valuable for their geodiversity and for their capacity to support diverse life forms today and into the future.

Geodiversity has been defined in several ways (see Table 1.2 in Gray 2013). Some definitions of geodiversity refer to variability in soil, geological, and geomorphological features and the processes that give rise to them (Gray 2013 and references therein). Other definitions tend to have a wider scope and also include topography, hydrology, and climate (Benito-Calvo et al. 2009; Parks and Mulligan 2010). These more inclusive definitions of geodiversity capture variability in the entire geosphere (Hjort et al. 2012) that link to important drivers of biodiversity (e.g., energy, water, and nutrients (Richerson and Lum 1980; Kerr and Packer 1997)). The geosphere includes the lithosphere, atmosphere, hydrosphere, and cryosphere (Williams 2012) and processes within and among them and encompasses the abiotic components of Earth's "Critical Zone," or the portion of Earth where biotic and abiotic processes support life on Earth's surface (NRC 2001). Just as the Critical Zone arises from interactions among abiotic and biotic processes, geodiversity is not separated from biotic influences and biodiversity. A key step in the prioritization of conservation areas using this approach is to understand the relationships between biodiversity and geodiversity. Remotely sensed biodiversity and geodiversity data have the potential to answer questions of scale to better inform conservation decisions because they can provide coverage at nearly continuous large spatial extents (i.e., regional to global) and at fine spatial and temporal resolutions (Fig. 10.1 for a spatial example). Here, we provide an overview of remotely sensed data sources that can be used to measure geodiversity and biodiversity to better understand biodiversity-geodiversity relationships, which is a key step in conserving nature's stage.

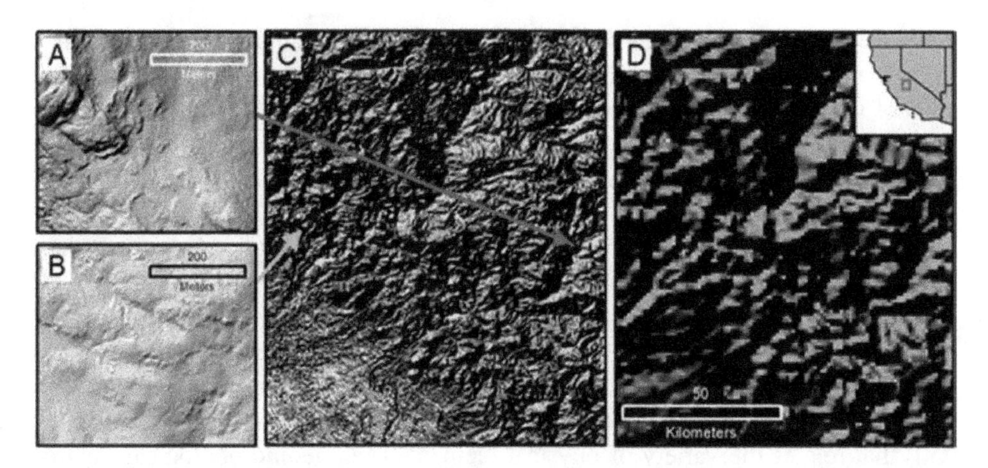

Fig. 10.1 Topography at different spatial grains. Hillshade maps calculated from digital elevation models (DEMs) at 1 m resolution (**a**) and (**b**), 90 m resolution (**c**), and 1 km resolution (**d**). The inset map in (**d**) shows the locations of panels (**c**) and (**d**) in California, which have the same extent. Data for panels (**a**) and (**b**) are from the National Ecological Observatory Network's (NEON) Airborne Observation Platform Light Detection and Ranging (LiDAR) system (Kampe et al. 2010). Data for panels (**c**) and (**d**) are from the Shuttle Radar Topography Mission (SRTM) via earthenv.org (Robinson et al. 2014)

10.2 Geodiversity Indices

Geodiversity represents an opportunity for habitat differentiation (Radford 1981) and available niche space (Dufour et al. 2006) that is thought to support biodiversity (Gray 2008). The continuous nature of remote sensing (RS) data enables exploration of novel measures of geodiversity. In this section we focus our discussion on metrics of variability, although absolute values (e.g., minimum and maximum thresholds) of some geographical features are also informative for understanding species' limits and ultimately species diversity. Studies have used two aspects of variability: the absolute range of conditions and the spatial configuration of these conditions (Spehn and Körner 2005; Dufour et al. 2006; Jackova and Romportl 2008; Serrano et al. 2009; Hjort and Luoto 2010; Hjort and Luoto 2012). The range in conditions is an estimate of the different elements in the area of interest. Given sampling units larger than the minimum pixel resolution, the proportional area covered by distinct geographical features could be used to calculate an evenness index of geodiversity. Categorical features have also treated geodiversity variables similarly to species with measured presences or abundances in various geodiversity metrics (Serrano et al. 2009; Tuanmu and Jetz 2015).

Alternatively, geodiversity could be quantified as variability in continuous observations such as elevation or climate. A focus on variability allows for different geological contexts (past and present) to be taken into account. One of the most common measures of environmental heterogeneity is elevational range (Stein et al. 2014), simply the absolute difference between elevation at two sites or sample units (i.e., among or within sites, respectively). Using elevation as an example, the average difference, squared, between the elevation in a focal cell and all other cells in a sample unit could be used as a measure of topographic heterogeneity. The coefficient of variation is a similar measure of heterogeneity, though it is standardized to the mean elevation of the sample unit. Pairwise site differences in multiple geographical features can be used as predictors in matrix regression such as generalized dissimilarity models (Ferrier et al. 2007) or more generally a Mantel test (Tuomisto et al. 2003; Legendre et al. 2005), though mechanistic interpretation is limited when geographical features are combined in this way.

Additional approaches include a geodiversity atlas that classifies areas as having very high, high, moderate, low, and very low geodiversity (Kozlowski 1999), quantifying geodiversity in terms of total component resource potential (i.e., energy, water, space, and nutrients; Parks and Mulligan 2010), and the geodiversity index (Gd) that relates the variety of physical elements (i.e., geomorphological, hydrological, soils) with the roughness and surface of the previously established geomorphological units according to the formula:

$$Gd = \frac{EgR}{lnS} \tag{10.1}$$

where Eg is the number of different physical elements, R is the coefficient of roughness of the unit, and S is the surface of the unit (km^2). The Gd is a semiquantitative scale that permits the establishment of five values of geodiversity, from very low to very high for each homogeneous unit. It is argued that use of Gd would allow easier comparison of units and aid suitable management of protected areas (Serrano et al. 2009; Hjort and Luoto 2010; Tukiainen et al. 2017).

With continuously measured remotely sensed geographical features, the sample unit (i.e., grain size) can be modified to examine within site and total site (and thus between sites) geodiversity. Additionally, RS data can uniquely address how relationships between geodiversity and biodiversity change across scales. Various combinations of changing grain and extent (change grain maintain extent, change extent maintain grain, change grain and extent) could be examined to explore scaling relationships (Barton et al. 2013).

10.3 Remote Sensing of Geodiversity

In the following sections, we describe the different components of geodiversity (Table 10.1), some of the ways they can be quantified, and the current state of technologies available to measure them remotely via airborne or satellite observations (Table 10.2). To match current interests in global biodiversity databases (e.g., the Global Biodiversity Information Facility, gbif.org), and because of the importance of scaling from local to much larger extents, we focus here on globally available data; however, we also mention some local scale RS applications. In particular, given that more and more remotely sensed data have been made publically available, we highlight open access remotely sensed geodiversity data.

Table 10.1 Elements of geodiversity

Lithosphere	Geology	Minerals	
		Rocks	
		Unconsolidated solids	
		Fossils	
	Geomorphology	Tectonics	
	Soils	Soil chemical properties	
		Soil physical properties	
	Topography	Elevation	
		Landforms (e.g., ridges, spurs)	
		Slope	
		Aspect	
		Energy	
		Roughness	
Atmosphere	Climate and weather	Temperature	Extreme events
		Precipitation	
		Wind	
Hydrosphere	Surface water		
	Groundwater		
Cryosphere	Ice		
	Snow		

Adapted from Serrano et al. (2009)

Table 10.2 Examples of remotely sensed geodiversity elements

Geosphere	Geodiversity element	RS data set
Lithosphere	Geology	Ground-penetrating radar (GPR)
	Topography	Advanced Spaceborne Thermal Emission and Reflection Radiometer (ASTER)
		Shuttle Radar Topography Mission (SRTM)
		Sentinel-2
Atmosphere	Surface temperature	MODIS (Moderate Resolution Imaging Spectroradiometer) surface temperature
		AVHRR (Advanced Very High Resolution Radiometer) surface temperature
		Sentinel-3
	Rainfall	Tropical Rainfall Measurement Mission (TRMM)
		Global Precipitation Measurement (GPM) mission
	Wind direction and speed	Quick Scatterometer (QuickSCAT)
		Rapid Scatterometer (RapidScat)
Hydrosphere	Soil moisture	ESA's Soil Moisture and Ocean Salinity (SMOS)
		NASA's Soil Moisture Active Passive (SMAP) observatory
	Gravity anomalies	Gravity Recovery and Climate Experiment (GRACE)
Cryosphere	Ice sheet mass balance	Geoscience Laser Altimeter System (GLAS) sensor onboard the Ice, Cloud, and land Elevation Satellite (ICESat)

10.3.1 Lithosphere

10.3.1.1 Lithosphere: Topography

Topographic barriers can influence geographic patterns of biodiversity by physically isolating populations of plants and animals (Janzen 1967). Topography also can be used as an indirect measure of microclimate, as topographic position can influence temperature and precipitation (e.g., Ollinger et al. 1995). Topography of the lithosphere crust is often represented by elevation (the height above sea level of a given point on the ground) or bathymetry (the depth to the bottom of a water body). In February 2000 the SRTM radar system flew on the US Space Shuttle Endeavour for 11 days collecting radar-derived elevation data from 60°N to 56°S. These data were originally released at 90 m resolution; however, in 2015, 30 m data (1 arc second) were released for the entire SRTM extent. There are many other sources of elevation data including NASA Advanced Spaceborne Thermal Emission and Reflection Radiometer ASTER (Fig. 10.2), *active* radar satellites designed for ice measurement (see the Cryosphere section), and more. NASA is currently working to develop a best available digital elevation model (DEM) for the planet, NASADEM. For this the entire SRTM data set will be reprocessed, Geoscience Laser Altimeter System (GLAS) data will be incorporated to remove artifacts, and the Advanced Spaceborne Thermal Emission and Reflection Radiometer Global Digital Elevation Model version 2 (ASTER) and Global Digital Elevation Map (GDEM) V2 DEMs will be used for refinement.

Elevation is only one of many products under the umbrella of topography. Slope (the angle between two elevation points) and aspect (the direction a slope is facing) are two of the many indices that can be derived from elevation data. Importantly, most of these indices are kernel-dependent, meaning they rely on data not just from an individual point but from surrounding points as well. For example, ArcGIS 10.3 (ESRI; Redlands, California) calculates the slope of a given pixel (elevation value) as the maximum slope between that center pixel and the eight surrounding pixels. The "terrain" function in the raster package in R statistical software (Hijmans and van Etten 2019) permits several different methods for calculating slope based on either a 4- or an 8-cell kernel, and these calculations differ slightly from those in the Geospatial Data Abstraction Library (GDAL; gdal.org). Environment for Visualizing Images (ENVI) software (Harris Geospatial Solutions, Broomfield, Colorado) allows the user to select any kernel size then fits a quadratic surface to the entire kernel, calculating slope and other parameters based on that surface (Wood 1996). These different methods could lead to somewhat different results; in particular, the selection of a small versus a large kernel could change the slope estimated. Imagine, for example, with fine-grained data, the inside of a tip-up pit on the side of a north-facing slope. The local aspect could be south facing, while a larger kernel could reveal that the landscape is north facing.

Beyond slope and aspect, there are many other kernel-dependent topographic measures. For instance, Topographic Position Index (TPI) is defined as the difference between a central pixel and the mean of its surrounding pixels. Terrain

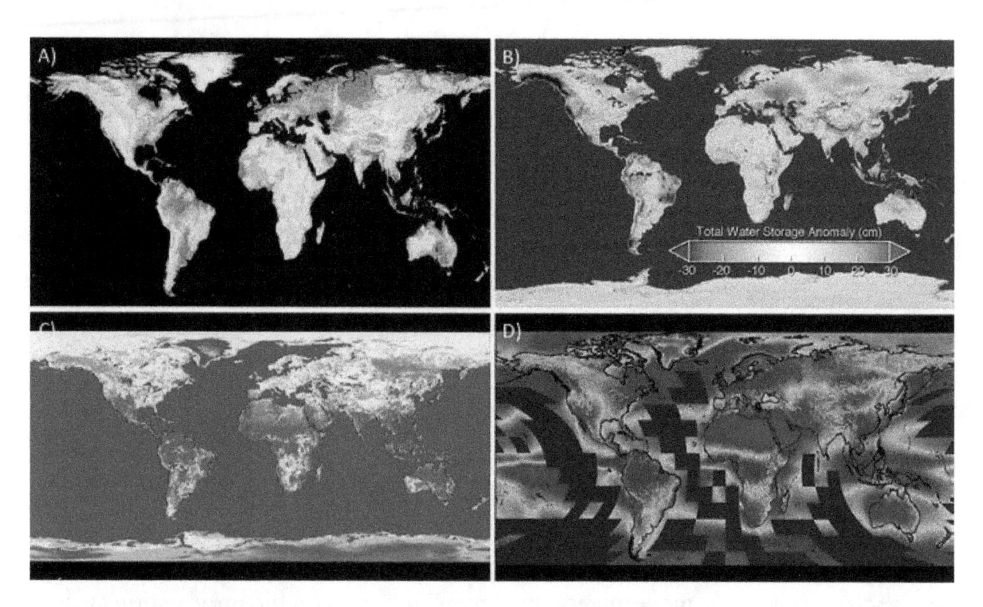

Fig. 10.2 Four examples of geodiversity variables derived from National Aeronautics and Space Administration (NASA) data products. (**a**) Earth's elevation, from which topographic diversity can be calculated, from 2009 imagery from the Advanced Spaceborne Thermal Emission and Reflection Radiometer (ASTER) instrument aboard NASA's Terra satellite (30 m spatial resolution). Image courtesy of NASA/JPL/METI/ASTER Team, NASA's Goddard Space Flight Center. https://svs.gsfc.nasa.gov/11734. Elevation in meters shown with yellows being lower in elevation than greens or reds. (**b**) Gravity Recovery and Climate Experiment's (GRACE) Terrestrial Water Storage Anomaly as of April 2015 relative to a 2002–2015 mean. Image courtesy of NASA's Scientific Visualization Studio (1° spatial resolution). (**c**) Soil Moisture Active Passive (SMAP) global radiometer map. Image courtesy of NASA (9 km spatial resolution). H-polarized brightness temperatures are shown in degrees Kelvin with warmer colors (reds and oranges) showing warmer temperatures and cooler colors (blues and yellows) showing cooler temperatures. (**d**) Mean annual cloud frequency (%; reds indicate higher cloud frequency than blues) over 2000–2014 derived from NASA's Moderate Resolution Imaging Spectroradiometer (MODIS) satellites (Wilson and Jetz 2016; 1 km spatial resolution)

Ruggedness Index (TRI), in contrast, is the mean of the difference between the central pixel and its surrounding pixels (Riley et al. 1999). Wood (1996) describes a number of convexity and curvature metrics based on the first and second derivatives of the quadratic surface described above. DEMs can also be classified into topographic features or landforms like peaks, ridges, channels, and pits, though these definitions depend on specific threshold values that may either be prescribed by software or defined by the user. Incident solar radiation can also be calculated for a given day or aggregated for a year based on a given point's elevation and latitude and the elevations of surrounding pixels. Although this section mainly describes DEM-derived morphometric landforms, it is also important to acknowledge that the genesis of landforms interacts with the ecology of a system. For instance, two hills with similar shapes may have very different associated vegetation if one is sandy (e.g., a dune) and the other is made of tills (e.g., end moraine).

While SRTM-derived products are typically used to produce "best available" topographic information, a challenge with SRTM is that the mission occurred only once. In geologically and tectonically active areas and areas where humans are influencing geology, satellite-derived data can be used to detect even very small changes over time. For example, Ge et al. (2014) used synthetic aperture radar (SAR) interferometry to detect subsidence in the Bandung Basin (Indonesia) likely due to groundwater extraction. Yun et al. (2015) used SAR data to map areas of change and potential damage after the 2015 Gorkha earthquake in central Nepal. Using SAR instruments in concert with LiDAR instruments on airborne flights has allowed for greater than 30 cm vertical accuracy (Corbley 2010). The launch of the Global Ecosystem Dynamics Investigation (GEDI) mission onboard the International Space Station has the potential to allow for improved global topographic data (Stavros et al. 2017).

10.3.1.2 Lithosphere: Geology and Soils

Geology consists of several subdisciplines, including lithology, tectonics, volcanology, and seismology. A modern geologic "map" in a geographic information system (GIS) framework may include polygons outlining the different substrate types and their ages, lines showing faults, and points identifying small outcrops or places where cores were collected. These static (unchanging through time) representations are developed through the painstaking work of geologists who gather in-situ records of rock type and estimates of geologic feature extents. Geologic maps vary in quality and access due largely to the density and biases of field technicians. When considering long-term evolutionary histories that generate deeper phylogenetic patterns, geological processes of uplift and erosion can become important (Cowling et al. 2009). Nevertheless, for more historically proximate species, community assembly, the available minerals, substrate structure, and topography are likely to play a more important role, especially in plants. For example, although all locations across the Mauna Loa environmental matrix in Hawai'i share a common parent material, differences in age, texture, and nutrient availability (due to variation in climate and weathering) lead to dramatically different vegetation patterns (Vitousek et al. 1992).

Similar to geologic maps, soil maps are typically developed through fieldwork and image interpretation for a single time period. Nevertheless, soils have higher spatial variability than bedrock and may change rapidly in response to natural or man-made disturbance. Recently there have been calls to improve the quality and dynamism of soil maps (Grunwald et al. 2011). The SoilGrids1km data product (Hengl et al. 2014) is one such example. It is a modeled product that relies on indirect remotely sensed variables, such as Moderate Resolution Imaging Spectroradiometer (MODIS), leaf area index (LAI), land surface temperature (LST), and topography from the SRTM to produce estimates at six depths of soil organic carbon, soil pH, sand, silt, and clay fractions, bulk density, cation-exchange capacity, coarse fragments, and depth to bedrock.

Imaging spectroscopy has been broadly applied for geologic mapping (Goetz et al. 1985; Gupta 2013). Multispectral imagery, like NASA's ASTER instrument and the European Space Agency's (ESA's) Sentinel-2 satellite, that is part of the

Copernicus program has long been used for mapping lithography in exposed surface environments (Rowan and Mars 2003; Hewson et al. 2005; Massironi et al. 2008; van der Werff and van der Meer 2016). Hyperspectral imagery has been used successfully to map minerals in many low-vegetation landscapes. For example, the Hyperion sensor, aboard the now decommissioned EO-1 satellite, was used to map mineralogy in Australia (Cudahy et al. 2001). The Airborne Visible/Infrared Imaging Spectrometer (AVIRIS) and new AVIRIS-Next Generation missions continue to push the boundary of imaging spectroscopy used in mineral mapping (Krause et al. 1993; Crowley 1993; Green et al. 1998). These instruments can also provide information on soil nutrient availability in areas dominated by vegetation cover via the influence of soils on foliar chemistry (e.g., Ollinger et al. 2002).

Ground-based RS has also provided insights for subsurface geologic mapping. For instance, ground-penetrating radar (GPR) uses radar pulses to map the relative densities of materials belowground and effectively maps soil and bedrock in layers (Davis and Annan 1989). Airborne GPR can greatly enhance the temporal and spatial resolution of geologic maps (Catapano et al. 2014; Campbell et al. 2018).

10.3.2 Atmosphere: Climate and Weather

Climate is an important control on mineral weathering, soil formation, and landforms (Jenny 1941). Surface temperature and cloud cover are readily observed with RS. The Advanced Very High Resolution Radiometers [AVHRR; National Oceanographic and Atmospheric Administration (NOAA)] have been collecting surface radiation data in the visible, infrared, and thermal spectra with twice-daily global coverage since 1981 that currently gathers data at ~1 km spatial resolution. AVHRR data can be used to map cloud cover and land and water surface temperatures; however, changes in satellite technology and the lack of onboard calibration in the AVHRR sensors have made the use of these data challenging due to a need for standardization of data across satellite technologies (Cao et al. 2008). The launch of the MODIS sensors on NASA's Terra (launched in 1999) and Aqua (launched in 2002) satellites and ESA's Sentinel-3 satellite as part of the Copernicus program (3-A launched in 2016 and 3-B launched in 2018) significantly improved global mapping capabilities. The two MODIS sensors map most of the planet twice a day with 36 bands ranging from the visible to the thermal infrared. The MODIS bands were selected to capture properties of the land surface but also ocean properties, atmospheric water vapor, surface temperature, and clouds (Fig. 10.2). Products from MODIS, such as surface temperature and cloud presence, have been used either to directly map climate variables for use in ecological research (e.g., Cord and Rödder 2011; Wilson and Jetz 2016) or to inform modeled climate products like Worldclim-2 (Fick and Hijmans 2017). Furthermore, surface temperature can better characterize plant ecological differences (Still et al. 2014) because it more accurately captures canopy temperature, which is not the same as air temperature, and because many air temperature products (such as Worldclim-2) are interpolated (see Pinto-Ledézma and Cavender-Bares, Chap. 9).

Satellite-derived rainfall products are estimated through a combination of measurements, including surface reflectance of clouds (i.e., cloud coverage, type, and top temperature), passive microwave (i.e., column precipitation content, cloud water and ice, rain intensity and type), and lightning sensors. The Tropical Rainfall Measurement Mission (TRMM) operated from 1997 to 2015, providing information on rainfall amount and intensity and lightning activity globally every 3 hours at 5 km resolution from 38°N to 38°S. As a follow-up to TRMM, the Global Precipitation Measurement (GPM) mission relies on a constellation of satellites, including a core GPM observatory, to produce 0.1° resolution data every 30 minutes from 60°N to 60°S. Initiated in 2014, GPM allows new explorations of extreme weather events. Like MODIS temperature measurements, TRMM and GPM precipitation measures have been directly incorporated into ecological research (e.g., Deblauwe et al. 2016) and used to inform modeled climate products like the Climate Hazards Group Infrared Precipitation with Station data product (CHIRPS; Funk et al. 2015).

There is also a broad set of efforts to generate reanalysis products that combine the history of Earth observations to develop temporally and spatially consistent global models of climatic and environmental variables. For instance, the NASA Modern-Era Retrospective Analysis for Research and Applications (MERRA) models close to 800 radiative and physical properties of the Earth's atmosphere at 3- to 6-hour time steps from 1979 to present at ~50 km spatial resolution (Rienecker et al. 2011). While this obviously sacrifices spatial resolution, these efforts open the door for longer-term analysis of climatic influence on biologic phenomena.

One commonly overlooked source of geologic substrate lies in the atmosphere. Airborne dust particles provide an essential source of nutrients in many environments and can originate from sources hundreds to thousands of miles away (Chadwick et al. 1999). Aeolian transport of phosphorus from North Africa to South America is thought to be an important driver of Amazonian productivity (e.g., Okin et al. 2004). Studies have mapped dust sources and rates using MODIS products (Ginoux et al. 2012) and produced 3-D models of dust transportation using LiDAR on the Cloud-Aerosol Lidar and Infrared Pathfinder Satellite Observation (CALIPSO) satellite (Yu et al. 2015). Furthermore, the SeaWinds instrument on the Quick Scatterometer (QuickSCAT) satellite and the subsequent Rapid Scatterometer (RapidSCAT) aboard the International Space Station measures wind speed and direction over the ocean's surface.

10.3.3 Hydrosphere

The hydrosphere consists of the water on, in, and above Earth's surface and is known to have a large influence in structuring riparian and aquatic communities of organisms (reviewed by Atkinson et al. 2017). The hydrosphere interacts with other types of geodiversity in the lithosphere, cryosphere, and atmosphere. Topography alone can be used to indirectly provide a crude estimate of many hydrological variables, including watershed size, soil water content (Moore et al. 1991), flow paths, and surface water. In addition, two types of satellite data can be used to estimate soil moisture and groundwater, which in some systems are important drivers of plant diversity because drought sensitivity may shape plant distributions (e.g., Engelbrecht et al. 2007). The ESA's Soil Moisture and Ocean Salinity (SMOS; launched 2009) and NASA's Soil Moisture Active Passive observatory (SMAP; launched 2015; Fig. 10.2) both use microwave radiometers to detect surface soil moisture globally in areas with low topographic variation and low-vegetation cover. The Gravity Recovery and Climate Experiment (GRACE; launched in 2002; Fig. 10.2) is a pair of satellites that measure gravity anomalies around the world, allowing researchers to estimate available groundwater reserves and their change over time.

Water quality is a critical driver of aquatic biodiversity across taxa, from plants to animals (Stendera et al. 2012). Watershed disturbance, sediment runoff, and nutrient pollution are major aquatic biodiversity stressors, affecting phytoplankton and aquatic and wetland vegetation abundance and diversity (Lacoul and Freedman 2006; Mouillot et al. 2013) and higher trophic levels (e.g., zooplankton, shrimps, larval fish, and birds (Thackeray et al. 2010). Optical RS can be used to retrieve a limited but important set of water quality variables, including particulate and dissolved organic and inorganic matter, chlorophyll-a, as well as other phytoplankton pigments like the phycocyanins common in potentially harmful cyanobacteria blooms. Surface or "skin" water temperature is measured from instruments with thermal bands (Giardino et al. 2018; Alcântara et al. 2010). The major limitation in RS of water quality is in sensor resolution. Sensors must have a fine enough pixel size to resolve water bodies, with high enough radiometric sensitivity to detect small changes in a dark target (10% or less of the total signal received by the sensor, Muller-Karger et al. 2018; Hestir et al. 2015). While some water quality products are publically distributed with limited spatial coverage [e.g., United Nations Educational, Scientific and Cultural Organization (UNESCO) regions], free data processors distributed by NASA (Sea-viewing Data Analysis System [SeaDAS]) and the ESA (Sentinel Application Platform) enable users to compute their own water quality products.

10.3.4 Cryosphere

Earth's fossil record illustrates how changes in glacial cover over time have governed the distribution of biodiversity (e.g., Veloz et al. 2012), and many aspects of the globe's biodiversity are influenced by snow, ice, and permafrost (reviewed by Vincent et al. 2011). The frozen parts of the Earth system, the cryosphere, can be detected with a number of different RS tools. The cryosphere can be divided into several different components—seasonally snow-covered land, permafrost, glaciers and ice sheets, lake ice, and sea ice. Because cloud cover is a frequent problem at high latitudes, cryosphere RS often relies on longwave techniques that can pass through clouds. A recent book, *Remote Sensing of the Cryosphere* (Tedesco 2014), describes these tools and methods in great detail; here we review some of the major techniques. In all of the discussion below, the importance of change over time is paramount; inter- and intra-annual variation in snow and ice cover are important drivers of physical and biological processes.

The 3-D extent of snow and ice can easily be mapped using optical techniques; snow reflects strongly in the visible and near-infrared (NIR) range but absorbs in the shortwave infrared (SWIR), making it spectrally distinct from other white objects such as rooftops and clouds. These distinctions may still be challenging with multispectral sensors, but hyperspectral sensors permit mapping of snow versus clouds and even some estimation of snow particle size (e.g., Burakowski et al. 2015). Passive microwave sensors can be used to estimate snow depth and snow water equivalent, while active microwave sensors can map liquid water content. Tools and techniques for mapping snow are reviewed by Dietz et al. (2011).

Ice and permafrost features can be mapped with many of the tools and methods described in preceding sections. Snow cover can be mapped using optical sensors and methods; subsidence of the cryosphere can be mapped with SRTM (near global extent, 30–90 m spatial resolution, single snapshot in time) and SAR (airborne, 2 m spatial resolution); and passive microwave radiometers such as SMOS (global extent, 50 km spatial resolution, 3-day temporal resolution) and SMAP (near global extent for low-vegetation areas, 9–36 km spatial resolution, 8-day temporal resolution) can be used to map frozen versus thawed ground surfaces (Entekhabi et al. 2014). Because glaciers and ice sheets are fundamentally a combination of snow, ice, and liquid water, many of the techniques described above, such as optical sensors and passive microwave radiometers, can be used to map their extent and status. In addition, the GLAS sensor onboard the Ice, Cloud, and land Elevation Satellite (ICESat; near global spatial extent, 70 m spatial resolution, 91-day temporal resolution from 2003 to 2009) permitted the mapping of ice sheet mass balance (Zwally et al. 2011). ICESat-2 is scheduled for launch in 2018 (global spatial extent, 14 km spatial resolution, 91-day temporal resolution). SAR has also been used to map ice flow on Antarctica (Rignot et al. 2011).

Sea, lake, and river ice cover can be mapped using optical techniques (Jeffries et al. 2005), while thickness has been measured using ICESat and passive microwave sensors (e.g., Kwok and Rothrock 2009). The difference between first-year sea ice and older sea ice can be identified by changes in salinity using multichannel passive microwave sensors like the Advanced Microwave Scanning Radiometer for Earth Observing System (AMSR-E) onboard NASA's Aqua satellite (global spatial extent, 474 km spatial resolution, 12-hour temporal resolution, operational 2002–2015). River ice mapping is critical for monitoring and predicting river habitat quality and duration for a variety of organisms (e.g., Charney and Record 2016; Pavelsky and Zarnetske 2017). The extent and duration of river icing types have been mapped with different polarizations of passive microwave data from Canada's RADARSAT-1 (1995–2013) and RADARSAT-2 (launched 2007) (Weber et al. 2003; Jeffries et al. 2005; Yoshikawa et al. 2007 for aufeis features) and with MODIS Terra (Pavelsky and Zarnetske 2017).

10.4 Remote Sensing of Biodiversity

Approaches for using RS to track biodiversity are reviewed in several chapters in this book (Fernandez and Pereira, Chap. 18; Serbin and Townsend, Chap. 4; Meireles et al., Chap. 7). Biodiversity has many forms—including taxonomic, functional, genetic, and phylogenetic diversity (Serbin and Townsend, Chap. 4; Meireles et al., Chap. 7). Each form may exhibit different relationships with both geophysical and biological drivers, owing to a variety of mechanisms (Gaston 2000; Lomolino et al. 2010). For example, reorganization of organisms in response to changing environments leads to species assemblages becoming more or less similar through biotic homogenization or differentiation (Baiser et al. 2012). Such biotic homogenization/differentiation is usually characterized taxonomically (Olden and Rooney 2006). However, functional traits (i.e., traits representing the interface between species and their environment) possessed by species are often more important to ecosystem functions valued by society (Baiser and Lockwood 2011) and may be more appropriate to use in assessing biodiversity-ecosystem function relationships (Flynn et al. 2011). Many functional traits may also exhibit a phylogenetic signal (Srivastava et al. 2012), so it is important to consider multiple measures of diversity (i.e., taxonomic, functional, and phylogenetic) when assessing patterns of biodiversity (Serbin and Townsend, Chap. 4; Meireles et al., Chap. 7; Lausch et al. 2016; Lausch et al. 2018).

One caveat to measures of biodiversity generated from high-resolution RS data is that as the spatial resolution of data increases, the spatial extent typically decreases (Turner 2014; Gamon et al., Chap. 16). This limitation hinders our ability to understand how biodiversity relates to different drivers (e.g., geodiversity) at different spatial scales to better inform conservation decisions. There have been recent calls from scientists for new satellite missions and data integration efforts to address this issue (Schimel et al., Chap. 19). For instance, Jetz et al. (2016) call for a Global Biodiversity Observatory to generate worldwide remotely sensed data on several plant functional traits. Petorelli et al. (2016) and Fernández and Pereira (Chap. 18) identify satellite RS data that, given technological and algorithmic developments in the near future, could be capable of meeting the criteria of EBVs for conservation outlined by the international Group on Earth Observations—Biodiversity Observation Network (GEO BON) at a global spatial extent.

Until finer resolution, remotely sensed biodiversity data exist at large spatial extents, data available from in-situ measurements of organisms can inform the relationships between biodiversity and geodiversity. Publically available biodiversity data with geographic locations include expert range maps of individual species from IUCN (IUCN 2017), occurrence data [e.g., Global Biodiversity Information Facility (GBIF, GBIF 2016); Botanical Information and Ecology Network (BIEN, Enquist et al. 2016)], citizen science networks [e.g., Invasive Plant Atlas of New England, IPANE, Bois et al. 2011], and national [e.g., US Forest Service Forest Inventory and Analysis (FIA), Bechtold and Paterson 2005)] and international inventory networks [e.g., the Amazon Forest Inventory Network (RAINFOR), Peacock et al. 2007].

Each of these data sets comes with its own uncertainties (e.g., observation errors) and user challenges. For instance, citizen science data require detailed metadata on the sampling process to ensure that citizen scientists are able to reduce error and bias as they collect data and to enable those analyzing the data to model potential uncertainty (Bird et al. 2014). Despite these sources of uncertainty and logistical hurdles, these data provide a useful starting point for understanding the relationship between biodiversity and geodiversity.

10.5 A Case Study Linking RS of Geodiversity to Tree Diversity in the Eastern United States

To motivate explorations of the relationship between biodiversity and geodiversity with remotely sensed data, we provide an example using biodiversity data from the FIA program of the US Forest Service (O'Connell et al. 2017) and geodiversity data on elevation from SRTM. We selected elevation as a covariate because patterns of tree diversity often vary with elevation (Körner 2012). While some studies promote the use of many geodiversity components (Serrano et al. 2009; Hjort and Luoto 2010; Bailey et al. 2017; Tukiainen et al. 2017), a great deal of the variation in geodiversity is captured by the standard deviation in elevation (Hjort and Luoto 2012), which is used in this analysis.

The FIA program uses a two-phase protocol to characterize the nation's forest resources. In phase one, all land in the United States is categorized as either "forested" or "not forested" using remotely sensed data. In phase two, in every 2428 ha of land classified as forested, one permanent FIA plot is placed for in-situ sampling. Each FIA plot consists of four 7.2-m-fixed-radius subplots wherein all trees >12.7 cm diameter at breast height are measured. FIA plot measurements began in the 1940s, but a consistent nationwide sampling protocol was not implemented until 2001. In the analysis presented, we used data from the most recent full plot FIA inventory from 2012–2016; the SRTM data were collected in 2009. Although there is not perfect temporal overlap in the geodiversity and biodiversity data used in this example, we do not expect that topography at a spatial resolution of 50 km would have changed much over the time period encompassed by both data sets for this part of the world.

We fixed the spatial extent of the analysis to the contiguous United States east of 100°W longitude (n = 90,250 plots total) and selected a grain size of 50 km for calculating alpha (within site), beta (turnover between sites), and gamma (total across all sites) diversities within a radius centered on each FIA plot. All biodiversity metrics were based on species abundances as quantified by the total basal area of each tree species in each plot. Alpha diversity was calculated as the median abundance-weighted effective species number of all plots falling within a 50 km radius of the focal FIA plot, including the focal plot. Beta diversity was calculated as the mean abundance-weighted pairwise Sørensen dissimilarity of all pairs of plots within a 50 km radius of the focal plot, including the focal plot. Gamma

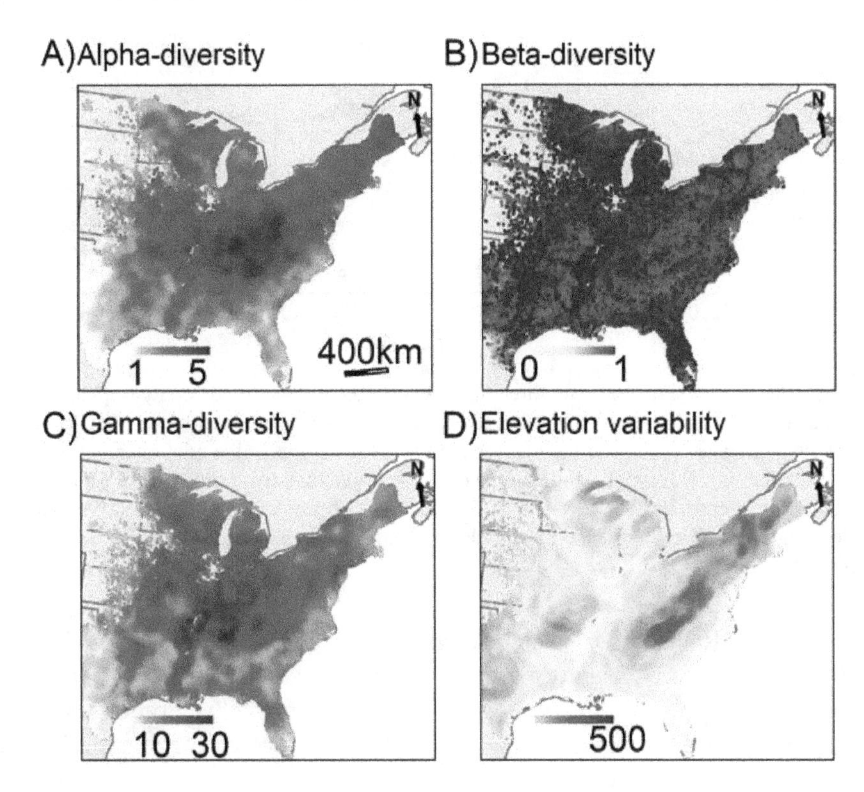

Fig. 10.3 Mapped variation in tree diversity calculated within 50 km radii. Tree data come from the Forest Inventory and Analysis of the US Forest Service (FIA, O'Connell et al. 2017. (**a**) Taxonomic alpha diversity. (**b**) Taxonomic beta diversity. (**c**) Taxonomic gamma diversity. (**d**) The standard deviation of all elevation pixels within the radius from 30 m Shuttle Radar Topography Mission (SRTM) data

diversity was calculated as the aggregated effective species number of all plots within a 50 km radius of the focal plot, including the focal plot. For each 50 km radius centered on a focal plot, we computed the standard deviation of elevation across pixels within the radius from 30 m SRTM data (Fig. 10.3). To avoid edge effects, all plots within 100 km of the political borders of the United States were excluded, retaining 80,411 plots. To avoid pseudo-replication, we generated 999 subsamples of plots separated by at least 100 km, yielding ~370 plots per subsample. Because of the saturating relationship between biodiversity and geodiversity, we fit natural splines with 3 degrees of freedom to relate all focal plots' univariate diversity to elevation standard deviation (SD) (linear regression for alpha and gamma diversity; beta regression for beta diversity), and goodness of fits of the models were assessed with r-squared (Fig. 10.4).

This example shows how the relationships between biodiversity and geodiversity for a subset of different biodiversity metrics vary depending on the metric of biodiversity calculated. Here beta and gamma diversity do not show a strong relationship (r^2 = 0.03 and r^2 = 0.07, respectively; Figs. 10.3 and 10.4) with geodiversity, but alpha diversity shows a stronger, positive relationship with elevation variability

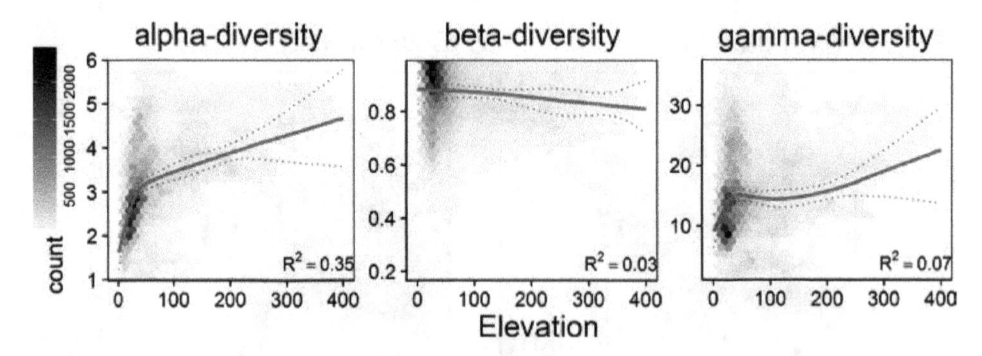

Fig. 10.4 The relationships between three measures of tree taxonomic diversity (alpha, beta, and gamma) and geodiversity (i.e., elevation standard deviation) at a spatial resolution of 50 km. Points indicate the aggregated plots, and the red line indicates the natural spline relationship fitted with a linear regression model for alpha and gamma diversity and a beta regression model for beta diversity. Dotted lines represent the 2.5% and 97.5% quantiles of predicted values across 999 spatially stratified random subsamples of the data, and the given r-squared value is the mean across all the subsamples

($r^2 = 0.35$; Figs. 10.3 and 10.4). Interestingly, in a sister study, Zarnetske et al. (2019) found that FIA tree diversity with a different spatial extent—in California, Washington, and Oregon—showed a different relationship with elevation variability. Furthermore, beta and gamma diversity showed a strong increasing relationship with elevation variability, whereas alpha diversity did not. This comparison between the case study illustrated in this chapter and the results of Zarnetske et al. (2019) highlights the importance of considering how the relationship between geodiversity and biodiversity may change with different spatial scales (Gamon et al., Chap. 16). Bailey et al. (2017) also showed that landforms detected with airborne RS at smaller spatial resolutions explained more of the variation in alpha diversity of alien vascular plants in Great Britain than did climate measured at larger spatial resolutions. While these examples do not provide an exhaustive exploration of the ways in which tree diversity responds to geodiversity, they clearly show how remotely sensed data may help us understand the relationships between geodiversity and biodiversity and how these relationships may be different in different geographic areas.

This example shows how the relationship between taxonomic biodiversity and geodiversity depends on the biodiversity metric chosen. There are various methodologies for calculating biodiversity metrics and various facets of biodiversity (e.g., functional, taxonomic, phylogenetic), and the theoretical pros and cons of each remain controversial (e.g., Jost 2007; Clark 2016), so it may not be obvious which metric is the best. Furthermore, different conclusions may be drawn depending on the types of taxa used in the analysis.

In a similar vein, the choice of an appropriate geodiversity metric may not be obvious. Here we use a single measure of geodiversity, standard deviation of elevation. However, different definitions of the term *geodiversity* include different components of geology, topography, and, in some instances, climate (Parks and Mulligan 2010; Gray 2013). The amalgamation of these different variables to characterize geodiversity as a whole is an area in need of development.

10.5.1 Challenges and Opportunities

10.5.1.1 The Interplay Between Biodiversity and Geodiversity over Time

Although we have focused thus far on the effects of geodiversity on biodiversity, biodiversity can also affect geodiversity. Ecosystem engineers (Jones et al. 1994) and foundation species (Record et al. 2018) can influence biodiversity through habitat formation (Hastings et al. 2007). Geodiversity can be modified by species impacting the structure and function of landscape features. For example, elephants dig, form trails, and trample (Haynes 2012), and vegetation and sediment interact to form streams and coastal dunes (Zarnetske et al. 2012; Atkinson et al. 2017). In turn, these species-modified features can feed back to mediate the strength and direction of biotic interactions among species and ultimately influence patterns of biodiversity (Zarnetske et al. 2017). Even climate can be influenced by biodiversity and biogeographic patterns. Forests directly affect Earth's climate through atmospheric exchange (Bonan 2008). If shrubs expand by 20% and continue to dominate in areas north of 60°N latitude, for example, Arctic annual temperature could increase by 0.66°C– 1.84°C, via decreased albedo and increased evapotranspiration (Bonfills et al. 2012).

Many of these feedbacks between biodiversity and geodiversity are not detectable given a single snapshot in time and require longer time series. RS with repeat samples taken as satellites orbit the Earth provide data with high spatial and deep temporal coverage that can be used to assess changes in the dominance of a species within a community (Pau and Dee 2016). Changes in the dominance structure of communities (or its counterpart, evenness) should be early indicators of global change because these changes occur before the complete loss or replacement of species (Hillebrand et al. 2008). Furthermore, tracking dominant species should be especially important for quantifying biomass or abundance-driven ecosystem functions and services (Pau and Dee 2016). For instance, Cavanaugh et al. (2013) used 28 years of Landsat imagery to map the poleward expansion of mangroves, which are important in preventing coastal erosion, in the eastern United States. Furthermore, the 45-year time series of Landsat data provide an excellent opportunity for detecting changes in habitat due to species, which may have extreme impacts on the abiotic stage.

10.5.1.2 Scale and Expertise Mismatches

The relationships between geodiversity and biodiversity are likely to change across spatial and temporal scales. For instance, a focused spotlight shining down on one part of the stage (e.g., the tip of a mountaintop) might exhibit different covariation between geodiversity and biodiversity than a broad swath of light on another portion of the stage (e.g., an expansive low-lying valley). Spatial patterns of biodiversity and geodiversity are each scale dependent (Rahbek 2005; Bailey et al. 2017; Cavender-Bares et al., Chap. 2; Gamon et al., Chap. 16), and it is well established that ecological processes influencing the assembly of communities of organisms are scale dependent (Levin 1992; McGill 2010). A spatially explicit framework for conceptualizing community assembly describes external filters (e.g., climate or soils) that sort species from a regional pool at a spatial scale larger than the community and internal filters that sort species into a community from a subset of the species that make it through the external filter (e.g., microenvironmental heterogeneity, biotic interactions; Violle et al. 2012; Fig. 10.5). These "assembly rules" about how communities form remain a controversial paradigm with uncertainty about which processes operate at which scales (McGill 2010; Belmaker et al. 2015). Observing and quantifying relationships between geodiversity and biodiversity and how these relationships change with scale, however, are essential for moving forward regardless of one's position on these controversies. To most effectively use geodiversity to help explain and predict patterns of biodiversity, we need a framework that addresses the scaling relationship between biodiversity and geodiversity.

Furthermore, there are important disconnects in both scale and expertise between biodiversity science and RS (Petorelli et al. 2014) that once addressed will aid in the development of such a framework. Whereas the availability of remotely sensed geodiversity data products has increased, many of the scales are too coarse to reflect the environmental and biological conditions that often drive more fine-scaled spatially heterogeneous biodiversity patterns (Nadeau et al. 2017) and thus may require complex post-processing techniques unfamiliar to most biodiversity scientists before they can be used appropriately in biodiversity models. Also, there are likely many important aspects of geodiversity that at this time can only be derived through in-situ measurements and cannot be remotely sensed. Determining how physical and biological drivers influence biodiversity across spatial and temporal scales is a central focus of ecology. However, most models predicting future patterns of biodiversity assume broad-scale climatic drivers—temperature and precipitation—are sole drivers and leave out important biological drivers (Zarnetske et al. 2012; Record et al. 2013). Biological drivers such as dispersal ability and biotic interactions (e.g., competition) are often mediated by the structure of the landscape, including geophysical feature configuration, topographic complexity, and habitat patch arrangement (Zarnetske et al. 2017). Yet a significant knowledge gap remains about how the relationships between biodiversity and its geophysical and biological drivers change with respect to space and time—perhaps owing to the scale mismatch between fine-scale point-level biodiversity data and many coarse-scale remotely sensed data products.

Many ecological questions are addressed at scales much finer than the grain size of MODIS or GPM, which makes statistical downscaling a necessity for remotely sensed products to be used. Yet the landscape of options for statistical downscaling is vast and complex (Pourmokhtarian et al. 2016). In addition, the increasing availability of airborne topographic data like LiDAR makes the possibility of finer-grain analysis even more viable, yet these data also bring another dimension of complexity and a lack of standardization across platforms and methods.

Open access analytical tools and training will provide ways forward given data downloading and processing challenges. The Application for Extracting and

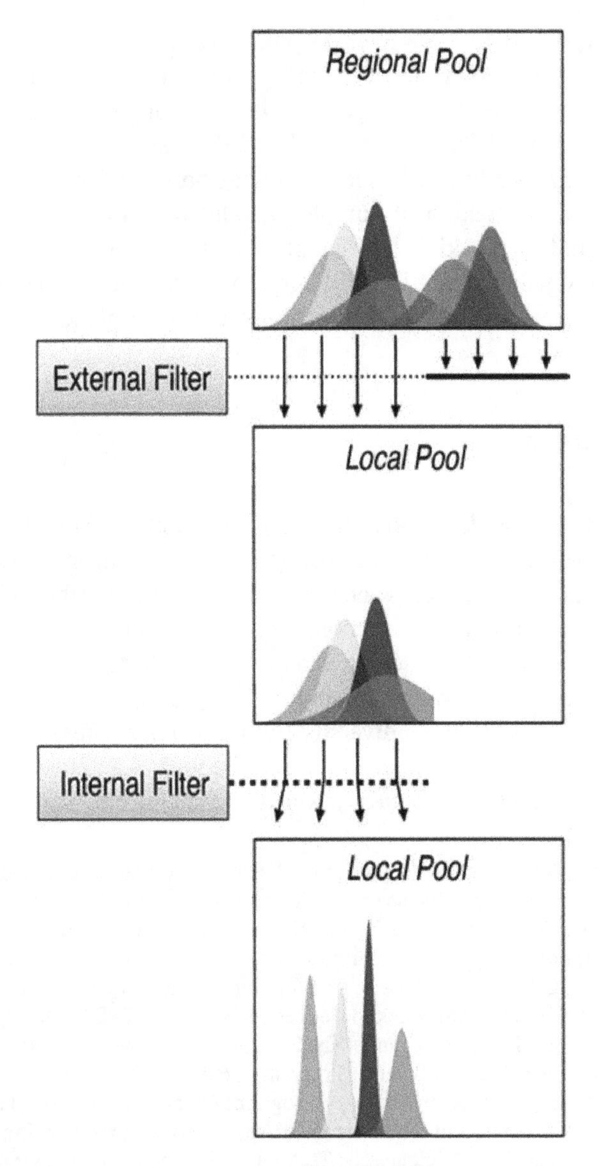

Fig. 10.5 Conceptual diagram adapted from Violle et al. (2012) showing how spatially explicit (i.e., local versus regional) filters influence the assembly of traits in an observed community (bottom schematic). The regional species pool (top schematic) contains all of the species capable of seeding into the local community. However, the observed local community may only contain a subset of the species in the regional pool after species have passed through a series of filters. Both internal and external filters encompass different aspects of the stage, whereas internal filters may also include the actors. Examples of external filters include broad-scale climate or soil types for which some species may not have physiological tolerances. Internal filters include microenvironmental heterogeneity and/or biotic interactions. In this schematic, the traits that passed through the external and internal filters partition in the observed local community, perhaps due to competitive effects between species for resources

Exploring Analysis Ready Samples (AppEEARS) offered by NASA and US Geological Survey (USGS) is a user-friendly tool that enables simple and efficient downloads and transformations of geospatial data from a number of federal data archives from the United States [e.g., the Land Process Distributed Active Archive Center (LP DAAC)]. Additionally, Geomorphons provides a user-friendly interface that automates the calculation of complex geodiversity features from topography data (Jasiewicz and Stepinski 2013). Training the next generation of ecologists and conservation biologists in RS will be integral to overcoming some of these hurdles and bridging the gaps between RS and ecology and conservation.

10.6 Conclusion

Cross-scale studies of relationships between geodiversity and biodiversity using RS and large field-based data sets hold promise for evaluating processes underlying biodiversity and identifying scales and methods for its monitoring and management. Realizing this potential will require more interaction among biodiversity scientists, geoscientists, RS experts, and statisticians to reconcile the challenges associated with differences in scales, available data products, disciplinary barriers, and available methods for connecting geodiversity to biodiversity. These challenges are far from trivial, but overcoming them has the potential to result in key ecological insights that will help us to be better stewards of the entire ecological theater.

Acknowledgments We thank J. Cavender-Bares, J. Gamon, A. Lausch, M. Madritch, F. Schrodt, P. Townsend, and S. Ustin for comments on the chapter while at NIMBioS. Funding for the NASA bioXgeo working group was provided by the National Aeronautics and Space Administration (NASA) Ecological Forecasting Program, Earth Science Division, Grant #NNX16AQ44G. Additional support for PLZ, KD, QR, JG, and AF came from Michigan State University. SR was supported by Bryn Mawr College KG Fund. AMW acknowledges support from NASA's Ecological Forecasting Program, Earth Science Division, Grant #NNX16AQ45G. MH was supported by the NASA Biodiversity Program and MODIS Science Program, Grant #NNX14AP07G. SO acknowledges support from the NSF Macrosystems Biology Program, Grant #638688. We thank the National Center for Ecological Analysis and Synthesis (NCEAS) for housing the working group meetings. KDG was supported by an AAAS Science and Technology Policy Fellowship served at NASA. The views expressed in this paper do not necessarily reflect those of NASA, the US Government, or the American Association for the Advancement of Science.

References

Alcântara EH, Stech JL, Lorenzzetti JA, Bonnet MP, Casamitjana X, Assireu AT, de Moraes L, Novo EM (2010) Remote sensing of water surface temperature and heat flux over a tropical hydroelectric reservoir. Remote Sens Environ 114:2651–2665

Anderson MG, Ferree CE (2010) Conserving the stage: climate change and the geophysical underpinnings of species diversity. PLoS One 5:e11554

Atkinson CL, Allen DG, Davis L, Nickerson ZL (2017) Incorporating ecogeomorphic feedbacks to better understand resiliency in streams: a review and directions forward. Geomorphology 305:123–140

Baiser B, Lockwood JL (2011) The relationship between functional and taxonomic homogenization. Glob Ecol Biogeogr 20:134–144

Baiser B, Olden JD, Record S, Lockwood JL, McKinney ML (2012) Pattern and process of biotic homogenization in the New Pangaea. P R Soc B 279:4772–4777

Barton PS, Cunningham SA, Manning AD, Gibb H, Lindenmayer DB, Didham RK (2013) The spatial scaling of beta diversity. Glob Ecol Biogeogr 22:639–647

Bechtold WA, Paterson PL (eds) (2005) The enhanced forest inventory and analysis program: national sampling design and estimation procedure. General technical report SRS-80. USA Department of Agriculture, Forest Service Southern Research Station, Asheville, NC, USA

Beier P, Brost B (2010) Use of land facets to plan for climate change: conserving the arenas, not the actors. Cons Biol 24:701–710

Beier P, Hunter ML, Anderson M (2015) Special section: conserving nature's stage. Cons Biol 29:613–617

Belmaker J, Zarnetske P, Tuanmu M-N, Zonneveld S, Record S, Strecker A, Beaudrot L (2015) Empirical evidence for the scale dependence of biotic interactions. Glob Ecol Biogeogr 24:750–761

Benito-Calvo A, Pérez-González A, Magri O, Meza P (2009) Assessing regional geodiversity: the Iberian Peninsula. Earth Surf Proc Land 34:1433–1445

Bird TJ, Bates AE, Lefcheck JS, Hill NA, Thomson RJ, Edgar GJ, Stuart-Smith RD, Wotherspoon S, Krkosek M, Stuart-Smith JF, Pecl GT (2014) Statistical solutions for error and bias in global citizen science datasets. Biol Conserv 173:144–154

Bois ST, Silander JA, Mehrhoff LJ (2011) Invasive plant atlas of New England: the role of citizens in the science of invasive alien species detection. Bioscience 61:763–770

Bonan GB (2008) Forests and climate change: forcings, feedbacks, and the climate benefits of forests. Science 320:1444–1449

Bonfills CJW, Phillips TJ, Lawrence DM, Cameron-Smith P, Riley WJ, Subin ZM (2012) On the influence of shrub height and expansion on northern high latitude climate. Environ Res Lett 7:015503

Burakowski EA, Ollinger SV, Lepine LC, Schaaf CB, Wang Z, Dibb JE, Hollinger DY, Kim J, Erb A, Martin ME (2015) Spatial scaling of reflectance and surface albedo over a mixed-use, temperate forest landscape during snow-covered periods. Remote Sens Environ 158:465–477

Campbell S, Affleck RT, Sinclair S (2018) Ground-penetrating radar studies of permafrost, periglacial, and near-surface geology at McMurdo Station, Antarctica. Cold Reg Sci Technol 148:38–49

Cao C, Xiong S, Wu A, Wu X (2008) Assessing the consistency of AVHRR and MODIS L1B reflectance for generating fundamental climate data records. J Geophys Res-Atmos 113:1–10

Catapano I, Antonio Affinito, Gianluca Gennarelli, Francesco di Maio, Antonio Loperte, Francesco Soldovieri, (2014) Full three-dimensional imaging via ground penetrating radar: assessment in controlled conditions and on field for archaeological prospecting. Applied Physics A 115 (4):1415–1422

Cavanaugh KC, Kellner JR, Forde AJ, Gruner DS, Parker JD, Rodriguez W, Feller IC (2013) Poleward expansion of mangroves is a threshold response to decreased frequency of extreme cold events. P Natl Acad Sci USA 111:723–727

Chadwick OA, Derry LA, Vitousek PM, Huebert BJ, Hedin LO (1999) Changing sources of nutrients during four million years of ecosystem development. Nature 397:491–497

Chapin FS, Walker BH, Hobbs RJ, Hooper DU, Lawton JH, Sala OE, Tilman D (1997) Biotic control over the functioning of ecosystems. Science 277:500–504

Charney ND, Record S (2016) Combining incidence and demographic modelling approaches to evaluate metapopulation parameters for an endangered riparian plant. AOB Plants 8:1–11

Charney ND, Babst F, Poulter B, Record S, Trouet VM, Frank D, Enquist BJ, Evans MEK (2016) Observed forest sensitivity to climate implies large changes in 21st century North American forest growth. Ecol Lett 19:1119–1128

Clark JS (2016) Why species tell more about traits than traits about species: predictive analysis. Ecology 97:1979–1993

Corbley K (2010) GEOSAR-making mapping the 'impossible' possible. GEO Inf 13:43–47

Cord A, Rödder D (2011) Inclusion of habitat availability in species distribution models through multi-temporal remote sensing data? Ecol Appl 21:3285–3298

Cowling RM, Proches S, Partridge TC (2009) Explaining the uniqueness of the Cape flora: incorporating geomorphic evolution as a factory for explaining its diversification. Mol Phylogenet Evol 51:64–74

Crowley JK (1993) Mapping playa evaporite minerals with AVIRIS data: a first report from Death Valley, California. Remote Sens Environ 44:337–356

Cudahy TJ, Hewson R, Huntington JF, Quigley MA, Barry PS (2001) The performance of the satellite-borne Hyperion hyperspectral VNIR-SWIR imaging system for mineral mapping at Mount Fitton, South Australia. In: Geoscience and remote sensing symposium (IGARSS 2001), pp 314–316

Davis JL, Annan AP (1989) Ground-penetrating radar for high-resolution mapping of soil and rock stratigraphy. Geophys Prospect 37:531–551

Deblauwe V, Droissart V, Bose R, Sonké B, Blach-Overgaard A, Svenning JC, Wieringa JJ, Ramesh BR, Stévart T, Couvreur TLP (2016) Remotely sensed temperature and precipitation data improve species distribution modelling in the tropics. Glob Ecol Biogeogr 25:443–454

Dietz AJ, Kuenzer C, Gessner U (2011) Remote sensing of snow—a review of available methods. Int J Remote Sens 33:4049–4134

Dufour A, Gadallah F, Wagner HH, Guisan A, Buttler A (2006) Plant species richness and environmental heterogeneity in a mountain landscape: effects of variability and spatial configuration. Ecography 29:573–584

Elith J, Leathwick JR (2009) Species distribution models: ecological explanation and prediction across space and time. Ann Rev Ecol Evol Syst 40:677–697

Engelbrecht BMJ, Comita LS, Condit R, Kursar TA, Tyree MT, Turner BL, Hubbell SP (2007) Drought sensitivity shapes species distribution patterns in tropical forests. Nature 447:80–83

Enquist BJ, Condit R, Peet RK, Schildhauer M, Thiers BM (2016) Cyberinfrastructure for an integrated botanical information network to investigate the ecological impacts of global climate change on plant biodiversity. Peer J 4:e2615v2

Entekhabi D, Yueh S, O'Neill PE, Kellogg KH, Allen A, Bindlish R, Brown M, Chan S, Colliander A, Crow WT (2014) SMAP handbook: soil moisture active passive

ESA. Endangered Species Act 1973 16 U.S.C. § 1531 et seq.

Evans MEK, Merow C, Record S, McMahon SM, Enquist BJ (2016) Towards process-based range modeling of many species. Trends Ecol Evol 31:860–871

Ferrier S, Manion G, Elith J, Richardson K (2007) Using generalized dissimilarity modelling to analyse and predict patterns of beta diversity in regional biodiversity assessment. Div Dist 13:252–264

Fick SE, Hijmans RJ (2017) WorldClim 2: new 1-km spatial resolution climate surfaces for global land areas. Int J Climatol 37:4302–4315

Flynn DFB, Mirotchnick N, Jain M, Palmer MI, Naeem S (2011) Functional and phylogenetic diversity as predictors of biodiversity-ecosystem function relationships. Ecology 92:1573–1581

Funk CP, Peterson P, Landsfeld M, Pedreros D, Verdin J, Shukla S, Husak G, Rowland J, Harrison L, Hoell A, Michaelsen J (2015) The climate hazards infrared precipitation with stations - a new environmental record for monitoring extremes. Sci Data 2:150066

Gaston KJ (2000) Global patterns in biodiversity. Nature 405:220–227

GBIF (2016) Global biodiversity information facility. Available at: https://gbif.org

Giardino C, Brando VE, Gege P, Pinnel N, Hochberg E, Knaeps E, Reusen I, Doerffer R, Bresciani M, Braga F, Foerster S, Champollion N, Dekker A (2018) Imaging spectrometry of inland and coastal waters: state of the art, achievements and perspectives. Surv Geophys 40:1–29

Ginoux P, Prospero JM, Gill TE, Hsu NC, Zhao M (2012) Global-scale attribution of anthropogenic and natural dust sources and their emission rates based on Modis deep blue aerosol products. Rev Geophys 50:1–36

Goetz AFH, Vane G, Solomon JE, Rock BN (1985) Imaging spectrometry for earth remote sensing. Science 228:1147–1153

Gray M (2008) Geoheritage 1. Geodiversity: a new paradigm for valuing and conserving geoheritage. Geosci Can 35:2

Gray M (2013) Geodiversity: valuing and conserving abiotic nature, 2nd edn. Wiley-Blackwell, London

Green RO, Eastwood ML, Sarture CM, Chrien TG, Aronsson M, Chippendale BJ, Faust JA, Pavri BE, Chovit CJ, Solis M, Olah MR (1998) Imaging spectroscopy and the airborne visible/infrared imaging spectrometer (AVIRIS). Remote Sens Environ 65:227–248

Grunwald S, Thompson JA, Boettinger JL (2011) Digital soil mapping and modeling at continental scales: finding solutions for global issues. Soil Sci Soc Am J 75:1201

Gupta RP (2013) Remote sensing geology, 2nd edn. Springer, New York

Hastings A, Byers JE, Crooks JA, Cuddington K, Jones CG, Lambrinos JG, Talley TS, Wilson WG (2007) Ecosystem engineering in space and time. Ecol Lett 10:153–164

Haynes G (2012) Elephants (and extinct relatives) as earth-movers and ecosystem engineers. Geomorphology 157:99–107

Hengl T, De Jesus JM, MacMillan RA, Batjes NH, Heuvelink GBM, Ribeiro E, Samuel-Rosa A, Kempen B, Leenaars JGB, Walsh MG, Gonzalez MR (2014) SoilGrids1km - global soil information based on automated mapping. PLoS One 9:e105992

Hestir EL, Brando VE, Bresciani M, Giardino C, Matta E, Villa P, Dekker AG (2015) Measuring freshwater aquatic ecosystems: the need for a hyperspectral global mapping satellite mission. Remote Sens Environ 167:181–195

Hewson RD, Cudahy TJ, Mizuhiko S, Ueda K, Mauger AJ (2005) Seamless geological map generation using ASTER in the Broken Hill-Curnamona Province of Australia. Remote Sens Environ 99:159–172

Hijmans RJ, Cameron SE, Parra JL, Jones PG, Jarvis A (2005) Very high resolution interpolated climate surfaces for global land areas. Int J Climatol 25:1965–1978

Hijmans R, van Etten J (2019) raster: Geographic data analysis and modeling. R Package version 3.0–7

Hillebrand H, Bennet DM, Cadotte M (2008) Consequences of dominance: a review of evenness effects on local and regional ecosystem processes. Ecology 89:1510–1520

Hjort J, Luoto M (2010) Geodiversity of high-latitude landscapes in northern Finland. Geomorphology 115:109–116

Hjort J, Luoto M (2012) Can geodiversity be predicted from space? Geomorphology 153:74–80

Hjort J, Heikkinen RK, Luoto M (2012) Inclusion of explicit measures of geodiversity improve biodiversity models in a boreal landscape. Biodivers Conserv 21:3487–3506

Hutchinson GE (1965) The ecological theater and the evolutionary play. Yale University Press, Connecticut

International Union for Conservation of Nature and Natural Resources (2001) IUCN red list categories and criteria (Int. Union Conserv. Nat. Nat. Resour. Species Survival Commission, Cambridge, U.K.) Version 3.1

IUCN (2017) The IUCN red list of threatened species. Version 2017–2. Available at: http://www.iucnredlist.org

Jackova K, Romportl D (2008) The relationship between geodiversity and habitat richness in Sumava National Park and Krivoklatsko PLA (Czech Republic): a qualitative analysis approach. J Landsc Ecol 1:23–28

Janzen DH (1967) Why mountain passes are higher in the tropics. Am Nat 101:233–249

Jasiewicz J, Stepinski TF (2013) Geomorphons—a pattern recognition approach to classification and mapping of landforms. Geomorphology 182:147–156

Jeffries MO, Morris K, Kozlenko N (2005) Ice characteristics and processes, and remote sensing of frozen rivers and lakes. In: Duguay CR, Pietroniro A (eds) Remote sensing in northern hydrology: measuring environmental change. American Geophysical Union, Washington D.C., pp 63–90

Jenny H (1941) Factors of soil formation. McGraw-Hill, New York

Jetz W, Cavender-Bares J, Pavlick R, Schimel D, Davis FW, Asner GP, Guralnick R, Kattge J, Latimer AM, Moorcroft P, Scaepman ME, Schildhauer MP, Schneider FD, Schrodt F, Stahl U, Ustin SL (2016) Monitoring plant functional diversity from space. Nat Plants 2:e3

John W. Williams, Stephen T. Jackson, (2007) Novel climates, no-analog communities, and ecological surprises. Frontiers in Ecology and the Environment 5 (9):475-482

Jones CG, Lawton JH, Shachak M (1994) Organisms as ecosystem engineers. In: Samson FB, Knopf FL (eds) Ecosystem management. Springer, New York, pp 130–147

Joseph J. Bailey, Doreen S. Boyd, Jan Hjort, Chris P. Lavers, Richard Field, (2017) Modelling native and alien vascular plant species richness: At which scales is geodiversity most relevant?. Global Ecology and Biogeography 26(7):763–776

Jost L (2007) Partitioning diversity into independent alpha and beta components. Ecology 88:2427–2439

Kampe TU, Johnson BR, Kuester M, Keller M (2010) NEON: the first continental-scale ecological observatory with airborne remote sensing of vegetation canopy biochemistry and structure. J Appl Remote Sens 4:043510

Kerr JT, Packer L (1997) Habitat heterogeneity as a determinant of mammal species richness in high-energy regions. Nature 385:252–254

Körner C (2012) Alpine treelines: functional ecology of the global high elevation tree limits. Springer, New York, pp 21–30

Kozlowski S (1999) Programme of geodiversity conservation in Poland. Pol Geol Inst Spec Papers 2:15–19

Krause FA, Lefkoff AB, Dietz JB (1993) Expert system-based mineral mapping in northern Death Valley, California/Nevada, using the airborne visible/infrared imaging spectrometer (AVIRIS). Remote Sens Environ 44:309–336

Kwok R, Rothrock DA (2009) Decline in Arctic Sea ice thickness from submarine and ICESat records. Geophys Res Lett 36:1958–2008

Lacoul P, Freedman B (2006) Environmental influences on aquatic plants in freshwater ecosystems. Environ Rev 14:89–136

Lausch A, Bannehr L, Beckmann M, Boehm C, Feilhauer H, Hacker JM, Heurich M, Jung A, Klenke R, Neumann C, Pause M, Rocchini D, Schaepman ME, Schmidtlein S, Schulz K, Selsam P, Settele J, Skidmore AK, Cord AF (2016) Linking earth observation and taxonomic, structural, and functional biodiversity: local to ecosystem perspectives. Ecol Indic 70:317–339

Lausch A, Borg E, Bumberger J, Dietrich P, Heurich M, Huth A, Jung A, Klenke R, Knapp S, Mollenhauer H, Paasche H, Paulheim H, Pause M, Schweitzer C, Schmulius C, Settele J, Skidmore A, Wegmann M, Zacharias S, Kirsten T, Schapeman M (2018) Understanding forest health with remote sensing, part III: requirements for a scalable multi-source forest health monitoring network based on data science approaches. Remote Sens 10:1120

Lawler JJ, Ackerly DD, Albano CM, Anderson MG, Dobrowski SZ, Gill JL, Heller NE, Pressey RL, Leidner AK, Böhm M, He KS, Nagendra H, Dubois G, Fatoyinbo T, Hansen MC, Paganini M, de Klerk HM, Asner GP, Kerr JT, Estes AB, Schmeller DS, Heiden U, Rocchini D, Pereira HM, Turak E, Fernandez N, Lausch A, Cho MA, Alcaraz-Segura D, McGeoch MA, Turner W, Mueller A, St-Louis V, Penner J, Petteri V, Belward A, Reyers B, Geller GN (2016) Framing the concept of satellite remote sensing essential biodiversity variables: challenges and future directions. Remote Sens Ecol Cons 2:122–131

Legendre P, Borcard D, Peres-Neto PR (2005) Analyzing beta diversity: partitioning the spatial variation of community composition data. Ecol Monogr 75:435–450

Levin SA (1992) The problem of pattern and scale in ecology. Ecology 73:1943–1967

Linlin Ge, Alex Hay-Man Ng, Xiaojing Li, Hasanuddin Z. Abidin, Irwan Gumilar, (2014) Land subsidence characteristics of Bandung Basin as revealed by ENVISAT ASAR and ALOS PALSAR interferometry. Remote Sensing of Environment 154:46-60

Lomolino MV, Brown JH, Sax DF (2010) Reticulations and reintegration of "A biogeography of the species" in Island Biogeography Theory Eds. Losos JB, Ricklefs RE. Princeton University Press, Princeton, NJ

Marcott KC, Shakun JD, Clark PU, Mix AC (2013) A reconstruction of regional and global temperature for the past 11,300 years. Science 339:1198–1201

Massironi M, Bertoldi L, Calafa P, Visona D, Bistacchi A, Giardino C, Schiavo A (2008) Interpretation and processing of ASTER data for geological mapping and granitoids detection in the *Saghro massif* (eastern anti-atlas, Morocco). Geosphere 4:736–759

McGill BJ (2010) Matters of scale. Science 328:575–576

Moore I. D., R. B. Grayson, A. R. Ladson, (1991) Digital terrain modelling: A review of hydrological, geomorphological, and biological applications. Hydrological Processes 5(1):3–30

Mouillot D, Graham NAJ, Villéger S, Mason NWH, Bellwood DR (2013) A functional approach reveals community responses to disturbances. Trends Ecol Evol 28:167–177

Muller-Karger FE, Hestir E, Ade C, Turpie K, Roberts DA, Siegel D, Miller RJ, Humm D, Izenberg N, Morgan F et al (2018) Satellite sensor requirements for monitoring essential biodiversity variables of coastal ecosystems. Ecol Appl 28:749–760

Nadeau CP, Urban M, Bridle JR (2017) Coarse climate change projections for species living in a fine-scaled world. Glob Change Biol 23:12–24

Nathalie Pettorelli, Martin Wegmann, Andrew Skidmore, Sander Mücher, Terence P. Dawson, Miguel Fernandez, Richard Lucas, Michael E. Schaepman, Tiejun Wang, Brian O'Connor, Robert H.G. Jongman, Pieter Kempeneers, Ruth Sonnenschein, Allison K. Leidner, Monika Böhm, Kate S. He, Harini Nagendra, Grégoire Dubois, Temilola Fatoyinbo, Matthew C. Hansen, Marc Paganini, Helen M. de Klerk, Gregory P. Asner, Jeremy T. Kerr, Anna B. Estes, Dirk S. Schmeller, Uta Heiden, Duccio Rocchini, Henrique M. Pereira, Eren Turak, Nestor Fernandez, Angela Lausch, Moses A. Cho, Domingo Alcaraz-Segura, Mélodie A. McGeoch, Woody Turner, Andreas Mueller, Véronique St-Louis, Johannes Penner, Petteri Vihervaara, Alan Belward, Belinda Reyers, Gary N. Geller, Doreen Boyd, (2016) Framing the concept of satellite remote sensing essential biodiversity variables: challenges and future directions. Remote Sensing in Ecology and Conservation 2(3):122–131

NRC. National Research Council (2001) Basic research opportunities in earth science, pp 14–15 The National Academies Press, Washington, DC, USA. https://doi.org/10.17226/9981

O'Connell BM, Conkling BL, Wilson AM, Burrill EA, Turner JA, Pugh SA, Christensen G, Ridley T, Menlove J (2017) The forest inventory and analysis database: database description and user's manual version 7.0 for phase 2. U.S. Department of Agriculture, Forest Service, p 830, Washington, DC, USA

Okin GS, Mahowald N, Chadwick OA, Artaxo P (2004) Impact of desert dust on the biogeochemistry of phosphorus in terrestrial ecosystems. Global Biogeochem Cy 18:2

Olden JD, Rooney TP (2006) On defining and quantifying biotic homogenization. Glob Ecol Biogeogr 15:113–120

Ollinger SV, Aber JD, Federer CA, Lovett GM, Ellis J (1995) Modeling physical and chemical climatic variables across the northeastern U.S. for a geographic information system. USDA Forest Service General Technical Report NE-191

Ollinger SV, Smith ML, Martin ME, Hallett RA, Goodale CL, Aber JD (2002) Regional variation in foliar chemistry and soil nitrogen status among forests of diverse history and composition. Ecology 83:339–355

Parks KE, Mulligan M (2010) On the relationship between a resource based measure of geodiversity and broad scale biodiversity patterns. Biodivers Conserv 19:2751–2766

Pau S, Dee LE (2016) Remote sensing of species dominance and the value for quantifying ecosystem services. Remote Sense Ecol Conserv 2:141–151

Pavelsky TM, Zarnetske JP (2017) Rapid decline in river icings detected in arctic Alaska: implications for a changing hydrologic cycle and river ecosystems. Geophys Res Lett 44(7):2016GL072397. https://doi.org/10.1002/2016GL072397

Peacock J, Baker TR, Lewis SL, Lopez-Gonzalez G, Phillips OL (2007) The RAINFOR database: monitoring forest biomass and dynamics. J Veg Sci 18:535–542

Petorelli N, Laurance WF, O'Brien TG, Wegmann M, Nagendra H, Turner W (2014) Satellite remote sensing for applied ecologists: opportunities and challenges. J Appl Ecol 51:839–848

Pourmokhtarian A, Driscoll CT, Campbell JL, Hayhoe K, Stoner AMK (2016) The effects of climate downscaling technique and observation dataset on modeled ecological responses. Ecol Appl 26:1321–1337

Radford AE (1981) Natural heritage: classification, inventory and information. University of North Carolina Press, Chapel Hill

Rahbek C (2005) The role of spatial scale and the perception of large-scale species-richness patterns. Ecol Lett 8:224–239

Record S, Charney ND (2016) Modeling species ranges. Chance 29:31–37

Record S, Fitzpatrick MC, Finley AO, Veloz S, Ellison AM (2013) Should species distribution models account for spatial autocorrelation? A test of model projections across eight millennia of climate change. Glob Ecol Biogeogr 22:760–771

Record S, McCabe T, Baiser B, Ellison AME (2018) Identifying foundation species in North American forests using long-term data on ant assemblage structure. Ecosphere 9:e01239

Richerson PJ, Lum K (1980) Patterns of plant species and diversity in California: relation to weather and topography. Am Nat 116:504–536

Rienecker MM, Suarez MJ, Gelaro R, Todling R, Bacmeister J, Liu E, Bosilovich MG, Schubert SD, Takacs L, Kim GK, Bloom S (2011) MERRA: NASA's modern-era retrospective analysis for research applications. J Clim 24:3624–3648

Rignot E, Mouginot J, Scheuchl B (2011) Ice flow of the Antarctic ice sheet. Science 333:1427–1430

Riley SJ, DeGloria SD, Elliot R (1999) A terrain ruggedness index that quantifies topographic heterogeneity. Intermountain J Sci 5:23–27

Robert L. Pressey, Mar Cabeza, Matthew E. Watts, Richard M. Cowling, Kerrie A. Wilson, (2007) Conservation planning in a changing world. Trends in Ecology & Evolution 22(11):583–592

Robinson N, Regetz J, Guralnick RP (2014) EarthEnv-DEM90: a nearly global, void-free, multi-scale smoothed, 90m digital elevation model from fused ASTER and SRTM data. ISPRS J Photogramm 87:57–67

Rowan LC, Mars JC (2003) Lithologic mapping in the Mountain Pass, California area using advanced spaceborne thermal emission and reflection radiometer (ASTER) data. Remote Sens Environ 84:350–366

Sala OE, Chapin FS, Armesto JJ, Berlow E, Bloomfield J, Dirzo R, Huber-Sanwald E, Huenneke LF, Jackson RB, Kinzig A, Leemans R, Lodge DM, Mooney HA, Oesterheld M, Poff NL, Sykes MT, Walker BH, Walker M, Wall DH (2000) Global biodiversity scenarios for the year 2100. Science 287:1770–1774

Sang-Ho Yun, Kenneth Hudnut, Susan Owen, Frank Webb, Mark Simons, Patrizia Sacco, Eric Gurrola, Gerald Manipon, Cunren Liang, Eric Fielding, Pietro Milillo, Hook Hua, Alessandro Coletta, (2015) Rapid Damage Mapping for the 2015 7.8 Gorkha Earthquake Using Synthetic Aperture Radar Data from COSMO–SkyMed and ALOS-2 Satellites. Seismological Research Letters 86(6):1549–1556

Serrano E, Ruiz-Flaño P, Arroyo P (2009) Geodiversity assessment in a rural landscape: Tiermes-Caracena area (Soria, Spain). Memorie desrittive della carta geologica d'Italia 87:173–180

Smith SJ, Edmonds J, Hartin CA, Mundra A, Calvin K (2015) Near-term acceleration in the rate of temperature change. Nat Clim Chang 5:333–336

Spehn EM, Körner C (2005) A global assessment of mountain biodiversity and its function. In: Huber UM, BugmannMel KM, Reasoner A (eds) Global change and mountain regions. Springer, Dordrecht, pp 393–400

Srivastava DS, Cadotte MW, MacDonald AAM, Marushia RG, Mirotchnick N (2012) Phylogenetic diversity and the functioning of ecosystems. Ecol Lett 15:637–648

Stavros EN, Schimel D, Pavlick R, Serbin S, Swann A, Ducanson L, Fisher JB, Fassnacht F, Ustin S, Dubayah R, Schweiger A (2017) ISS observations offer insights into plant function. Nature Ecol Evol 1:0194

Stein A, Gerstner K, Kreft H (2014) Environmental heterogeneity as a universal driver of species richness across taxa, biomes, and spatial scales. Ecol Lett 17:866–880

Stendera S, Adrian R, Bonada N, Cañedo-Argüelles M, Hugueny B, Januschke K, Pletterbauer F, Hering D (2012) Hydrobiologia 696:1–28

Still CJ, Pau S, Edwards EJ (2014) Land surface temperature captures thermal environments of C3 and C4 grasses. Glob Ecol Biogeogr 23:286–296

Tedesco M (2014) Remote sensing of the cryosphere. Wiley-Blackwell, Hoboken

Thackeray SJ, Sparks TH, Fredericksen M, Burthe S, Bacon PJ, Bell JR, Botham MS, Brereton TM, Bright PW, Carvalho L, Clutton-Brock T, Dawson A, Edwards M, Elliott JM, Harrington R, Johns D, Jones ID, Jones JT, Leech DI, Roy DB, Scott WA, Smith M, Smithers RJ, Winfield IJ, Wanless S (2010) Trophic level asynchrony in rates of phenological change for marine, freshwater and terrestrial environments. Glob Change Biol 16:3304–3313

Tuanmu M-N, Jetz W (2015) A global, remote sensing-based characterization of terrestrial habitat heterogeneity for biodiversity and ecosystem modeling. Glob Ecol Biogeogr 24:1329–1339

Tukiainen H, Bailey JJ, Field R, Kangas K, Hjort J (2017) Combining geodiversity with climate and topography to account for threatened species richness. Cons Biol 31:364–375

Tuomisto H, Ruokolainen K, Aguilar M, Sarmiento A (2003) Floristic patterns along a 43-km long transect in an Amazonian rainforest. J Ecol 91:743–756

Turner W (2014) Sensing biodiversity. Science 346:301–302

van der Werff H, van der Meer F (2016) Sentinel-2A MSI and Landsat 8 OLI provide data continuity for geological remote sensing. Remote Sens 8(11):883

Veloz SD, Williams JW, Blois JL, Feng H, Otto-Bliesner B, Liu Z (2012) No-analog climates and shifting realized niches during the late quaternary: implications for 21st-century predictions by species distribution models. Glob Change Biol 18:1698–1713

Vincent WF, Callaghan TV, Dahl-Jensen D, Johansson M, Kovacs KM, Michel C, Prowse T, Reist JD, Sharp M (2011) Ecological implications of changes in the Arctic cryosphere. Ambio 40:87–99

Violle C, Enquist BJ, McGill BJ, Jiang L, Albert CH, Hulshof C, Jung V, Messier J (2012) The return of the variance: intraspecific variability in community ecology. Trends Ecol Evol 27:244–252

Vitousek PM, Aplet G, Turner D, Lockwood JJ (1992) The Mauna Loa environmental matrix: foliar and soil nutrients. Oecologia 89:372–382

Walther G-R, Post E, Convey P, Menzel A, Parmesan C, Beebee TJC, Fromentin J-M, Hoegh Guldberg O, Bairlein F (2002) Ecological responses to recent climate change. Nature 416:389–395

Weber F, Nixon D, Hurley J (2003) Semi-automated classification of river ice types on the Peace River using RADARSAT-1 synthetic aperture radar (SAR) imagery. Can J Civil Eng 30:11–27

Williams RS (2012) Introduction—changes in the Earth's cryosphere and global environmental change in the earth system. In: Williams RS, Ferrigno JG (eds) State of the earth's cryosphere at the beginning of the 21st century—glaciers, global snow cover, floating ice, and permafrost and periglacial environments: U.S. Geological Survey Professional Paper 1386-A. U.S. Geological Survey, Reston, p 23

Williams JW, Jackson ST (2007) Novel climates, no-analog communities, and ecological surprises. Front Ecol Environ 5:475–482

Wilson AM, Jetz W (2016) Remotely sensed high-resolution global cloud dynamics for predicting ecosystem and biodiversity distributions. PLoS Biol 14:1–20

Wood J (1996) Scale-based characterisation of digital elevation models. Innovat GIS 3:163-175

Yachi S, Loreau M (1999) Biodiversity and ecosystem productivity in a fluctuating environment: the insurance hypothesis. P Natl Acad Sci USA 96:1463–1468

Yoshikawa K, Hinzman LD, Kane DL (2007) Spring and Aufeis (icing) hydrology in Brooks Range, Alaska. J Geophys Res-Biogeosci 112:G04S43

Yu H, Chin M, Bian H, Yuan T, Prospero JM, Omar AH, Remer LA, Winker DM, Yang Y, Zhang Y, Zhang Z (2015) Quantification of trans-Atlantic dust transport from seven-year (2007-2013) record of CALIPSO LIDAR measurements. Remote Sens Environ 15:232–249

Zarnetske PL, Skelly DK, Urban MC (2012) Biotic multipliers of climate change. Science 336:15–30

Zarnetske PL, Baiser B, Strecker A, Record S, Belmaker J, Tuanmu M-N (2017) The interplay between landscape structure and biotic interactions. Curr Landscape Ecol Rep 2:12–29

Zarnetske PL, Record S, Dahlin K, Read Q, Grady JM, Costanza J, Finley AO, Gaddis K, Hobbi M, Latimer A, Malone S, Ollinger S, Pau S, Turner W, Wilson A (2019) Connecting biodiversity and geodiversity with remote sensing across scales. Glob Ecol Biogeog 28:548-556.

Zwally HJ, Li J, Brenner AC, Beckley M, Cornejo HG, Di Marzio J, Giovinetto MB, Neumann TA, Robbins J, Saba JL, Yi D, Wang W (2011) Greenland ice sheet mass balance: distribution of increased mass loss with climate warming; 2003-07 versus 1992-2002. J Glaciol 57:88–102

11

Remote Sensing-Based Predictions of Plant Biodiversity

Andrea Paz, Marcelo Reginato, Fabián A. Michelangeli, Renato Goldenberg, Mayara K. Caddah, Julián Aguirre-Santoro, Miriam Kaehler, Lúcia G. Lohmann, and Ana Carnaval

11.1 Introduction

The spatial distribution of species is unquestionably tied to environments, particularly temperature and precipitation (Hutchinson 1957). By exploring this correlation, multiple studies have demonstrated that environmental descriptors are able to predict geographic patterns of biological diversity reasonably well (Peters et al. 2016;

A. Paz (✉) · A. Carnaval
Department of Biology, City College of New York, New York, NY, USA

Biology Program, The Graduate Center, City University of New York, New York, NY, USA

M. Reginato
Departamento de Botânica, Instituto de Biociências, Universidade Federal do Rio Grande do Sul, Porto Alegre, RS, Brazil

F. A. Michelangeli
Biology Program, The Graduate Center, City University of New York, New York, NY, USA

Institute of Systematic Botany, The New York Botanical Garden, The Bronx, NY, USA

R. Goldenberg
Universidade Federal do Paraná, Curitiba, PR, Brazil

M. K. Caddah
Universidade Federal de Santa Catarina, Florianopolis, SC, Brazil

J. Aguirre-Santoro
Biology Program, The Graduate Center, City University of New York, New York, NY, USA

Instituto de Ciencias Naturales, Facultad de Ciencias, Universidad Nacional de Colombia, Bogotá, Colombia

M. Kaehler · L. G. Lohmann
Departamento de Botânica, Instituto de Biociências, Universidade de São Paulo, São Paulo, SP, Brazil

Zellweger et al. 2016). Temperature, for example, has been repeatedly shown to be a good predictor of the species that inhabit a given area (the taxonomic dimension of biodiversity, e.g., Peters et al. 2016). However, the power to predict the distinct dimensions of biodiversity varies within and across groups of organisms. For instance, the contribution of different measures of temperature and precipitation appears to be idiosyncratic when multiple taxa are compared (Rompré et al. 2007; Laurencio and Fitzgerald 2010; Peters et al. 2016; Zellweger et al. 2016). Moreover, and in contrast to species richness (SR), the relationships between climate and the geographic distribution of evolutionary diversity in a region (i.e., the phylogenetic dimension of biodiversity), as well as the relationships between climate and endemism, have been less explored. Still, those relationships appear weaker due to the relatively larger contribution of history, biogeography, and contingency in the spatial distribution of lineages (da Silva et al. 2012; Barratt et al. 2017).

Most of those advances have relied on the use of climatic data sets that are interpolated from weather station data (Hijmans et al. 2005), summarizing spatial patterns of temperature and precipitation. These include the widely used WorldClim data set (Hijmans et al. 2005), country-specific data sets (e.g., Cuervo-Robayo et al. 2014), and the hybrid CHELSA database (Karger et al. 2017). The ease by which biodiversity scientists can access and download these databases, and the fact that they provide global-scale climatic information at biologically relevant scales (up to 1 km), have resulted in a sharp increase in the number of studies that explore the correlations between climate and biodiversity patterns. Yet the accuracy and the effectiveness of these global climatic descriptors have been questioned (Soria-Auza et al. 2010). Because the distribution of weather stations around the world is unequal, the confidence in those data sets is reduced in undersampled areas, which frequently correspond to the most biodiverse areas on Earth (see Pinto-Ledézma and Cavender-Bares, Chap. 9).

In this chapter, we explore the use of bioclimatic variables built from long-term climatologies derived from remote sensing (RS) as predictors of biodiversity patterns. We focus in a megadiverse region, with high topographic complexity: the Brazilian Atlantic Forest hotspot. We evaluate whether climate, inferred from RS sources, predicts which areas accumulate the highest diversity of species, evolutionary lineages, and endemism. For that, we use distribution and phylogenetic data from three plant clades representing different life forms, that are commonly found in the Brazilian Atlantic Forest: melastomes (178 species of shrubs and trees), bromeliads (43 species of epiphytes), and bignones (131 species of lianas). We also evaluate what (if any) gains emerge from the use of climatic descriptors based on RS, rather than weather stations, for this area. Given the sharp altitudinal changes observed in the Brazilian Atlantic Forest hotspot, it has been proposed that interpolated weather station data may perform more poorly than variables derived from RS (Waltari et al. 2014).

11.2 Study System

The Brazilian Atlantic Forest harbors one of the highest levels of endemism and threat globally, representing one the world's hotspots of biodiversity (Myers et al. 2000). Although only about 16% of the original forest persists (Ribeiro et al. 2009), the Atlantic Forest is topographically and environmentally complex, spanning more than 1,700 m in altitude and about 25° of latitude (Ribeiro et al. 2009). Climatic analyses of the forest, along with molecular studies of its biota, suggest that it encompasses multiple environmental spaces and associated species pools. More specifically, the northern (mostly lowland) and southern (mostly montane) elements are largely different in species composition and have responded differently to past climatic changes (Carnaval et al. 2014; Leite et al. 2016).

Melastomes represent the first clade selected for our study. The tribe Miconieae (Melastomataceae) is exclusively Neotropical, with ca. 1,900 species, mostly shrubs and small trees, but also herbs, lianas, epiphytes, and large trees (Michelangeli et al. 2004, 2008; Goldenberg et al. 2008). In the Atlantic Forest, the tribe is represented by ca. 310 species, 70% of which are endemic ("Flora do Brasil 2020"; Goldenberg et al. 2009). These species are largely grouped into three clades: the *Leandra* clade with ca. 215 species (Caddah 2013; Reginato and Michelangeli 2016), the *Miconia* section *Chaenanthera* clade (Goldenberg et al. 2018), and the *Miconia* sect. *discolor* clade. Most of these species are small trees and shrubs (although the *Pleiochiton* clade contains 12 species of shrubby epiphytes; Reginato et al. 2010, 2013), and the great majority are bee pollinated and have berry fruits that are dispersed by birds. In the Atlantic Forest, species of Miconieae are found throughout most environments and at all elevations, with species ranges varying from widely distributed within the domain and beyond, to microendemics found in a single mountain top (Michelangeli et al. 2008).

Bromeliads represent the second clade included in this investigation. The Bromeliaceae is an almost exclusively Neotropical family, with ca. 3,300 species of terrestrial or epiphytic rosette-forming herbs. In the Atlantic Forest, the Bromeliaceae is represented by 816 species, over 75% of which are endemic (Martinelli et al. 2009). The data set used here represents a clade of 70 species belonging to the *Ronnbergia-Wittmackia* alliance (Aguirre-Santoro et al. 2016; Aguirre-Santoro 2017). With the exception of one species, the basal grade of 26 species of *Wittmackia* is composed of species restricted to the Atlantic Forest. All of them are tank-forming epiphytes found in forested environments, many with very restricted distributions (Aguirre-Santoro 2017). In the Atlantic Forest, *Wittmackia* is found predominantly in the central and northern states.

Bignones are the third plant clade used in this analysis. The tribe Bignonieae (Bignoniaceae) originated at around 50 million years ago (mya) in the Brazilian Atlantic Forest and subsequently occupied Amazonia and the dry areas of Central Brazil (Lohmann et al. 2013). The group is very diverse ecologically, including species pollinated by hummingbirds, butterflies, bees, and bats (Gentry 1974; Alcantara and Lohmann 2010). Ant-plant interactions are extremely common and play an important role in herbivore defense (Nogueira et al. 2015). Most species in the family are dispersed by wind or water (Lohmann 2004).

Former drying of Neotropical climates, and the Andean orogeny, seems to have represented key diversification drivers for tribe Bignonieae (Lohmann et al. 2013). Today, it includes 383 species and 21 genera (Lohmann and Taylor 2014), representing the most diverse and abundant clade of lianas in Neotropical forests (Lohmann 2006). All species of the tribe are distributed among three main clades: (i) the "multiples of four clade" (referring to the multiples of four phloem wedges), with ca. 135 species (Lohmann 2006); (ii) the "*Fridericia* and Allies clade," with around 132 species (Kaehler et al. 2019); and (iii) the "*Adenocalymma-Neojobertia*" clade, with ca. 75 species (Fonseca and Lohmann 2018). The remaining species of the Tribe are distributed among eight small genera (Lohmann 2006).

11.3 Methods

To investigate the relationships between climate and biodiversity patterns in the Atlantic rainforest of Brazil, we selected three clades of angiosperms with different life forms, i.e., shrubs and small trees (tribe Miconieae, Melastomataceae), epiphytic herbs (the *Ronnbergia*/*Wittmackia* alliance, Bromeliaceae), and lianas (the "*Fridericia* and Allies" clade of tribe Bignonieae, Bignoniaceae).

For each group, we combined geo-referenced occurrence data from each species with information about its evolutionary relationships. Using personal field data, published records, and geo-referenced herbarium information, we gathered locality information for 352 species and 22,338 unique locality points vetted by experts for spatial and taxonomic accuracy as follows: (i) melastomes, 178 species and 10,253 records of members of tribe Miconieae; (ii) bromeliads, 43 species and 4,606 records of members of the *Ronnbergia*/*Wittmackia* alliance; and (iii) Bignones, 132 species and 7,480 records of members of the "*Fridericia* and Allies" clade of tribe Bignonieae (Lohmann, unpublished data; see Meyer et. al 2008 for further details on this data set).

For each species, we used the locality data to generate a multiple convex polygon representing its range, which was then converted into a gridded map (~5 km resolution). Maps of the individual species were then stacked, allowing us to compute the total number of species per pixel. Information about the species composition at each grid cell was then combined with published and novel data on the phylogenetic relationships among species of melastomes (Caddah 2013; Reginato and Michelangeli 2016; Goldenberg et al. 2018), bromeliads (Aguirre-Santoro et al. 2016), and bignones (Kaehler et al. 2019), to provide a measurement of phylogenetic diversity (PD) per pixel, using Faith's phylogenetic diversity index (Faith 1992). This metric quantifies the evolutionary history included in every community by adding the branch lengths leading to each taxon present in the community (Faith 1992).

We also identified pixels holding high or low levels of phylogenetic endemism (PE) by including information about the range of each species' sister taxon

(PE, Rosauer et al. 2009). This metric takes into account the evolutionary history (as branch lengths) and spatial restriction (here as range estimates; Rosauer et al. 2009). To allow for comparisons across plant groups with distinct life histories and environmental envelopes, we performed these analyses separately for each clade (melastomes, bromeliads, and bignones). We used the Biodiverse software (Laffan et al. 2010) to map the geographical patterns of SR, the PD, and the phylogenetic endemism of each group.

We then gathered climatic descriptors for the Atlantic Forest region using two sources of climatic data, both at a 0.05° resolution (~5 km): one derived from RS instruments (Deblauwe et al. 2016) and another derived from interpolated weather station data (WorldClim database; Hijmans et al. 2005). Both databases describe environmental variation in the form of 19 bioclimatic variables that reflect spatial and temporal differences in precipitation and temperature (Bio1-19, as defined in the WorldClim database). These data were estimated with the same formulae, across data sets. While the WorldClim data reflect bioclimatic conditions estimated from interpolated weather station information, the database of Deblauwe et al. (2016) was built based on temperature information from NASA's Moderate Resolution Imaging Spectroradiometer (MODIS) and precipitation from the Climate Hazards Group InfraRed Precipitation with Station (CHIRPS) data. To reduce collinearity between the 19 bioclimatic variables, we employed a variance inflation factor (VIF), retaining only those variables with VIF < 5 in both datasets in all analyses. This left us with seven variables from each source, in both cases bio 3, 8, 9 13,18, and 19, plus bio 2 for the dataset based on weather station data (WorldClim), and bio 7 for the RS-based (Deblauwe et al. 2016) dataset (see Table 11.1 for bioclimatic variable descriptions).

To investigate how much of the spatial patterns of SR, PD, and phylogenetic endemism can be explained by each set of climatic descriptors, we ran conditional autoregressive (CAR) models on the pooled data from each group. CAR models

Table 11.1 Bioclimatic variables used as predictors for analyses, after removing variables with high variance inflation factor (VIF)

Variable	Description
Bio 2	Mean diurnal range [mean of monthly (max temp–min temp)]
Bio 3	**Isothermality (Bio 2/Bio 7) *100**
Bio 7	Temperature annual range
Bio 8	**Mean temperature of the wettest quarter**
Bio 9	**Mean temperature of the driest quarter**
Bio 13	**Precipitation of the wettest month**
Bio 18	**Precipitation of the warmest quarter**
Bio 19	**Precipitation of the coldest quarter**

In bold, variables used in both the RS- and weather station-derived data sets

Bio 2 was used only in the weather station-based analysis; bio 7 was used only in the RS-based analysis implement a multiple spatial regression in which the covariance among residuals considers the neighborhood of each evaluated cell (Rangel et al. 2006). To evaluate model fit, we computed pseudo-R^2 for our models, including both the full model and the predictor-only effect (i.e., removing the effect of space and spatial autocorrelation). Also, to detect areas of potential concentration or overdispersion of the regression residuals, we generated residual maps for models of SR and PD.

11.4 Results and Discussion

The spatial patterns of SR, PD, and phylogenetic endemism vary across groups. While melastomes show higher PD and SR along the east coast of Brazil, bromeliads accumulate PD and SR in the northern region of the Atlantic Forest, and bignones in the central portion of the domain. Phylogenetic endemism concentrates in the coastal mountains for melastomes, in the northern coast for bromeliads, and toward the west and northwest for bignones (Fig. 11.1).

Bioclimatic variables derived from RS products have excellent predictive power for both SR and PD. The full CAR models built from RS sources performed well, irrespective of plant group ($R^2 > 0.89$, Table 11.2). Prediction of phylogenetic ende-

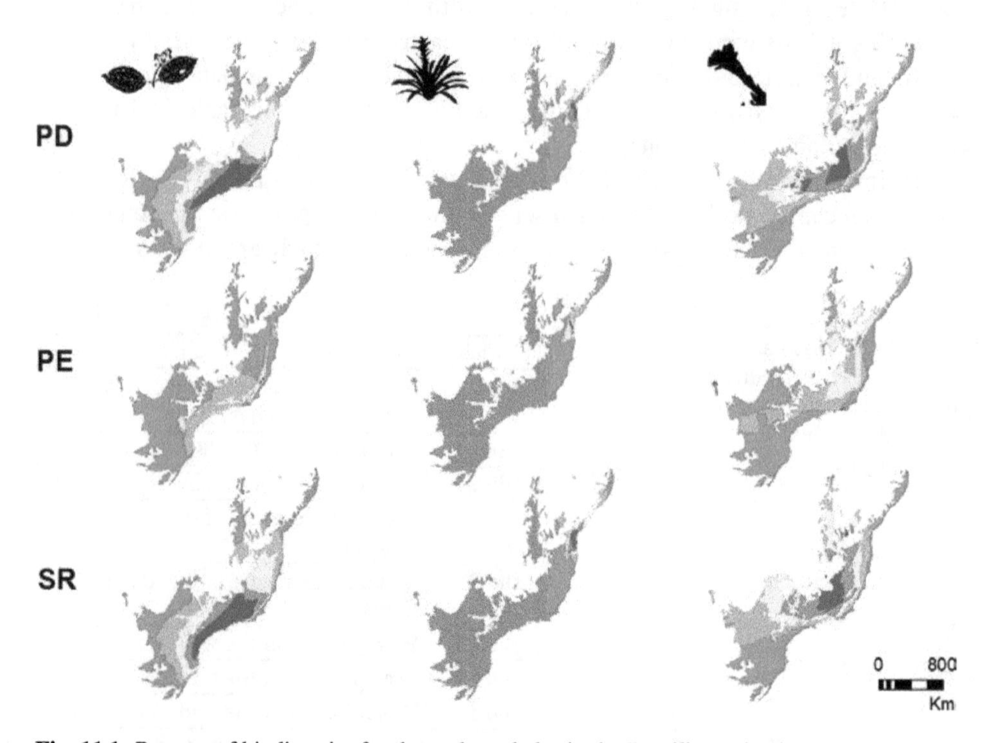

Fig. 11.1 Patterns of biodiversity for three plant clades in the Brazilian Atlantic Forest, from left to right: melastomes, bromeliads, and bignones. Each row corresponds to one biodiversity index, top to bottom: phylogenetic diversity (PD), phylogenetic endemism (PE), and species richness (SR). Warmer colors represent higher values of each of the indices

Table 11.2 Predictive power of models using either RS-based variables (RS) or weather station-derived variables (WC) as predictors of phylogenetic diversity (PD), phylogenetic endemism (PE), and species richness (SR) in three plant clades from the Brazilian Atlantic Forest: melastomes, bromeliads, and bignones

Clade	Predictors	Predicted	Full model R^2	Non-space R^2
Melastomes	RS	PD	0.96	**0.61**
		PE	0.62	0.01
		SR	**0.97**	**0.61**
	WC	PD	0.96	0.57
		PE	0.62	0.02
		SR	0.96	0.55
Bromeliads	RS	PD	**0.91**	**0.37**
		PE	**0.75**	**0.03**
		SR	**0.92**	**0.37**
	WC	PD	0.88	0.19
		PE	0.74	0.02
		SR	0.89	0.19
Bignones	RS	PD	0.94	0.58
		PE	0.57	0.10
		SR	0.96	0.72
	WC	PD	0.94	**0.59**
		PE	0.57	0.10
		SR	0.96	**0.74**

Numbers in bold have higher predictive power when comparing RS and WC for a single group

mism based on RS sources was not as successful: the performance of the CAR models was fair to good, irrespective of plant clade (R^2 0.57–0.75, Table 11.2). These observed differences in predictive power are not surprising and stand in agreement with the expectation that phylogenetic endemism may be more strongly impacted by historical processes and former climates (e.g., Late Quaternary) than by contemporary descriptors (Rosauer and Jetz 2015).

The predictive power of the models built with weather station data (WorldClim; Hijmans et al. 2005) was comparable to that of models based on satellite information, similar to Pinto-Ledézma and Cavender-Bares (Chap. 9), showing only slightly lower R^2 values overall (within 0.01, Table 11.2). This difference tended to increase (i.e., with models based on RS data performing better than those built with weather station data) when spatial autocorrelation effects were removed from the analyses (Table 11.2).

Climatic descriptors derived from both RS information (Deblauwe et al. 2016) and weather station data (Hijmans et al. 2005) failed to predict spatial patterns of phylogenetic endemism (PE) when decoupled from space (R^2 0.01–0.1, Table 11.2). Geography is naturally expected to impact maps of endemism because this analysis of geographical restriction of evolutionary history explicitly incorporates space in its calculations (Rosauer et al. 2009). Still, when this spatial imprint is removed from the data, we notice that contemporary climates are unable to predict the distribution of lineage restriction—in agreement with previous suggestions that historical climates, or the stochasticity of the processes associated with colonization or extinction, may have an important role in determining phylogenetic endemism (Carnaval et al. 2014; Rosauer and Jetz 2015).

Although the performance of the CAR models varied across diversity measures and taxa, the analyses decoupled from space recovered consistently lower predictive power in bromeliads, irrespective of diversity measures (Table 11.2). Unlike the other two groups, this clade is composed mainly of microendemics and represented only in a relatively small region of the Atlantic Forest (Fig. 11.1). We hypothesize that the larger influence of space, history, and chance events (particularly related to local extinctions) may be responsible for the lower correspondence between spatial patterns of biodiversity and climatic descriptors in groups of species that are narrowly distributed. This being true, it is expected that the predictive power of correlative models of biodiversity such as those presented here—including the use of RS data—will perform best in tropical groups in which most species have relatively large ranges.

The spatial distribution of the residuals of the correlation between biodiversity metrics and climate data differed across clades. In melastomes, they were homogeneously distributed across the forest, while for both bromeliads and bignones large residuals of SR and PD were observed in areas with low overall diversity. In bignones, residuals were especially concentrated in the south—where large geographic extensions showed more or less PD than expected (Fig. 11.2). Particularly the southern portion of the forest shows higher PD of bignones than expected, given the models based on climate data. This may be related to these plants' sensitivity to altitude, which limits their growth (Lohmann, pers. obs.). Bignoniaceae species are sensitive to temperature and precipitation, with abundance and species richness responding positively in warmer climates, and strongly negative in wetter climates. Thus, the species richness, or phylogenetic diversity, is increased in warmer and drier dry-season habitats (Punyasena et al. 2008). The south of the Atlantic forest is montane and the areas with larger residuals in South Brazil have dry, though cool winters. Given that no topographic variables were included in our models, this overprediction appears reasonable.

11.5 Conclusions and Future Directions

Community-level data from three representative tropical plant groups that include lianas, shrubs, and trees demonstrate that the use of RS data describing temperature and precipitation accurately predicts the spatial distribution of two essential biodiversity metrics (SR and PD) in a biodiversity hotspot. This predictive power is reduced when the approach is applied to a clade of spatially restricted (narrow endemic) species, such as bromeliads. Across all plant groups, predictive power is lower for diversity indices highly influenced by historical contingency and spatial configuration, such as phylogenetic endemism. For predictive purposes, and at the

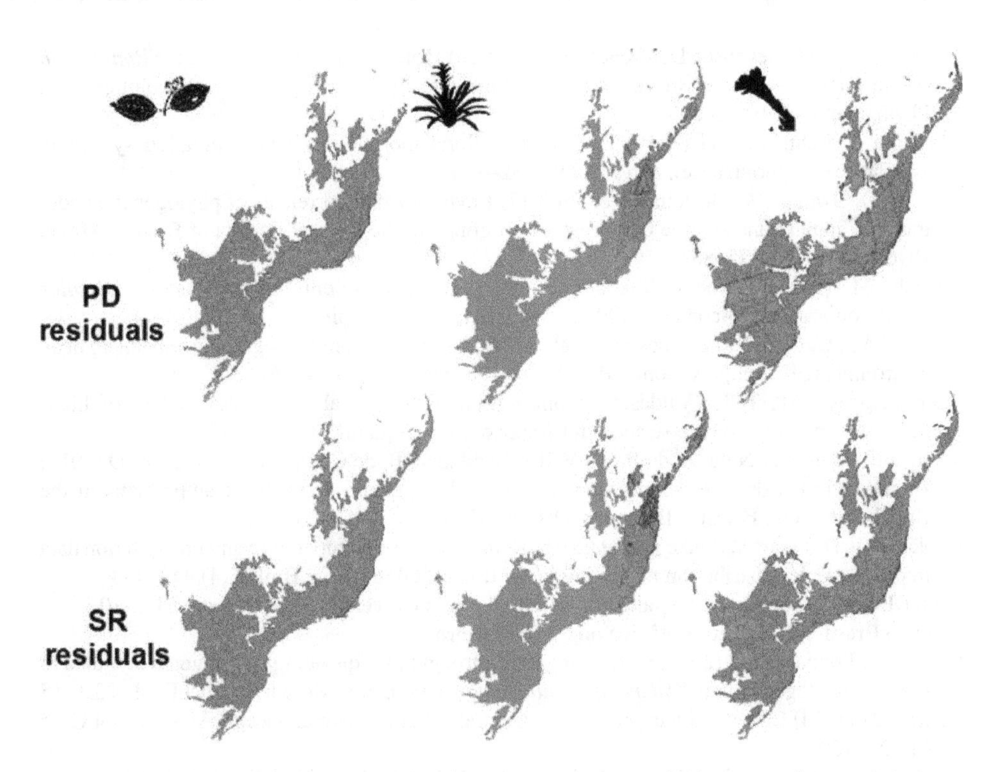

Fig. 11.2 Residuals of the CAR models using remote sensing variables as predictors of phylogenetic diversity (PD) and species richness (SR) for three groups of plants from the Brazilian Atlantic Forest, from left to right: melastomes, bromeliads, and bignones. Blue denotes areas with negative residuals (sites where less diversity is observed than expected under the model). Areas with positive residuals are shown in red. Darker shades of blue and red represent larger residuals

spatial scale of the Atlantic Forest, the performance of RS-based climate descriptors is comparable to, or slightly better than, that of weather station-based databases. These results show promise for predicting different dimensions of diversity in the tropics, based on RS data, especially for widely distributed groups. This approach may be particularly relevant in groups or regions for which direct or indirect species identification through RS (e.g., hyperspectral images) is feasible or available. It also may be extended to other groups of plants, and to animals. Future directions of this work include testing whether RS-based predictions of biodiversity work similarly well in other biological groups, biomes, and geographical areas, while also potentially including additional variables of interest, such as topography and historical climates.

References

Aguirre-Santoro J (2017) Taxonomy of the *Ronnbergia* Alliance (Bromeliaceae: Bromelioideae): new combinations, synopsis, and new circumscriptions of Ronnbergia and the resurrected genus Wittmackia. Plant Syst Evol 303:615–640

Aguirre-Santoro J, Stevenson D, Michelangeli F (2016) Molecular phylogenetics of the *Ronnbergia* Alliance (Bromeliaceae, Bromelioideae) and insights about its morphological evolution. Mol Phylogenet Evol 100:1–20

Alcantara S, Lohmann LG (2010) Evolution of floral morphology and pollination system in Bignonieae (Bignoniaceae). Am J Bot 97:782–796

Barratt CD, Bwong BA, Onstein RE et al (2017) Environmental correlates of phylogenetic endemism in amphibians and the conservation of refugia in the Coastal Forests of Eastern Africa. Divers Distrib 23:875–887

Caddah M (2013) Estudos Taxonomicos e filogenéticos em *Miconia* sect. *Discolor* (Melastomataceae, Miconieae). PhD thesis. UNICAMP, Campinas

Carnaval AC, Waltari E, Rodrigues MT et al (2014) Prediction of phylogeographic endemism in an environmentally complex biome. Proc R Soc B Biol Sci 281:20141461

Cuervo-Robayo AP, Téllez-Valdés O, Gómez-Albores MA et al (2014) An update of high-resolution monthly climate surfaces for Mexico. Int J Climatol 34:2427–2437

da Silva FR, Almeida-Neto M, do Prado VHM, Haddad CFB, de Cerqueira Rossa-Feres D (2012) Humidity levels drive reproductive modes and phylogenetic diversity of amphibians in the Brazilian Atlantic Forest. J Biogeogr 39:1720–1732

Deblauwe V, Droissart V, Sonke B et al (2016) Remotely sensed temperature and precipitation data improve species distribution modelling in the tropics. Glob Chang Biol 25(4):443–454

Faith DP (1992) Conservation evaluation and phylogenetic diversity. Biol Conserv 61:1–10

Flora do Brasil 2020 Jardim Botânico do Rio de Janeiro

Fonseca L, Lohmann LG (2018) Combining high-throughput sequencing and targeted loci data to infer the phylogeny of the "*Adenocalymma-Neojobertia*" clade. Mol Phylogenet Evol 123:1–15

Gentry AH (1974) Coevolutionary patterns in Central American Bignoniaceae. Ann Mo Bot Gard 61:728–759

Goldenberg R, Penneys DS, Almeda F, Judd WS, Michelangeli F (2008) Phylogeny of *Miconia* (Melastomataceae): initial insights into broad patterns of diversification in a megadiverse neotropical genus. Int J Plant Sci 169:963–979

Goldenberg R, Guimarães P, Kriebel R, Romero R (2009) Melastomataceae. In: Stehmann JR, Forzza R, Salino A et al (eds) Plantas da Floresta Atlantica. Instituto de Pesquisas Jardim Botânico do Rio de Janeiro, Rio de Janeiro, pp 330–343

Goldenberg R, Reginato M, Michelangeli F (2018) Disentangling the infrageneric classification of megadiverse taxa from Mata Atlantica: phylogeny of *Miconia* section *Chaenanthera* (Melastomataceae. Miconieae). Taxon 67:537–551

Hijmans RJ, Cameron SE, Parra JL, Jones PG, Jarvis A (2005) Very high resolution interpolated climate surfaces for global land areas. Int J Climatol 25:1965–1978

Hutchinson (1957) Concluding remarks. Cold Spring Harb Symp Quant Biol 22:415–427

Kaehler M, Michelangeli FA, Lohmann LG (2019) Fine-tuning the circumscription of Fridericia (Bignonieae, Bignoniaceae). Taxon 68(3). https://doi.org/10.1002/tax.12121

Karger DN, Conrad O, Böhner J et al (2017) Data descriptor: climatologies at high resolution for the earth's land surface areas. Scientific Data 4:1–20

Laffan SW, Lubarsky E, Rosauer DF (2010) Biodiverse, a tool for the spatial analysis of biological and related diversity. Ecography 33:643–647

Laurencio D, Fitzgerald LA (2010) Environmental correlates of herpetofaunal diversity in Costa Rica. J Trop Ecol 26:521–531

Leite YLR, Costa LP, Loss AC et al (2016) Neotropical forest expansion during the last glacial period challenges refuge hypothesis. Proc Natl Acad Sci 113:1008–1013

Lohmann LG (2004) Bignoniaceae. In: Smith N, Mori S, Henderson A, Stevenson SH (eds) Flowering plants of the neotropics. Princeton University Press, Princeton, pp 51–53

Lohmann LG (2006) Untangling the phylogeny of neotropical lianas (Bignonieae, Bignoniaceae). Am J Bot 93:304–318

Lohmann LG, Taylor CM (2014) A new generic classification of tribe Bignonieae (Bignoniaceae). Ann Mo Bot Gard 99:348–489

Lohmann LG, Bell CD, Calió MF, Winkworth RC (2013) Pattern and timing of biogeographical history in the Neotropical tribe Bignonieae (Bignoniaceae). Bot J Linn Soc 171:154–170

Martinelli G, Vieira C, Leitman P, Costa A, Forzza R (2009) Bromeliaceae. In: Stehmann JR, Forzza R, Salino A et al (eds) Plantas da Floresta Atlantica. Instituto de Pesquisas Jardim Botânico do Rio de Janeiro, Rio de Janeiro, pp 186–204

Meyer L, Diniz-Filho JAF, Lohmann LG (2008) A comparison of hull methods for estimating species ranges and richness maps. Plant Ecol Diversity 10:389–401

Michelangeli F, Penneys D, Giza J et al (2004) A preliminary phylogeny of the tribe Miconieae (Melastomataceae) based on nrITS sequence data and its implications on inflorescence position. Taxon 53:279–290

Michelangeli F, Judd WS, Penneys DS et al (2008) Multiple events of dispersal and radiation of the tribe Miconieae (Melastomataceae) in the Caribbean. Bot Rev 74:53–77

Myers N, Mittermeier RA, Mittermeier CG, da Fonseca GAB, Kent J (2000) Biodiversity hotspots for conservation priorities. Nature 403:853–858

Nogueira A, Rey P, Alcántara J, Feitosa R, Lohmann LG (2015) Geographic mosaic of plant evolution: Extrafloral nectary variation mediated by ant and herbivore assemblages. PLoS One 10:e0123806

Peters MK, Hemp A, Appelhans T et al (2016) Predictors of elevational biodiversity gradients change from single taxa to the multi-taxa community level. Nat Commun 7:13736

Punyasena SW, Eshel G, McElwain JC (2008) The influence of climate on the spatial patterning of Neotropical plant families. J Biogeogr 35:117–130

Rangel TFLVB, Diniz-Filho JAF, Bini LM (2006) Towards an integrated computational tool for spatial analysis in macroecology and biogeography. Glob Ecol Biogeogr 15:321–327

Reginato M, Michelangeli F (2016) Untangling the phylogeny of *Leandra* sensu str. (Melastomataceae, Miconieae). Mol Phylogenet Evol 96:17–32

Reginato M, Michelangeli F, Goldenberg R (2010) Phylogeny of *Pleiochiton* A. Gray (Melastomataceae, Miconieae): total evidence. Bot J Linn Soc 162:423–434

Reginato M, Baumgratz J, Goldenberg R (2013) A taxonomic revision of *Pleiochiton* (Melastomataceae, Miconieae). Brittonia 65:16–41

Ribeiro MC, Metzger JP, Martensen AC, Ponzoni FJ, Hirota MM (2009) The Brazilian Atlantic Forest: how much is left, and how is the remaining forest distributed? Implications for conservation. Biol Conserv 142:1141–1153

Rompré G, Robinson WD, Desrochers A, Angehr G (2007) Environmental correlates of avian diversity in lowland Panama rain forests. J Biogeogr 34:802–815

Rosauer DF, Jetz W (2015) Phylogenetic endemism in terrestrial mammals. Glob Ecol Biogeogr 24:168–179

Rosauer D, Laffan SW, Crisp MD, Donnellan SC, Cook LG (2009) Phylogenetic endemism: a new approach for identifying geographical concentrations of evolutionary history. Mol Ecol 18:4061–4072

Soria-Auza RW, Kessler M, Bach K et al (2010) Impact of the quality of climate models for modelling species occurrences in countries with poor climatic documentation: a case study from Bolivia. Ecol Model 221:1221–1229

Waltari E, Schroeder R, McDonald K, Anderson RP, Carnaval A (2014) Bioclimatic variables derived from remote sensing: assessment and application for species distribution modelling. Methods Ecol Evol 5:1033–1042

Zellweger F, Baltensweiler A, Ginzler C et al (2016) Environmental predictors of species richness in forest landscapes: abiotic factors versus vegetation structure. J Biogeogr 43:1080–1090

Permissions

All chapters in this book were first published in RSPB, by Springer; hereby published with permission under the Creative Commons Attribution License or equivalent. Every chapter published in this book has been scrutinized by our experts. Their significance has been extensively debated. The topics covered herein carry significant information for a comprehensive understanding. They may even be implemented as practical applications or may be referred to as a beginning point for further studies.

The contributors of this book come from diverse backgrounds, making this book a truly international effort. We would like to thank all the contributing authors for lending their expertise to make the book truly unique. They have played a crucial role in the development of this book. Without their invaluable contributions this book wouldn't have been possible. They have made vital efforts to compile up to date information on the varied aspects of this subject to make this book a valuable addition to the collection of many professionals and students.

This book was conceptualized with the vision of imparting up-to-date and integrated information in this field. To ensure the same, a matchless editorial board was set up. Every individual on the board went through rigorous rounds of assessment to prove their worth. After which they invested a large part of their time researching and compiling the most relevant data for our readers.

The editorial board has been involved in producing this book since its inception. They have spent rigorous hours researching and exploring the diverse topics which have resulted in the successful publishing of this book. They have passed on their knowledge of decades through this book. To expedite this challenging task, the publisher supported the team at every step. A small team of assistant editors was also appointed to further simplify the editing procedure and attain best results for the readers.

Apart from the editorial board, the designing team has also invested a significant amount of their time in understanding the subject and creating the most relevant covers. They scrutinized every image to scout for the most suitable representation of the subject and create an appropriate cover for the book.

The publishing team has been an ardent support to the editorial, designing and production team. Their endless efforts to recruit the best for this project, has resulted in the accomplishment of this book. They are a veteran in the field of academics and their pool of knowledge is as vast as their experience in printing. Their expertise and guidance has proved useful at every step. Their uncompromising quality standards have made this book an exceptional effort. Their encouragement from time to time has been an inspiration for everyone.

The publisher and the editorial board hope that this book will prove to be a valuable piece of knowledge for students, practitioners and scholars across the globe.

Index

Printed in the USA
CPSIA information can be obtained
at www.ICGtesting.com
JSHW052127021123
51365JS00005B/20